Specialized Plumbing Technologies

Specialized Plumbing Technologies presents a unique compilation of specialized plumbing systems rarely covered in general plumbing literature. Moving beyond basic water supply and drainage systems, it explores the diverse nature of building services dealing with fluid flow for special purposes.

The book aims to provide detailed guidance on the design, installation, usage, and maintenance of specialized plumbing systems, and the content covers 11 specialized plumbing systems, including dual water supply, chilled water supply, fire standpipe and sprinkler systems, fountains and waterfalls, sprinkler irrigation, steam supply, gas supply, medical gas supply, healthcare plumbing, and swimming pools. For each system, the book provides clear explanations of functionality, component sizing, major elements, and maintenance requirements. This comprehensive approach makes it valuable for those seeking to master plumbing and fluid services engineering beyond the more common applications, and the hundreds of illustrations break down complex components and systems to aid understanding.

The book serves anyone who needs detailed knowledge of specialized plumbing systems and is designed for a diverse readership spanning both academic and professional spheres. Students of civil, mechanical, and architectural engineering will find it highly educational, while professionals, researchers, and academics will appreciate its specialized technical content. Technical universities, educational institutions, and building development authorities will find it a valuable addition to their reference collections.

Syed Azizul Haq, PEng, a registered professional engineer, a patent holder of the "mechanical seal trap" and author of *Harvesting Rainwater from Buildings*, *Plumbing Principles and Practice*, and *Water Supply and Drainage for Buildings*, is the president of the RAiN Forum—a nonprofit voluntary platform for knowledge sharing and promotion of rainwater among scientists, engineers, architects, planners, academicians, practitioners, administrators, students, and others who are directly or indirectly concerned about sustainable water management—and, following 35 years as assistant engineer to additional chief engineer with the Public Works Department in Bangladesh, also served as a part-time faculty member and a guest lecturer at a number of universities.

Specialized Plumbing Technologies

Syed Azizul Haq

Routledge
Taylor & Francis Group
LONDON AND NEW YORK

Designed cover image: © the author

First published 2027
by Routledge
4 Park Square, Milton Park, Abingdon, Oxon OX14 4RN

and by Routledge
605 Third Avenue, New York, NY 10158

Routledge is an imprint of the Taylor & Francis Group, an informa business

British Library Cataloguing-in-Publication Data
A catalogue record for this book is available from the British Library

ISBN: 978-1-041-20766-5 (hbk)
ISBN: 978-1-041-19240-4 (pbk)
ISBN: 978-1-003-72405-6 (ebk)

DOI: 10.1201/9781003724056

Typeset in Times New Roman
by SPi Technologies India Pvt Ltd (Straive)

Dedicated to all sanitation workers and plumbers—especially the septic tank cleaners who lost their lives while serving mankind. May this book stand as a humble tribute to their dedication.

Contents

List of Figures xxi
List of Tables xxvi
Foreword 1 xxix
Foreword 2 xxx
Preface xxxi
Acknowledgment xxxiv
List of Abbreviations xxxv

1 Introduction **1**

Introduction 1
Specialized plumbing system 1
Chilled water 2
Non-potable water 2
Fire suppression 2
Irrigation 3
Amusement 3
Steam 4
Fuel gas 4
Medical gas 4
Health care 5
Swimming pool 5
Specialties of specialized plumbing 6
Challenges of specialized plumbing 6

2 Chilled water supply system **7**

Introduction 7
Chilled water 7
Use of chilled water 7
Chilled water supply for space cooling 8
Open chilled water supply system 9
Closed chilled water supply system 10
Components of a chilled water supply system 11

Chiller 12
Types of chillers 13
Condenser 14
Evaporator 16
Compressor 17
Expansion valve 20
Capacity range of chillers 21
Cooling tower 22
Open-circuit or wet cooling tower 22
Closed-circuit or fluid cooling tower 23
Cooling tower size 23
Cooling tower materials 24
Valves and appurtenances 24
Three-way valve 25
Balancing valve 25
Buffer tank 27
Pumps for chilled water supply 28
Chilled water circulation pump 28
Condenser water circulation pump 29
Chilled water distribution piping 29
Constant-rate chilled water distribution system 29
Primary-secondary chilled water distribution system 30
Variable primary flow chilled water distribution system 31
Venting chilled water supply piping 32
Chilled water system for drinking 32
Drinking fountain 35
Chilled water distribution system for drinking 35
Chilled water supply piping material 35
Design of major components of chilled water supply system 35
Sizing a chiller 36
Sizing a buffer tank 36
Cooling (refrigerant) load 37
Determining chilled water flow rate 37
Sizing chilled water distribution pipes 38
Sizing a cooling tower 39
Operation and maintenance 39

3 Dual water supply system **41**
Introduction 41
Potable and non-potable water 41
Quality parameters of potable and non-potable water 42
Non-potable water use 43

Dual water supply system 43
Urban dual-water supply system 43
A dual water supply system in a building 44
Advantages and disadvantages of a dual water supply system 44
Guidelines for a dual water supply system 45
Storage of non-potable water 46
Supply of potable and non-potable water in a building 46
Back flow 46
Back-siphonage 47
Backflow hazard rating 47
Backflow preventer 48
Types of backflow prevention assemblies 48
Separation of potable and non-potable water main 50
Switching over 50
Cross-connection between potable and non-potable water piping 50
Temporary cross-connection 50
Permanent cross-connection 51
Avoiding cross-connection 51

4 Fire standpipe system 53

Introduction 53
Fire 53
Causes of a fire in a building 54
Fire suppressing 54
Fire hydrant 54
Fire hydrant types 55
Fire standpipe system 55
Types of fire standpipe systems 56
Wet standpipe system 57
Dry standpipe system 57
Water reservoir 58
Fire flow 58
Fire pump 59
Types of fire pump 60
Fire-pump specialty 62
Suction pipe requirement 62
Fire pump driver 63
Electric motor for a fire pump 63
Diesel engine for a fire pump 63
Jockey pump 64
Fire pump controller 66
Stand pipe 67

Types of standpipe systems 67
Fire hose station 68
Fire station coverage 68
Fire hose station requirements 68
Location of the fire station 69
Fire station hose cabinet 69
Classes of fire hose stations 69
Landing valve 70
Fire hose 73
Hose reel 73
Hose rack 74
Friction loss of the fire hose 74
Fire hose nozzle 75
Types of fire hose nozzles 75
Flow through a nozzle 77
Fire department connection 78
Locating the FDCs 79
Quick coupling 79
Valves and appurtenances 80
Pressure-reducing valve 80
Air valve 80
Drain cock 81
Check valve 81
Designing fire hydrant elements 82
Sizing a water reservoir 82
Sizing a hose reel or pipe 83
Sizing fire pumps 84
Fire hydrant system maintenance 84

5 Fire sprinkler system 87

Introduction 87
Fire sprinkler system 87
Major components 88
Water reservoir 88
Pressure tank 89
Fire pump for fire sprinkler system 90
Flow rate 91
Occupancy hazard classification 91
Design area 91
Required pressure 92
Head and flow for fire pump 93
Fire pump performance curve 94
Fire pump testing 95
Jockey pump for fire sprinkler system 95

Types of fire sprinkler systems 96
Wet-pipe fire sprinkler 96
Dry-pipe fire sprinkler 97
Deluge fire system 98
Pre-action fire system 101
Sprinkler head 102
Thermal sensitivity 103
Temperature rating 104
Types of sprinkler heads 105
Automatic fire sprinkler system development 107
Building hazard classification 107
Sprinkler head requirement 109
Number of sprinkler heads 109
Sprinkler heads spacing 109
Pressure of sprinkler head 110
Sprinkler piping 110
Piping configuration 111
Branch or lateral pipe 111
Piping material 112
Combined firefighting system 113
Stand pipe size for combined system 114
Designing fire sprinkler system components 114
Determining number of sprinkler heads 114
Sprinkler pipe sizing 114
Velocity in pipe 116
Fire pump sizing 117
Jockey pump sizing 117
Reservoir sizing 118
Drainage system 118
Flushing 118
Installation, testing and commissioning 119

6 Fountain and waterfall **121**

Introduction 121
Technological considerations 121
Structural components of a fountain and waterfall 122
Water fountain 122
Types of water fountains 123
Jumping jet 124
Musical fountain 124
Dancing fountain 124
Indoor fountain 125
Outdoor fountain 126
Dry fountain 127

Floating fountain 127
Water jet fountain 128
Single-jet water display 128
Multi-jet water displays 130
Crystalline jet water display 130
Frothy jet water display 131
Mist jet water display 131
Nozzle 131
Lance jet nozzle 131
Palm tree jet nozzle 131
Fan jet nozzle 132
Finger jet nozzle 132
Crown jet nozzle 133
Bell or mushroom jet nozzle 134
Diffuser sheet nozzle 134
Geyser jet nozzle 134
Foam jet nozzle 136
Mist or spray jet nozzle 136
Fountain development planning 137
Water fall 139
Clear-sheet waterfall 139
Aerated waterfall 142
Cascading 143
Pool or basin 144
Water pool below a water fall 145
Pool basin accessories 145
Pump 146
Piping 147
Filtration system 147
Lighting and illumination 147
Designing fountain elements 148
Sizing a pump 148
Pipe sizing 149
Maintenance of fountain 150

7 Sprinkler irrigation system **151**

Introduction 151
Irrigation 151
Sprinkler irrigation 151
Types of sprinkler irrigation systems 152
Sprinkler irrigation system components 152
Accessories and appurtenances 153
Source of water 153
Pump 154

Centrifugal pump 154
Deep-well turbine pump 155
Propeller pump 156
Submersible pump 157
Selection of pump 158
Piping 158
Irrigation sprinkler head 159
Fixed head sprinkler 160
Rotary head sprinkler 160
Bubbler 161
Spot sprinkler 162
Pop-up sprinkler 162
Sprinkler sizes 163
Sprinklers arrangement pattern 163
Square pattern 165
Triangular pattern 165
Accessories and appurtenances 165
Valve 165
Pressure gauge 166
Quick coupler 167
Water meter 167
Timer controller 167
Drain 167
Material of items 167
Planning sprinkler system 168
Designing sprinkler system components 169
Precipitation rate 170
Application rate 170
Slope of land 170
Nozzle size and pressure 171
Irrigation duration 171
Wind effect 171
Sprinkler head sizing 172
Pipes designing 172
Pump sizing 174
Operation and maintenance of sprinkler system 178
Operation of sprinkler system 178
Maintenance of sprinkler components 178

8 Steam supply system **180**

Introduction 180
Steam 180
Steam condensation 180
Steam dryness 181

Use of steam in building 181
Steam distribution method 182
Steam supply pressure 182
Steam pressure management 184
Steam distribution system components 184
Steam boiler 185
Basic working principle 185
Types of boilers 185
Fire-tube boiler 186
Types of fire-tube boilers 186
Water-tube boiler 187
Types of water-tube boilers 188
Major components of a steam boiler 189
Combustion chamber 189
Heat exchanger 189
Burner 189
Shell 190
Exhaust stack 190
Feed-water tank 190
Flash steam from heat recovery system 191
Steam injection 191
Overflow 191
Feed pump take-off 191
Inspection opening 192
Mountings and accessories of a boiler 192
Water level indicator 192
Steam stop valve 193
Blow-off valve 193
Feed check valve 194
Steam strainer 195
Steam filter 195
Steam trap 196
Steam separator 197
Economizer 197
Air pre-heater 197
Super-heater 198
Safety measures for a boiler 198
Safety valve 199
Low-water cutoff switch 199
Pressure switch 201
Fusible plug 202
Expansion tank 202
Flash tank 202

Pressure range of boilers 204
Low-pressure boiler 204
Medium-pressure boiler 204
High-pressure boiler 204
Steam boiler capacity 204
Velocity of steam flow 205
Pipes for steam supply 205
Pumps in steam supply 206
Feed-water pump 206
Condensate pump 207
Circulation pump 207
Water heating by steam 208
Steam flow requirement for water heating 208
Space heating by steam 209
Comfortable room temperature 210
Major components of steam supply for space heating 211
Steam boiler for space heating 211
Radiator 211
Air vent valve 212
Steam distribution systems for space heating 212
One-pipe system 213
Two-pipe system 214
Steam supply for steam bathing 214
Steam bathing stall 215
Steam generator for bathing 216
Steam shower-head 217
Controller 217
Steam for indoor humidification 218
Steam for sterilization 218
Piping for steam supply 218
Design of major components of a steam supply system 219
Sizing a boiler 219
Sizing feed-water pump 220
Sizing steam radiator 221
Determining steam flow for radiator 222
Sizing feed-water tank 222
Sizing pipes for steam supply 224

9 Gas supply system 227

Introduction 227
Natural fuel gas 227
Liquefied petroleum gas 227

Uses of natural gas 228
Gas use in building 228
Gas supply system 229
Gas supply main 230
Types of gas supply in a building 230
Gas supply system from the main 231
Service connection to gas-main 231
Gas service piping requirements 232
Natural gas supply pressure in building 232
Pressure testing of gas piping 233
Independent gas supply system 233
Gas cylinder 233
Individual cylinder gas supply system 235
Central gas supply system 235
Components of the central gas supply system 235
Cylinder bundle 236
Changeover valve or regulator 237
Bubble tester 237
Gas supply pipe 237
Gas supply pipe installation 237
Gas piping-related issues 238
Gas pipe installation guidelines 238
Requirements of gas pipe shaft or duct 240
Gas flow-controlling devices 240
Stop valve 240
Gas pressure-controlling devices 241
Pressure regulatory valve 241
Pressure-reducing valve 241
Back-pressure regulator 242
Installing pressure-reducing valve 244
Gas meter 244
Types of gas meters 244
Sizing of gas meter 250
Installation of gas meter 251
Gas service line sizing 252
Gas supply pipe sizing 252
Maximum gas demand 253
Gas flow 254
Length of pipe 254
Gas supply pressure loss 256
Allowable pressure drop 256
Diversity factor 257
Change of gas pressure corresponding to height 257
Velocity of gas flow 258

10 Medical gas supply system **260**

Introduction 260
Medical gas 260
Types of medical gases 260
Nitrogen 261
Medical gas demand 262
Medical gas supply system 263
Types of medical gas supply systems 263
Centralized piped medical gas supply and vacuum system 263
Advantages of a centralized medical gas delivery system 263
Components of medical gas supply system 264
Medical oxygen plant 264
Nitrous oxide plant 265
Vacuum plant 266
Medical compressed air plant 267
Gas bank manifold 268
Manifold room 269
Control panel 270
Alarm system 271
Medical gas supply piping 271
Pipe installation 271
Medical gas supply pressure management 272
Medical gas supply pressure regulator 273
Terminal units or outlets or inlets 273
Wall-mounted terminal unis (outlet) 275
Trunking-type terminal unit 275
Vacuum inlet unit 276
Color coding usage of medical gases 276
Sizing of medical gas supply components 276
Medical gas pipe sizing 276
Particular medical gas supply pipe sizing 278
Medical gas supply system testing 283

11 Healthcare plumbing **284**

Introduction 284
Healthcare plumbing 284
Special fixtures and appliances 284
Hand-washing clinical sink 285
Laboratory glass washer 285
Emergency shower 286
Mop-service basin 287
Bedpan washer 287
Floor drain 289

Medical equipment of plumbing concern 290
Dialysis machine 290
Heart-lung machine 292
Electron microscope 293
Still 293
Sterilizer 295
Dental lavatory 296
Special valves and accessories 298
Trap primer 298
Atmospheric vacuum breaker 298
Drain tempering valve 299
Areas where plumbing is installed 300
Patient room 300
Ward room 301
Nurse station 301
Intensive-care room 301
Emergency (triage) room 301
Examination and treatment room 301
Physical therapy treatment room 302
Cystoscopic room 302
Nourishment station 302
Operating room 303
Recovery room 304
Birthing room 304
Anesthesia workroom 304
Fracture room 304
Pathology laboratory 304
Pharmacy and drug room 305
Autopsy room 305
Kitchen 305
Laundry 306
Water supply systems 306
Various qualities of waters supplied 306
Water service connection 306
Water source 307
Water storage 307
Water distribution systems 307
Hot water supply system 308
Chilled water supply system 309
Pure water supply system 309
Pure-water supply piping materials 312
Non-potable water supply system 314
Disinfecting water supply system 314

Drainage system 314
Special drainage system 314
Corrosive-waste handling pipe materials 315
Precautionary measures 316
Acidic waste neutralization 316
Acid-waste metering detail 316
Indirect wastewater drainage 319
Air gap and air break 319
Efficient healthcare plumbing system planning 320
Accessibility 320
Maintainability 320
Redundancy 320
Serviceability 321

12 Swimming pool plumbing 322

Introduction 322
Swimming pool 322
Types and sizes 323
Components of swimming pool 323
Basin of swimming pool 324
Materials of construction 325
Basin construction 326
Skimmer 326
Skimmer number 327
Balance tank 327
Piping and accessories of balance tank 328
Balance tank sizing 329
Pool pump 329
Types of pool pumps 330
Pool water flow 331
Turnover 331
Flow velocity 332
Pump flow 332
Head of pump 332
Power of pump 334
Pool water quality and condition 334
Filtering pool water 335
Types of filters used 336
Six-way valve 338
Sizing a pressure filter 339
Backwashing 340
Heating pool water 340
Types of pool water heaters 340
Sizing a pool heater 341

Disinfecting pool water 343
Inline chlorination 343
Water supply and circulation 344
Water supply 344
Water recirculation 345
Water inlets 345
Multiport valves 345
Swimming pool drainage 346
Drain suction inlet 346
Swimming pool plumbing 347
Swimming pool plumbing principles 347
Piping of swimming pool 348
Swimming pool maintenance 349

Appendices 351
Index 363

Figures

2.1 Basic chilled water cooling system using an air-cooled chiller 8
2.2 Open-type chilled water supply system for space cooling, using a water-cooled chiller 10
2.3 Chilled water supply using an air-cooled chiller for air conditioning 11
2.4 A chiller unit 12
2.5 Functional diagram of an air-cooled chiller 13
2.6 Functional diagram of a water-cooled chiller 14
2.7 A finned-tube condenser 15
2.8 An evaporator 16
2.9 A centrifugal compressor 17
2.10 A scroll compressor 18
2.11 A Twin-screw compressor 19
2.12 A thermal expansion valve or thermostatic expansion valve (TEV or TXV) 20
2.13 A thermal expansion valve with a sensing bulb 21
2.14 A water-cooled cooling tower 22
2.15 A closed-circuit cooling tower 23
2.16 Three-way valves in three operating conditions 25
2.17 A static balancing valve 26
2.18 A buffer tank with its accessories 27
2.19 Buffer tank connection with chillers 28
2.20 A schematic diagram of the constant-flow rate primary chilled water distribution system 30
2.21 A schematic diagram of the primary-secondary chilled water distribution system 31
2.22 A schematic diagram of the variable primary flow chilled water distribution system 31
2.23 Air-cooled chilled water supply systems piping with air vent 33
2.24 A chilled water supply (down-feed) system developed for drinking purposes 33
2.25 A drinking fountain including chiller inside 34
3.1 Schematic diagram of an urban dual water supply system 44
3.2 Down-feed dual water supply system in a building 45
3.3 Example of back-pressure and back-siphonage 47
3.4 A cut section of the reduced pressure principle assembly 48
3.5 A double check (DC) valve assembly 49
3.6 A cross-connection supported by double check valve 51

4.1	Triangle of fire	54
4.2	A schematic diagram of a fire hydrant system	56
4.3	Electric motor-driven end suction fire pump (side view)	60
4.4	Diesel engine-driven horizontal split fire pump (front view)	61
4.5	Anti-vortex plate assemblies	62
4.6	Fire pump driven by a diesel engine (side view)	63
4.7	A jockey pump	64
4.8	Jockey pump connections	65
4.9	Fire pumping including main electric pump, diesel pump and jockey pump and piping connections	66
4.10	Installation of a fire hose connection for a Class-I fire station	70
4.11	Installation of a fire hose reel for a Class-II fire hose station	71
4.12	Installation of a Class-III fire hose station	71
4.13	Various types of landing valves	72
4.14	Various fire hose pipes	73
4.15	Various types of nozzles for firefighting	76
4.16	Various types of fire departments connections (FDC) and the stand pipe sizes	78
4.17	A quick coupling (cut section)	79
4.18	A pressure reducing valve on a pipe	80
4.19	Air valve on top of standpipe	81
4.20	Check valve installed on a pump delivery and fire department connection pipe	82
4.21	Fire flow area of a multistory building	83
5.1	Major components of a fire sprinkler system	89
5.2	Partitioning of a large firewater reservoir	90
5.3	Pressurized tanks for fire-sprinkler water supply	90
5.4	Density-per-area curves for sprinkler heads, based on occupancy hazards	92
5.5	Fire pump performance curve	94
5.6	The combined supply curve	95
5.7	Typical layout of wet pipe sprinkler systems	96
5.8	Alarm check valve for wet-riser	97
5.9	Dry-riser and its valve	98
5.10	Dry-pipe sprinkler valve assembly	98
5.11	The deluge fire sprinkler system	99
5.12	A manually operated deluge sprinkler valve	100
5.13	Sprinkler head, pendent type	102
5.14	Temperature rating of the sprinkler bulb and the corresponding color	104
5.15	An open fire sprinkler head	105
5.16	An upright sprinkler head	106
5.17	A sidewall sprinkler head	106
5.18	The tree-piping configuration	111
5.19	The loop-piping configuration	112
5.20	The grid-piping configuration	112
5.21	Density/Area curves for various hazard classifications	116
6.1	Dancing fountain and its nozzle	125
6.2	An indoor water fall	126
6.3	A floating fountain	127

6.4	Fountain of upright single lance jet nozzles arranged in a circular way	129
6.5	Spout fountain paired with a sculpture	130
6.6	A Palm-tree jet fountain and the nozzle	132
6.7	A Fan-jet fountain and the nozzle	132
6.8	A Finger-jet fountain and the nozzle	133
6.9	A crown spray fountain and the nozzle	133
6.10	A mushroom jet fountain and the nozzle	134
6.11	A water film fountain and a diffuser sheet nozzle	135
6.12	A geyser jet fountain and the nozzle	135
6.13	A foam jet fountain and the nozzle	136
6.14	A mist fountain and the misting-nozzle	137
6.15	A simplest fountain	137
6.16	Fountain having the pump in a concealed space in the basin	138
6.17	Fountain having the pump outside the basin	138
6.18	Fountain pump in an underground chamber outside the basin	139
6.19	Various natures of clear water sheet falls	140
6.20	Clear sheet waterfall: a. Curtain type and b. wall type	141
6.21	Typical aerated waterfall weirs	142
6.22	A cascade waterfall	143
6.23	Fountain splash coverage	144
6.24	Fountain drain including overflow armature	146
7.1	Schematic diagram of a simplified fixed-type sprinkler irrigation system	153
7.2	A Centrifugal pump	154
7.3	A deep well turbine pump	155
7.4	A deep well propeller pump	156
7.5	A deep tube well submersible pump	157
7.6	Fixed spray irrigation sprinkler heads and their coverage	160
7.7	A rotating irrigation sprinkler head	161
7.8	Watering by a bubbler	162
7.9	A pop-up irrigation sprinkler head	162
7.10	Sprinkler heads installed above ground: a. fixed type and b. rotor type	163
7.11	Wetting pattern for a single sprinkler (top view)	164
7.12	Sprinkler arrangement patterns	164
7.13	Layout of irrigation sprinkler heads and their spraying circle	168
7.14	Sprinkler coverage pattern under strong wind	172
7.15	Cumulative flow in different sections of a lateral pipe	174
7.16	Pump elevation and suction head	176
7.17	Friction losses in plastic pipe	177
8.1	A typical steam supply system in a building	183
8.2	Pressure and specific volume relationship for dry saturated steam	184
8.3	Pressure-reducing station of a steam distribution piping	184
8.4	A horizontal fire-tube boiler	187
8.5	A horizontal water-tube boiler	188
8.6	A feed-water tank with major accessories	190
8.7	A water level indicator	193
8.8	A steam stop valve in closed condition	193
8.9	A blow-off valve	194
8.10	A feed check valve	194

8.11 Y-type steam strainer 195
8.12 A steam filter 195
8.13 An inverted bucket steam trap 196
8.14 A steam separator 197
8.15 Tube type air preheater 198
8.16 Tubular steam super-heater 199
8.17 A spring-loaded safety valve 200
8.18 Probe type low water level electrode assembly 200
8.19 A pressure switch 201
8.20 A fusible plug 202
8.21 An expansion tank 203
8.22 A vertical steam flash tank 203
8.23 A floor-mounted steam heat radiator 211
8.24 Schematic of a one-pipe gravity return steam supply system for space heating 213
8.25 A one-pipe radiator with a vent valve on the side 213
8.26 Schematic of a two-pipe steam supply system for space heating 214
8.27 Steam bath enclosure and the system components 215
8.28 Bathing steam generator 216
8.29 Gravity displacement type autoclave 219
8.30 Chart for sizing steam supply pipes for flow velocity 25 mps 225
9.1 Block diagram of a gas distribution pipe line 229
9.2 Building gas service pipe connection 231
9.3 Gas cylinder head assembly 234
9.4 A cylinder bundle for supplying gas 236
9.5 A changeover valve regulator 237
9.6 A gas bubble tester 238
9.7 An electro-pneumatic gas shutoff valve 240
9.8 Single-stage pressure regulator valve 242
9.9 Two-stage pressure regulator valve 243
9.10 A back-pressure regulator 243
9.11 A diaphragm gas flow meter 245
9.12 A rotary gas flow meter 246
9.13 A turbine gas flow meter 246
9.14 An Orifice gas flow meter 247
9.15 An ultrasonic gas flow meter 247
9.16 A thermal mass gas flow meter 248
9.17 A vortex gas flow meter 249
9.18 A Coriolis gas flow meter 249
9.19 Gas meter location and installation indoors 251
9.20 Capacity of steel pipe of Schedule 40 255
10.1 Schematic diagram of an oxygen plant 265
10.2 Schematic diagram of a nitrous oxide plant 265
10.3 Schematic diagram of a central vacuum plant 266
10.4 Schematic diagram of a central compressed medical air plant 267
10.5 Two stage medical oxygen manifolds 269
10.6 Schematic diagram of a manual control panel for cylinder gas cylinders manifold 270

10.7 Medical gas pipe installation 272
10.8 Various types of medical gas outlets 274
11.1 Mixture faucet with wrist-blade handle 285
11.2 A laboratory glass washer 286
11.3 Combined emergency shower and eye wash 287
11.4 Floor-mounted mop service basin 288
11.5 Plumbing for manual bedpan washer 288
11.6 Automatic bedpan washer 289
11.7 Funnel floor drain 290
11.8 A dialysis machine including its major components 292
11.9 A heart-lung machine and its water supply and drainage system 293
11.10 Chilling water recirculating system and a laboratory chiller 294
11.11 The components and process of a water distillation unit 294
11.12 Autoclave drainage incorporating a drain-tempering valve 295
11.13 A dental care area with a supporting plumbing system 296
11.14 A trap-primer connected to a floor-drain 298
11.15 Atmospheric vacuum breakers and their installation 299
11.16 A drain tempering valve (DTV) 300
11.17 A specimen-collector 302
11.18 Hydrotherapy whirlpool immersion baths 303
11.19 A solids interceptor 305
11.20 RO membrane details 311
11.21 Ion-exchange processes 311
11.22 Acid-neutralizing tank installation detail 317
11.23 Acid-waste solids interceptor detail 318
11.24 Sampling manhole 318
11.25 Air break and air-gap 320
12.1 A schematic diagram of a skimmer type swimming pool incorporating a skimmer and other major components 324
12.2 A skimmer on the side wall of a swimming pool 326
12.3 Detail of a pool skimmer 327
12.4 A swimming pool incorporating a balancing tank 328
12.5 A swimming pool pump with strainer 329
12.6 Recirculation flow of a swimming pool 331
12.7 Nomograph for Hazen William's formula, based on C = 100 333
12.8 A pressure sand filter for swimming pools 337
12.9 A multi-grade media gravity water filter 338
12.10 A Six-way valve 339
12.11 Automatic chlorine dispenser (chemical feeder) for a swimming pool 344
12.12 Two-way and a three-way valve 346
12.13 Drain inlets 347
12.14 Schematic diagram of plumbing of a swimming pool incorporating balancing tank 348
12.15 Schematic diagram of plumbing of a swimming pool incorporating skimmer 348

Tables

2.1	Main chilled water distribution pipes: maximum allowable flow, velocity, and pressure drop	38
2.2	Lateral chilled water distribution pipes: maximum allowable flow, velocity, and pressure drop	38
3.1	Drinking water quality standards as recommended by the FEPA and WHO	42
3.2	Non-potable (rain) water quality standards as prescribed by the Uniform Plumbing Code	42
3.3	Safe distance of installing non-potable storage tank from some elements	46
4.1	Water delivery rate for firefighting in a building	58
4.2	Required fire flow and flow duration	59
4.3	The approximate size of a fire hose rack cabinet	74
4.4	Coefficients of friction loss for various hose sizes	75
4.5	The k-factor of different nozzle orifice	77
4.6	Fire department connection number based on targeted fire flow	78
4.7	Sizing of the drain cocks fire hydrant stand pipe	81
4.8	Minimum fire flow based on exposure distance	83
5.1	The k factor for different sprinkler heads	93
5.2	Response Time Index (RTI) value for different categories of sprinkler heads	103
5.3	Temperature rating of the sprinkler bulb and the corresponding color code	104
5.4	Types of occupancies under various classes of fire hazards	108
5.5	Protection area and maximum spacing (Extended coverage for pendent and upright spray sprinklers	110
5.6	Important differences between the most popular piping materials recommended for fire sprinkler systems	113
5.7	Sprinkler pipe sizing based on pipe schedule method	115
5.8	Maximum flows through pipes for EN 12845 fire sprinkler systems	117
5.9	Water flow rate and duration requirements for hydraulically calculated systems	118
5.10	Drain size for flushing	119
7.1	Preferred pump types for sprinkler irrigation, based on range of flow rates and the delivery heads	159
7.2	Sprinkler flow rate with respect to pressure	164
7.3	Suggested maximum application rate for average soil, slope, and tilt	170
7.4	Available water-holding capacity of various types of soil and typical crop rooting depths	171
7.5	Maximum allowable spacing of irrigation sprinklers, based on wind speeds	172

7.6 Pipe sizing table based on velocity limitation for Schedule 40 plastic pipes 173
7.7 Water requirement for various vegetables 175
7.8 Equivalent length (meter) of straight pipe for fittings like bends, returns, tees, and valves by making screw joints 177
8.1 Comparative advantages and disadvantages of fire-tube and water-tube boilers 189
8.2 Steam table (part). Properties of steam at varying pressures and temperatures 209
8.3 The comfortable room temperature in different rooms of particular use and the corresponding wattage required 210
8.4 Steam generator rating for various volume of steam bathing enclosure 217
8.5 Sizing steam pipes based on velocity 224
9.1 Amount of gas content in a 20-liter cylinder at various pressures 234
9.2 Sizes of different diaphragm gas meters for various gas flows 250
9.3 Sizes of gas service connection plastic pipe based on meter capacity and length of service pipe from the connection on main to the meter, operating at various pressure level and allowable pressure drop 253
9.4 Approximate natural gas input for typical appliances 254
9.5 Capacity of Natural Gas Supplying Pipe (MBH ≈ CFH) for different sizes of pipes of Steel Pipe - Schedule 40 255
9.6 Allowable pressure drop for various supply pressures 256
9.7 Multi-family dwellings diversity factor in percent of total load for domestic gas ranges with connected load ratings of 1.78 cum/h or more per range 257
9.8 Increase of pressure with altitude (measured from sea level) 257
9.9 Specific gravity and other properties of fuel gases 258
10.1 Maximum intervals between supports for gas pipes 272
10.2 Color coding of various medical gases 276
10.3 Diversity factor for probable flow of gas in a pipe supplying multiple pipes leading to terminal outlets or inlets 277
10.4 Maximum permitted pressure loss in medical gas and medical vacuum systems 278
10.5 Size of medical gases and vacuum piping 280
10.6 Medical gas pipe sizing for supplying pressure at 4 bar, based on pipe length and flow rate at various pressure loss 281
10.7 Medical gas pipe sizing for supplying pressure at 7 bar, based on pipe length and flow rate at various pressure loss 282
11.1 AAMI water standards for preparing dialysate for dialysis 291
11.2 The daily water consumption for various types of healthcare facility with respect to number of beds 307
11.3 Acidic-waste neutralization tank sizing 317
12.1 Sizes of various types of pools 323
12.2 Skimmer sizing, based on surface area of pool 327
12.3 Turnover rate (hours) for different types of pools 332
12.4 Head loss for flowing water through various elements of swimming pool plumbing 334
12.5 Recommended chemical concentrations in swimming pool water 335
12.6 Pressure sand filter dimension 339
A1.1 Technical specifications of domestic type air-cooled water chiller 351

A2.1 Pipe or tube materials and dimensions 352
A3.1 Friction loss in meter-head, per 30-meter fire hose 352
A3.2 Solid stream flow through nozzles 353
A4.1 Water supply requirements for sprinkler systems 353
A4.2 Pressure losses in water meters (bar) 353
A5.1 Fuel and gas properties 354
A6.1 Display height, flow rate and pressure for geyser nozzles of various diameter 355
A7.1 Properties of saturated steam (part-1) 356
A7.2 Properties of saturated steam (part-2) 357
A8.1 Optimum pump type for specific flow rate and head applications 357
A9.1 Medical fixtures water demand and corresponding fixture unit 358
A10.1 Chlorine dose in fluid ounce, to reach desired chlorine level in water 358
A10.2 Super chlorination ounce (Amount needed to introduce 10 ppm) 359
A11.1 Conversion factors, mixed units 359

Foreword 1

It is with great honor and deep appreciation that I write this foreword for *Specialized Plumbing Technologies*, an authoritative and timely contribution to the field of modern building services engineering by Mr. Syed Azizul Haq, PEng. Having served as the Director of the Engineering Staff College, Institution of Engineers, Bangladesh (IEB), I have witnessed first-hand the growing complexity and demands of plumbing systems in today's built environment. This book arrives as a much-needed resource for engineers, architects, consultants, and practitioners navigating this evolving landscape.

Mr. Syed Azizul Haq, PEng, a respected author of *Plumbing Principles and Practice* (Taylor and Francis, UK), has long been a dedicated educator and technical expert in this field. As a regular trainer at the Engineering Staff College, he has enriched the knowledge of countless engineers and architects with his practical insights and clear instruction. In *Specialized Plumbing Technologies*, he goes further to address some of the most critical and specialized systems shaping modern infrastructure—ranging from chilled water and dual water supply systems to fire hydrants, sprinkler irrigation, medical gas supply systems, and the intricacies of healthcare and swimming pool plumbing.

Each chapter reflects his deep technical understanding and practical experience, bridging theory with application. Of particular importance are the chapters on firefighting systems, medical gas supply, and water features—topics increasingly essential in public health, safety, and sustainable design. The inclusion of both traditional and emerging systems offers a comprehensive reference that is as educational as it is practical.

I believe this work will serve not only as a valuable academic text but also as a field guide for professionals striving for excellence in plumbing design and implementation. It is a commendable effort by Mr. Syed Azizul Haq, PEng, to elevate the standards of engineering education and practice in Bangladesh and beyond.

I congratulate the author for this significant achievement and commend this book to all who seek to advance their expertise in this critical domain of engineering.

Professor Abdul Hannan
Former Director
Engineering Staff College
Institution of Engineers, Bangladesh (IEB)
Dhaka.

Foreword 2

It is my great pleasure to write this foreword for the book *Specialized Plumbing Technologies*, authored by Syed Azizul Haq, PEng, and published by Taylor & Francis, UK. In an era when building services engineering is rapidly evolving in response to technological advancements, sustainability imperatives, and public health concerns, this book arrives at a particularly opportune time.

Syed Azizul Haq is a seasoned engineer whose professional career reflects both depth of experience and a sustained commitment to excellence. Having served the Public Works Department (PWD) of Bangladesh from 1981 to 2016, retiring as an Additional Chief Engineer, he has been closely involved in the planning, design, implementation, and supervision of a wide range of building projects across the country. His long and distinguished public service has provided him with rare practical insight into plumbing systems as they operate under real-world conditions—insight that is clearly reflected throughout his writing.

The author is already well known in professional and academic circles for his authoritative publications, including *Plumbing Practices* (2006), *Harvesting Rainwater from Buildings* (Springer), *Plumbing Principles and Practice* (2nd Edition) (Taylor & Francis), and *Water Supply and Drainage for Buildings* (Nova, USA). These works have served as valuable references for engineers, architects, planners, and students alike. With *Specialized Plumbing Technologies*, Engr. Haq advances further by addressing contemporary and specialized aspects of plumbing that are increasingly relevant to modern buildings and infrastructure.

What distinguishes this book is its balanced and thoughtful approach—combining sound engineering principles with modern technologies, prevailing codes, and recognized best practices. The discussions are practical, systematic, and firmly grounded in the author's extensive field experience, while remaining aligned with international standards. As such, the book is expected to be highly useful not only to practicing engineers and consultants, but also to academics, researchers, and postgraduate students seeking a deeper understanding of advanced plumbing systems.

As a Chartered Engineer (CEng) and Fellow of the Institution of Civil Engineers (FICE), UK, and a Fellow of the American Society of Civil Engineers (FASCE), and as the Managing Director of Design Planning and Management Consultants Ltd., Dhaka, Bangladesh, I have observed first-hand the growing importance of specialized plumbing technologies in achieving sustainability, efficiency, and public health protection in the built environment. I am confident that this book will make a meaningful contribution to the profession and will be well received by the global engineering community.

I commend Syed Azizul Haq for his continued commitment to knowledge dissemination and professional development, and I am pleased to recommend *Specialized Plumbing Technologies* as a valuable addition to the technical literature in the field of building services engineering.

M. A. Sobhan, PEng, CEng, FICE, FASCE
Managing Director, Design Planning and Management Consultants Ltd.

Preface

Plumbing—the lifeline of a building—is commonly misunderstood by bounding its domain within the development of water supply—generally normal water, occasionally hot water, and drainage system in a building. However, those devoted to mastering this subject must go further, as they will encounter the diverse nature of building services involving specialized fluid flows for specific purposes—a domain known as specialized plumbing. Having such understanding, it is realized by heart that knowledge of plumbing would remain ever-partial unless the knowledge in specialized plumbing is dealt in.

After having primary knowledge of plumbing, during my professional life I encountered several jobs that, in many respects, appeared special to me in terms of technology know-how, for which I had to go for reviewing literature related to those subjects of new interest. In the majority of cases, I had to face difficulties in finding that knowledge of specialized plumbing in one, or rather, found sporadic in books on to-some-extent related subjects.

Based on understanding that like me many of this profession are also having the same difficulties in having diverse dimensional knowledge of special plumbing in particular plumbing books, I—after authoring books focusing basics of plumbing practice—planned to write another book on plumbing discussing exclusively the rarely discussed subjects of special plumbing. So, this book is an endeavor to fulfill this objective which is harnessed with following major topics identified as specialized plumbing.

To make a comprehensive book on the special technologies of plumbing, some special topics related to the services of buildings and beyond have been included in this book, which are rarely found accumulated under a cover-title of one book of concerned subjects. Among all the aspects of plumbing practice, only the special technologies are selected as the content of this book. Finally, this book has been furnished with the following 12 chapters.

Chapter 1—Introduction: This chapter predominantly introduces all the chapters that the book explains in the following chapters. It also provides a brief overview of the specialties of the plumbing technologies that are being introduced. At the end, the challenges of developing specialized plumbing systems have been underscored.

Chapter 2—Chilled Water Supply System: Generally, in hot climates, buildings' indoor environments are tried to keep cool for their inhabitants, and in some cases provisions for chilled water are made available for drinking purposes inside a building. For these purposes, a chilled water supply system is developed in the buildings. For these system chillers, pumps, piping, tanks, valves, etc. are used. So, a description of all the components and their sizing, installation maintenance, etc. have been discussed in this chapter.

Chapter 3—Dual Water Supply System: In areas where there is a crisis of having sufficient potable water to meet the total demand of water, in some cases the use of potable water can be replaced by the use of non-potable water. In this situation, a dual water supply system incorporating potable and non-potable water is to be developed. This chapter is written to discuss all the methodologies of planning and design to develop this system, along with the various issues to be addressed in the development.

Chapter 4—Fire Standpipe System: Buildings are supposed to be prone to being caught by unwanted fire that causes loss of life and property. Suppressing of such unwanted fire in time is the prime concern of safety of the building, for which various measures are taken; the development of fire standpipe is one of the most dependable systems to be considered in this regard. This chapter is dedicated to discussing all the concerns of this firefighting system in a building.

Chapter 5—Fire Sprinkler System: The firefighting systems using water as an extinguishing agent may be automatic or manually operated. A fire sprinkler system is a very common automatic firefighting system using water as an extinguishing agent. This chapter elaborately discusses all the technical aspects of developing, operating, and maintaining the fire sprinkler system.

Chapter 6—Fountain and Waterfall: Water is not only for the survival of life; it is also for beautifying and amusing life. Fountains and waterfalls are the water features designed and developed inside or outside of buildings for these purposes. Knowledge of developing water fountains and waterfalls fall under the purview of two major fields of science and technology: architecture and hydro-electro-mechanical. This chapter is designed to discuss all the hydro-electro-mechanical aspects of water fountain and waterfall development and maintenance as well.

Chapter 7—Sprinkler Irrigation System: Watering for irrigation can be done in many ways, both manually and automatically. Automatic irrigation systems can be developed in many ways; of them, the sprinkler irrigation system is considered to be the most effective one. Special hydro-electro-mechanical knowhow is to be applied in developing this irrigation system. This chapter is therefore harnessed with the topics related to the development of an efficient sprinkler irrigation system.

Chapter 8—Steam Supply System: Not the liquid state of water is the only fluid needed in a building, but steam is also needed in many building occupancies. Steam is used for various needs in those buildings for which a different supply system is to be developed side by side of a water supply system. It also needs generation of steam in the same building. Steam supply systems need special knowledge of fluid mechanics and knowhow of supply methodology for their efficient performance and safety as well. This chapter is articulated with major aspects of a steam supply system in a building.

Chapter 9—Gas Supply System: There is a need for another gaseous fluid that is fuel gas. So, a supply system of fuel gas is needed in a building when the requirement of this gas exists at different multiple locations of a grand building, particularly in high-rise buildings. For this purpose, a supply system is to be developed incorporating pipes and pressure-developing mechanisms. Based on the properties of gases, various precautionary and safety measures are incorporated in the system. Including all these and other concerned topics, this chapter has been organized.

Chapter 10—Medical Gas Supply System: In a healthcare establishment, particular types of gases are required for several reasons, primarily to support patient care, surgeries, and medical procedures. Special technology and measures are adopted for supplying these gases from the source to the points of use, mostly for patients or for their treatment process. With utmost care and accuracy, these gases are supplied, maintaining accurate flow, pressure, and purity. Special plumbing is therefore needed to be developed. This chapter deals with those topics related to the supply of all types of medical gases in a healthcare establishment.

Chapter 11—Healthcare Plumbing: In the ever-developing world of healthcare services, in addition to efficient and easy-to-maintain plumbing systems, more reliable, flexible, and disease-spreading protected plumbing systems are required, for which acquiring additional knowledge in this regard is the key to supporting sustainable healthcare services. With a view to achieving such an objective, this chapter has been purposefully included in this book.

Chapter 12—Swimming Pool Plumbing: People bound to remain in buildings desire more relaxation, movement, and amusement in this midst. So, they opt for having a bath, swimming, or playing in a pond-like, hydro-mechanized structured water body, generally a swimming pool inside or outside the building. In developing a swimming pool, three major aspects of technology are involved: architectural, structural, and hydro-electro-mechanical. In this chapter, the hydro-electro-mechanical components of swimming pools have been discussed elaborately.

Last but not least, I am confident that this book will be favored not only by students, professionals, academia, and researchers but also by building developers, traders, manufacturers, and policymakers.

Syed Azizul Haq, PEng

Acknowledgment

In the name of Allah (SWT), the Most Gracious, the Most Merciful. I begin by humbly acknowledging the boundless blessings of my Lord, Allah (SWT), who granted me the strength, guidance, and opportunity to write this book and ultimately bring it to publication for a global audience.

I am deeply indebted to my beloved family for their patience, prayers, and sacrifices. Their constant supplications to Almighty Allah (SWT) for my success in this life and the Hereafter and for enabling me to perform righteous deeds have been my greatest source of strength. Writing this book with the intention of serving humanity is, I hope, a small endeavor toward *Sadaqah jariyah*—a continuing charity.

I also wish to sincerely thank my peers, students, and trainees, whose encouragement and curiosity constantly inspired me to embark on new writings. Many of them would often ask, "What will your next book be?"—a question that motivated me to take on the challenge of addressing specialized plumbing issues rarely found in existing literatures under one cover.

I also wish to acknowledge with gratitude the support of **Taylor & Francis, UK**, for accepting this manuscript for publication and for their professional guidance throughout the editorial and production process. Their efforts have been instrumental in bringing this book to a wider readership.

Last but not the least, I extend my apologies to my family members and relatives for the time and attention I was unable to give them while absorbed in this project. I am also grateful to the many individuals who, in ways big and small, supported me with kind words, encouragement, and practical help along this long journey. Though I cannot name each one here, their contributions are deeply appreciated.

Finally, I warmly invite readers to share their thoughts, corrections, comments, and suggestions with me via email at azizulhaqsyed@yahoo.com. Their feedback will, Insha'Allah, help improve and refine future editions of this book.

Syed Azizul Haq, PEng

Abbreviations

Ω-cm	Ohm-centimeter
°C	Degree centigrade
μS/cm	Micro-Siemens per centimeter
AAMI	Association for the advancement of medical instrumentation
ABS	Acrylonitrile butadiene styrene
accm	Actual cubic centimeter per minute
AFF	Above floor finish
AGSS	Anesthetic gas scavenging system
AHU	Air handling unit
ANSI	American national standards institute
AS	Australian standard
ASME	American society of mechanical engineers
ASPE	American society of plumbing engineers
ASTM	American society of testing material
Bar g	Bar gauge pressure
BHP	Brake horsepower, also Boiler horse power
BOD	Biological oxygen demand
BS	British standards
BS EN	British standards European norm
Cal/gram°C	Calorie per gram per celcius
CBT	Chilled-water buffer tank
CFC	Chlorinated fluorocarbon
cfu	Colony-forming unit
CFU/ml	Colony-forming unit per milliliter
CGS	Central gas supply
cm	Centimeter
CNG	Compressed natural gas
CO_2	Carbon dioxide gas
CP	Circulation pump
CPVC	Chlorinated Poly Vinyl Chloride,
CSST	Corrugated stainless steel tubing
CT	Computed Tomography
cum/h	Cubic meter per hour
cum/sqm/h	Cubic meter per square meter per hour
cum/s	Cubic meter per second

D	Dimensional
DC	Direct current, Double check
DCVA	Double check valve assembly
DE	Diatoms or diatomaceous earth
deg	Degree
DI	Deionized
DN	Diameter nominal
DPCV	Differential pressure control valves
DRV	Double regulating valve
DTV	Drain tempering valve
EC	Electrical conductivity
EN	European norm
ESFR	Early suppression fast response
EU	Endotoxin Unit
FCU	Fan coil unit
FD	Fire department
FDC	Fire department connection
FEPA	Food and environment protection act
FM	Factory mutual, Fire marshal
FRP	Fiber reinforced plastics
GWP	Global warming potential
H+	Hydrogen ions
HCFC	Hydro-chlorofluorocarbon
HCU	Heater-cooler unit
HD	Hemodialysis
HDPE	High-density poly ethylene
HDT	Heat distortion temperature
HFC	Hydrofluorocarbon
hfg	Latent Heat of evaporation
Hg	Mercury
Hp	Horsepower
HOCl	Hypochlorous acid
ICU	Intensive Care Unit
ID	Internal diameter
IPC	International plumbing code
J/(mol·K)	Joule per mole kelvin
K	Kelvin
kcal/hr	Kilocalorie per hour
kcal/hr	Kilocalorie per hour
Kcal/kg/°C	Kilocalorie per kilogram per degree centigrade
kg/cm	Kilogram per centimeter
kg/cum	Kilogram per cubic meter
kg/k/mole	kilogram per Kelvin per mole
kg/mole	kilogram per mole
kg/sq.cm	Kilogram per square centimeter
kg/sqcm	Kilogram per square centimeter
kg/sqcm g	Kilogram per square centimeter gauge pressure
kJ/kg	Kilo Joule per kilogram

kJ/kg °C	Kilo Joule per kilogram centigrade
kJ/kg/K	Kilojoule per Kilogram per Kelvin
kJ/kgK	Kilojoules per kilogram per kelvin
kJ/s	Kilojoule per second
km/hr	Kilometer per hour
kN/sqm	Kilo Newton per square meter
kPa	Kilo Pascal
kW	Kilowatt
l/d/b	liter per day per bed
LED	Light Emitting Diode
LPG	Liquefied petroleum gas
LNG	Liquefied natural gas
Log	Logarithm
lpd	Liter per day
lph	Liter per hour
lpm	Liter per minute
lpm/acre	Liter per minute per acre
lpm/sqm	Liter per minute per square meter
lpm/√bar	Liters per minute per square root of bar
lps	Liter per second
LWCO	Low water cutoff
m	Meter
MΩ	Mega Ohm
MΩ-cm	Mega Ohm-centimeter
μS/cm	Micro Siemens per centimeter
MDPE	Medium-density poly Ethylene
m/s^2	Meter per second squared or Meter per square second
mg/l	Milligram per liter
mg/l	Milligram per liter
MBTU	Million British thermal units
MJ/cum	Megajoule per cubic meter
ml	Milliliter
mm/h	Millimeter per hour
mlps	Milliliter per second
mmHg	Millimeter mercury
MP	Melting point
mPa	megapascal
mps	Meter per second
MRI	Magnetic resonance imaging
MS	Mild steel
mV	millivolts
N/sqm	Newton per square meter
N_2	Nitrogen gas
N_2O	Nitrous oxide gas
NaCl	Sodium chloride
NFPA	National fire protection association
NICU	Neonatal intensive care unit
NOWL	Normal operating water level

NPSH	Net positive suction head
NPT	National pipe thread (American standard)
NTU	Nephelometric turbidity units
O_2	Oxygen gas
OCl^-	Hypochlorite
ODP	Ozone depletion potential
OH−	Hydroxide ions
OPSO	Over pressure shut off
OR	Operating room
OS&Y	Outside screw and yoke
PE	Poly ethylene
PEX	Cross-linked polyethylene
pH	Potential of hydrogen
PICU	Pediatric Intensive Care Unit
PLC	Programmable logic controller
PMGV	Piped medical gas and vacuum
PMP	Pressure motive pump
PP	Poly Propelling
ppm	Parts per million
PSA	Pressure swing adsorption
PVC	Poly Vinyl Chloride
PVDF	Polyvinylidene fluoride
PVR	Precision vacuum regulator
RCC	Reinforced cement concrete
REDOX	Oxidation-reduction
RH	Relative humidity
RPM	Revolution per minute
RO	Reverse osmosis
RP	Reduced pressure
RPZD	Reduced pressure zone device
RTI	Response time index
sccm	Standard cubic centimeter per minute
scmh	Standard cubic meter of gas per hour
SDR	Standard dimension ratio
SI unit	System international unit
sqm	Square meter
SS	Stainless steel
TDH	Total dynamic head
TDS	Total dissolved solids
TEV	Thermostatic expansion valve (TXV)
TOH	Total operating head
TVC	Total viable count
UFC/l	Unité formant colonie par liter of water
UL	Underwriters laboratories
UP	Uniform plumbing
UPSO	Under pressure shut off
USA	United States of America
uPVC	Unplasicized poly vinyl chloride

UV	Ultraviolet
VPF	Variable primary flow
WAGD	Waste anesthetic gas disposal
wc	Water column
WHO	World Health Organization

1 Introduction

Introduction

Plumbing is one of the building services that deals with the supply of water predominantly and gas where needed, along with drainage of various types of wastewaters and storm water, in a building and its premise. In a building, supply of water might be needed in various conditions for various purposes in various building occupancies having special equipment, appliances, etc. The system of supplying water would therefore differ in planning, designing, and maintenance depending upon the nature of use and system of supply, in comparison to normal water supply systems generally developed in the buildings. So, plumbing needed for these special supply systems of water and gas, and the drainage for special types of wastewaters other than the common nature of wastewater generally discharged from the drains of common buildings, may be termed as specialized plumbing. In this book, each chapter discusses the special nature of plumbing for different types and nature of water and gas supply along with drainage systems for the special nature of various types of wastewaters.

Specialized plumbing system

Theoretically, plumbing deals with fluid flow in a building. Fluid is that which can flow. In a building the fluids that are mostly used are water and fuel gas. So, plumbing technology focuses generally and mostly on the technologies concerned with water supply, hot water supply, sanitary, and storm drainage system development. In addition to common and normal use of water and waste disposal in a building, there are various other uses of water for various special needs inside and outside of a building, for which special water supply and drainage system in terms of safety and security might be required. The plumbing system to be developed for such special needs of water along with the necessary drainage system might be termed as specialized plumbing technology. Specialized plumbing technology includes the plumbing system for the following uses of water and gas:

1 Chilled water
2 Non-potable water
3 Fire suppression
4 Irrigation
5 Amusement
6 Swimming
7 Health care

DOI: 10.1201/9781003724056-1

8 Steam
9 Fuel gas
10 Medical gas

Chilled water

Chilled water refers to water maintained at a low temperature—typically between 5 °C and 15 °C—which is above its freezing point but below typical indoor ambient temperatures. It is generally used for cooling indoor environments. Potable chilled water is used for drinking.

Water is chilled by using a device called chiller which is fed normal water to produce chilled water. From the chiller chilled water is supplied to various locations, as desired, through a piping network including some controlling system. The technology of the chilled water supply system for cooling indoor environments, and the supply of potable chilled water for drinking, falls under the purview of plumbing technology. Chilled water supply systems are generally installed in buildings of developed areas that experience very hot climates.

The specialty of chilled water supply systems as compared to normal water supply systems lies in its purpose of use, temperature maintained, and area of usage, particularly in building systems and industrial applications.

Chilled water supply piping follows a closed-loop system and requires insulation to prevent heat gain and condensation; it uses special equipment like cooling towers, air handling units, fan coils, drinking fountains, etc. The system requires careful monitoring of temperature and pressure; and needs to manage energy loss in compressors and circulation systems.

Non-potable water

Non-potable water is water that is not suitable for drinking, cooking, or bathing, but is not so objectionable in appearance, use, or odor as to prevent its use for other non-potable purposes. Non-potable water is generally used for flushing toilets, irrigation, and cleaning of floors, yards, vehicles, etc.

Where there is limited availability of potable water or cost is too high to fulfill the total potable water need, but where non-potable water serves the purpose, the recycled wastewater known as gray water or rainwater is used. In this case a dual water supply system is developed: one pipe network is for potable water supply and the other one is for non-potable gray or rainwater supply.

Adoption of dual water supply greatly promotes water conservation and thereby retards imbalance in available water resources. It is more eco-friendly and supports sustainable water management. So, this system is primarily demanded in green smart buildings and cities.

Dual water supply system involves comparatively higher initial installation cost but saves money in the long term. In a dual water supply system some precautionary measures are taken to prevent contamination of potable water by the adjacent non-potable water supply piping.

Fire suppression

Water plays very effective role in fire suppression or as a fighting agent in major fire incidences in a building. To throw sufficient water at every corner of a building susceptible to fire occurrence, a fire hydrant or sprinkler system, or a combination thereof is developed in that building. In fire hydrant system a riser pipe is fitted with hose station boxes in every floor, which is supported by a pump. Hose station boxes house hose reel of desired length and size having a nozzle at its end. The pump draws water from a reservoir to supply it through the standpipes and particular hose pipe when activated by opening a landing valve on it, to cause flow of water through the hosepipe, and at the end of it, a water jet flow through a nozzle to throw water on a distant fire.

In another system known as a sprinkler system, the stand or hydrant pipe supplies water to a pipe network on which sprinkler heads are installed that are activated automatically to spray water when a fuse inside bursts out due to a rise in temperature caused by any fire under it.

The specialty of developing a water supply system for firefighting is all about its emergency readiness, high pressure and flow, reliability during fire events, and the coverage of all the areas of a building under the system. The system is dedicated to fire protection only, using basically a centrifugal pump but having the special feature of auto-starting when pressure drops in the system. Sufficient water is kept stored only for firefighting for a considerable time. The fire pump requires frequent testing, as it remains mostly non-driven.

Irrigation

Buildings might have some built green spaces inside the building and landscape outside. These areas are in need of water for irrigation for the nursing of plants, herbs, or even grass. For this purpose an automatic or manual irrigation system is developed. In this irrigation system a pipe network is developed to supply water that is thrown in various forms of spraying using special nozzles, either moving or stationary, placed at suitable locations to cover the designated area to be irrigated. Pump is needed to draw water from a source and to deliver through the pipe network.

The specialty of the water supply system in sprinkler irrigation compared to the regular water supply system in an area lies particularly in its design, purpose, operation, etc. In a water supply system in sprinkler irrigation, it is to be ensured that even distribution of water over crops is more or less uniform, like rainfall. This system uses special sprinkler heads or nozzles at the pipe ends, which break water into small droplets and spray over an area covered according to the size and pressure designed. This system may require filters to prevent nozzle clogging, especially if using water from open sources like canals or ponds.

Amusement

We also use water for amusement purposes. In a building or on its premises, various water structures are built primarily for the amusement of the onlookers to the various forms of dynamic water features in the state of flowing, falling, or spreading. Water features are created in different types of structures and so named differently as water fountains, waterfalls, water cascades, sprinklers, etc. For all these water structures having particular water features, a supply of water is needed through pipes and thereby requires the provision of proper drainage in those water structures. To maintain water quality, inclusion of a filtration system becomes essential. For making the special feature of displaying water, a special pumping arrangement is to be designed, and additional special measures are to be included in the supply system.

The specialty of a water supply system for a fountain or waterfall is that the system must be a recirculating one. Initially water is filled in a basin, drawing from a water source, and then the supply is made self-contained through recirculation. Fountains or falls use special nozzles or outlets at the supply pipe end, according to the feature of displaying water. To maintain desired water quality, the system includes filters, UV systems, or chemicals, particularly to keep water clear and algae-free.

Steam

In a building, steam supply may be required for its different purposes of uses like space heating, steam bathing, etc., which need the production of steam by heating water. Steam, being a gaseous form of water and too hot, is required in healthcare facilities for disinfection or sterilization of various items used and for generating hot water. Steam supply needs a special plumbing system incorporating some special control devices and accessories for an efficient steam supply system and so falls in the group of special plumbing systems.

The specialty of a steam supply plumbing system lies mostly in its design, purpose of use, and materials tailored to safely and efficiently handle high temperatures and pressures, as this system operates at much higher pressures and temperatures. So all the elements, including pipes, valves, appurtenances, and their joints, must be rated for these extremes. All component materials must withstand thermal expansion, corrosion, and pressure stress. Pipes are properly insulated to prevent heat loss and protect personnel from burns. Ideal slope in pipes is important to allow proper drainage of condensate. Steam supply systems include pressure relief valves, check valves, and emergency shut-offs to prevent accidents.

Fuel gas

Buildings that are generally used for dwelling need gas for cooking, hot water generation, space heating, cleaning laundry services, etc. In healthcare facilities, fuel gas is not required only for kitchen areas; it is also required in other areas like boiler rooms, water heaters, incinerators, etc. Those are special requirements in such establishments. Fuel gas being prone to fire, its supply system involves special requirements compared to other non-flammable gases.

The specialty in fuel gas supply system involves the design, installation, maintenance, and giving additional importance of careful inspection of the systems, particularly in leak detection and pressure maintenance, using special detectors to ensure system integrity. A special quality of having deep understanding of fuel gas properties and pressure requirements is required for the plumbers.

Medical gas

Medical gases are those required for life saving and relieving from ailment particularly from breathing problems. So, a supply system for medical gases is required in healthcare facilities of large capacity in terms of bed number and types of service or treatment provided. Medical gas supply system is to be developed with utmost preciseness in all respects, as it is concerned with the life and death of the patients.

The medical gas supply system can be characterized with several specialties that distinguish it from other gas supply systems. These specialties are mainly due to the critical nature of its use in healthcare facilities only where ensuring utmost safety, purity, and reliability standards is the prime concern.

Medical gases must meet pharmaceutical-grade purity standards. Prevention of contaminants, filtration and monitoring systems are more stringent. So, regular testing, maintenance and certification are mandatory.

Medical gas supply system must be designed with redundancy to ensure uninterrupted supply, particularly during maintenance or even while system fails. So the system requires alarm systems, manifold switchover units, emergency backup sources. Piping should be copper-based to avoid corrosion and bacteria growth, and must include safety features like color-coded piping, non-interchangeable connectors, and pressure regulators. At the ends there are terminals installed at patient bedsides, operating rooms, ICUs, etc., in a well-organized and accessible manner.

Health care

Healthcare establishments are very much concerned with reviving the good health of the patient and maintaining the good health of the service providers. For various types of special healthcare services, there might be a need for water of various conditions, qualities, and states to satisfy the special needs of the respective services. Additionally, a special type of drainage system would be required for the safety of both the patient and their caretakers. So, healthcare services need a special plumbing system, somewhat special in comparison to the common plumbing system.

Healthcare plumbing systems are quite unique often, compared to common plumbing systems developed generally in residential or commercial buildings, due to the sensitive and high-stakes environment they support.

Healthcare plumbing systems primarily aim at infection control by preventing growth and spread of harmful pathogens, avoiding water stagnation, preventing backflow of contaminated water supplies, etc. Healthcare plumbing systems include additional purified water supply systems like RO, deionized water, etc. There is also a need for a strictly regulated hot water supply for ensuring patient safety and hygiene.

Plumbing systems for any water or gas supply must incorporate backup systems and looped circuits for ensuring uninterrupted service. Plumbing materials shall be of high grade in corrosion resistance and antimicrobial growth. Special faucets and fixtures like hands-free faucets and laminar flow devices are used to avoid aerosol spread and contamination.

A special drainage system is required for bio-hazardous waste that needs special treatment before draining into a common drainage system. Healthcare plumbing may need to incorporate a vacuum system, particularly for surgical suction, dental clinics, etc.

Swimming pool

Ancient people used to take baths in ponds, lakes, and other surface water bodies, of which ponds were the smallest in size. With the development of building structures, ponds started disappearing, replicas of which started appearing in the form of a water pool constructed inside or outside of the buildings, termed as swimming pools. Large swimming pools are constructed as containers of water to swim in. Water from various sources is filled in, and there is provision for draining water from the pool. All these piping systems for supplying water and draining wastewater fall under the purview of plumbing.

The specialty of swimming pool plumbing compared to general plumbing for buildings lies in its unique requirements for water circulation, maintaining balanced hydraulic flow through filtration, chemical disinfection, and temperature management.

Pool plumbing is designed based on recirculating and conditioning the same water to keep it clean, comfortable, and safe for swimming, incorporating special pumps, filters, water heaters, and disinfectors. So, pipes, valves, and fittings must be of chemical-resistant material. Pool plumbing is often buried and pressurized, requiring a specialized leak detection system.

Specialties of specialized plumbing

Plumbing for water supply and drainage systems is considered to be well taught and trained; furthermore, information related to these services are available profusely, both in print and electronic media. So, all the concerned people are very much acquainted with all the related problems and solutions. But, in the cases of special plumbing, related information are to be collected from different sources, hardly found in one domain of particular major subjects. The special information that is very much pertinent for developing specialized plumbing includes the following points:

1 Physical and chemical properties of the fluids to be handled
2 Properties and quality of wastewater to be generated and the precautionary measures to be taken therefore
3 Limitations regarding pressure and velocity of flow to be maintained
4 Information regarding pipes of special materials, strength and jointing system, where demanded
5 Information regarding pros and cons, sizing, installation, safety, health and environmental issues, space requirement, etc., of special accessories and appurtenances to be installed
6 Guidelines for operation and maintenance of the system

Challenges of specialized plumbing

Development of special plumbing may not be as easy as the common plumbing practice of water supply and drainage for buildings. As the name implies, special plumbing needs special preparation, attention, and effort for its development, subject to availability of special plumbing items, experienced professionals, and skilled manpower. So, while developing a specialized plumbing system some challenges to be faced are as follows:

1 In underdeveloped areas special plumbing items, experienced professionals, and skilled manpower may not be available; if available those may be too costly.
2 Operation, repair, and maintenance cost may be comparatively too high.
3 Vendors offer consultancy services which may raise the cost of the system, making it higher, due to incorporation of some items that could be avoided.
4 Imported special items for a particular project may cause delay in project completion, increase in cost due to breakage, missing, or non-functioning of the items.
5 Absence of local codes for development of those special plumbing works and effective regulatory authority to monitor and approve the work done properly.

2 Chilled water supply system

Introduction

For comfortable living in a building built in hot climate area, it may need to dehumidify and keep the indoor environment cool. So, the building's indoor temperature might need to be brought down to a comfortable level. There are various ways of cooling the indoor environment; among them, mostly air conditioning is done, and the incorporation of ventilation systems for this purpose also helps bring down the temperature to some extent. Indoor air temperature can also be brought down by a system incorporating chilled water supply, which falls under the subject of plumbing and in the group of specialized plumbing technology. A chilled water supply is also made for various other needs in a building. For drinking purposes, chilled water supply systems are also developed in buildings.

In this chapter, only the chilled water supply system developed for space cooling and drinking purposes in a building is discussed.

Chilled water

Chilled water is colder water of very low temperature, but above the freezing temperature of water and below the room temperature. So, the temperature range for chilled water can be considered between 5°C and 15°C.

The chilled water is normally created at a particular location of the building for its use and consumption at various locations, although it may be supplied by a private vendor or a public utility service.

Use of chilled water

Chilled water is commonly used in buildings to provide cooling for indoor spaces. Chilled water is sometimes supplied to drinking fountains to draw water for drinking purposes.

For cooling a confined space, chilled water is flown through chilled beams, which are basically copper-pipe coils bonded to aluminum fins, thereby cooling the surrounding air by convection. Typically, chilled beams are distributed regularly across the ceiling of a compartment in a building. Chilled water is pumped through the pipes to provide cooling to the internal spaces, where warmer air is accumulated from under the compartments of the buildings.

For drinking purposes at various locations in a building, water is chilled centrally in a chiller, and then the chilled water is supplied to the drinking fountains installed at various locations in the building to supply chilled water.

DOI: 10.1201/9781003724056-2

Chilled water supply for space cooling

A chilled water supply for air conditioning represents a supply of water at a temperature generally no greater than 7.22°C, which is measured in one million British thermal units (MBTU).

Regardless of how it is provided, the chilled water is pumped through an air handler, which captures the heat from the surrounding air and then disperses the air throughout the area to be cooled. Chilled water is typically provided by compression refrigeration or absorption refrigeration, as introduced below.

In compression-type chilled water systems, a liquid refrigerant with a low boiling point is used, which absorbs heat from the return water and subsequently boils in an evaporator to form a gas. The resulting gas is then compressed, which increases its temperature further. The gas is then condensed, releasing its latent heat, which is then rejected. The process is then repeated.

The absorption-type chilled water system works on a similar basis, but in this case, the refrigerant gas is absorbed in a solution, which is then heated in a "generator", so that the refrigerant evaporates again, but this time at a higher pressure and temperature. The gas is then condensed, releasing its latent heat, which is then rejected. The process is then repeated.

Chilled water cooling system configurations can be developed in a very simple way, while they may be designed in a more complex form, as hinted in the following points:

1 Simple or basic chilled water cooling system configuration can consist of a single chiller and a single pump, as shown in Figure 2.1.
2 Complex chilled water cooling system configurations can consist of multiple chillers, multiple pumps, cooling towers, heat exchangers, and controllers like valves to redirect flow according to the indoor heat load.

A basic chilled water cooling system consists of the following functional parts:

1 Chillers that cool the water (or fluid)
2 Coils that transfer heat from air to water
3 Chilled water distribution pump
4 Cooling towers or condenser fans that reject heat from the chiller into ambient air

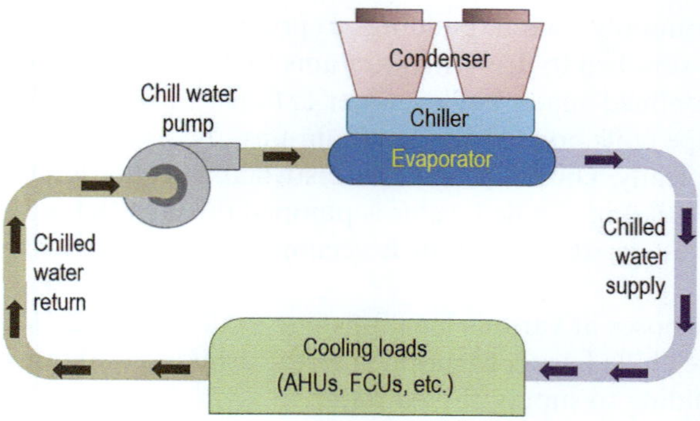

Figure 2.1 Basic chilled water cooling system using an air-cooled chiller.

5 Piping that includes supply pipes that send chilled water to the loads and return pipe to return warmer water to the chiller
6 Controls that coordinate the operation of the mechanical components together as a system.

The temperature of the chilled water will depend on the purpose for which it is being used. Chilled water supplying air handling units might be in the range of 5°C–10°C, the average being 7°C, with the return temperature about 12°C higher than the supply temperature by 5°C. The US standard rating of chillers is 7.22°C supply and 12.78°C return. Supply and return water temperatures need to be monitored.

In short, chilled water is pumped out at 7°C from the chillers and supplied around the building to the Air Handling Units (AHUs) and Fan Coil Units (FCUs), which extract the undesired excess heat from the building indoors and transfer it into the chilled water loop. The chilled water will enter the AHUs or FCUs at around 7°C, and by the time it leaves the heat exchanger within the AHUs or FCUs, it will have risen to around 12°C and will then find its way back to the air-cooled or water-cooled chiller to dump excess 5°C heat into the air or water to produce chilled water at 7°C. This cycle of flow is repeated in the system to continuously absorb heat from a confined space.

In a chilled water system, some makeup water is needed when leakage has occurred, generally at pump packing, when water is drained to allow system repair, and for evaporation loss. High-quality water should be used for makeup, so that scale deposits are not a problem. Piping should be well insulated to maintain efficient operation and prevent condensation.

Chilled water can be stored in insulated tanks for use when chiller units have spare capacity, such as during the night or during peak load periods.

A chilled water supply system is composed of two main loops or circuits: the refrigeration loop and the chilled water loop. The refrigeration loop is the sub-system that causes the chilling of water, while the chilled water loop is a distribution system by which chilled water is supplied to the points of use.

There are two types of chilled water supply systems used for space cooling:

1 Open chilled water supply system
2 Closed chilled water supply system

Open chilled water supply system

An open-type recirculating chilled water supply system uses the same water repeatedly to cool process equipment. Heat absorbed in the process is dissipated to allow reuse of the water. For this purpose, cooling towers, spray ponds, and evaporative condensers are used.

This system saves a tremendous amount of fresh water compared to the other method. In the open recirculating method, the quantity of water discharged as wastewater is greatly reduced, so chemical treatment is more economical. However, compared to other systems, open recirculating cooling systems are inherently more susceptible to the following treatment-related problems:

1 Cooling by evaporation can increase the presence of solid matter in the water, raising the corrosion and deposition potential.
2 The relatively higher temperatures significantly increase the corrosion potential.
3 The longer is the retention time of warmer water, the greater is the possibility of biological growth.

4 Airborne gases such as sulfur dioxide, ammonia, or hydrogen sulfide can be absorbed from the air, causing higher corrosion rates.
5 Microorganisms, nutrients, and potential fouling elements can also be absorbed into the water across the tower.

Figure 2.2 illustrates an open-type chilled water supply system for space cooling, using a water-cooled chiller.

Closed chilled water supply system

In the closed recirculating chilled water supply system, water circulates in a closed cycle and is subjected to alternate cooling and heating without air contact. In this system, generally chilled water is used as the heat transfer medium, though there are other water-based coolants to use.

Closed systems are adopted in air conditioning chilled water systems to transfer the refrigerant cooling to air washers, in which the air is chilled. Closed-water cooling systems provide a reliable method of temperature control. Figure 2.3 illustrates a closed system chilled water supply using an air-cooled chiller for air conditioning.

The chilled water supply system at 825 kPa and the condenser water circulation system are closed-loop systems.

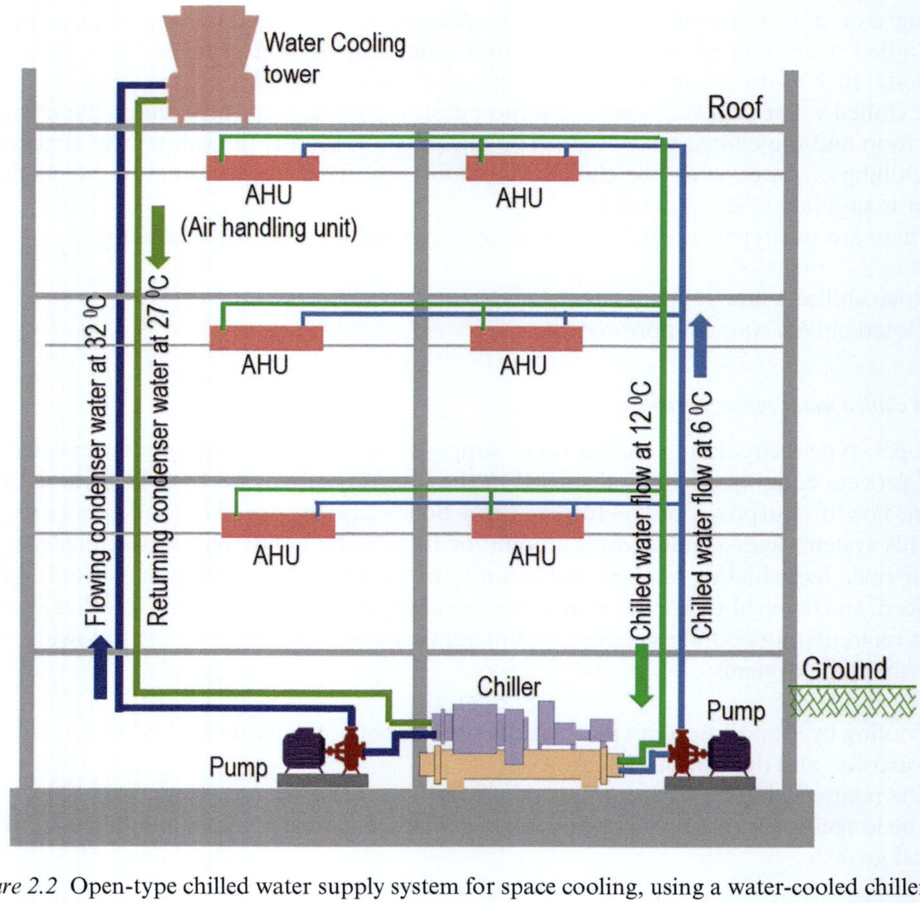

Figure 2.2 Open-type chilled water supply system for space cooling, using a water-cooled chiller.

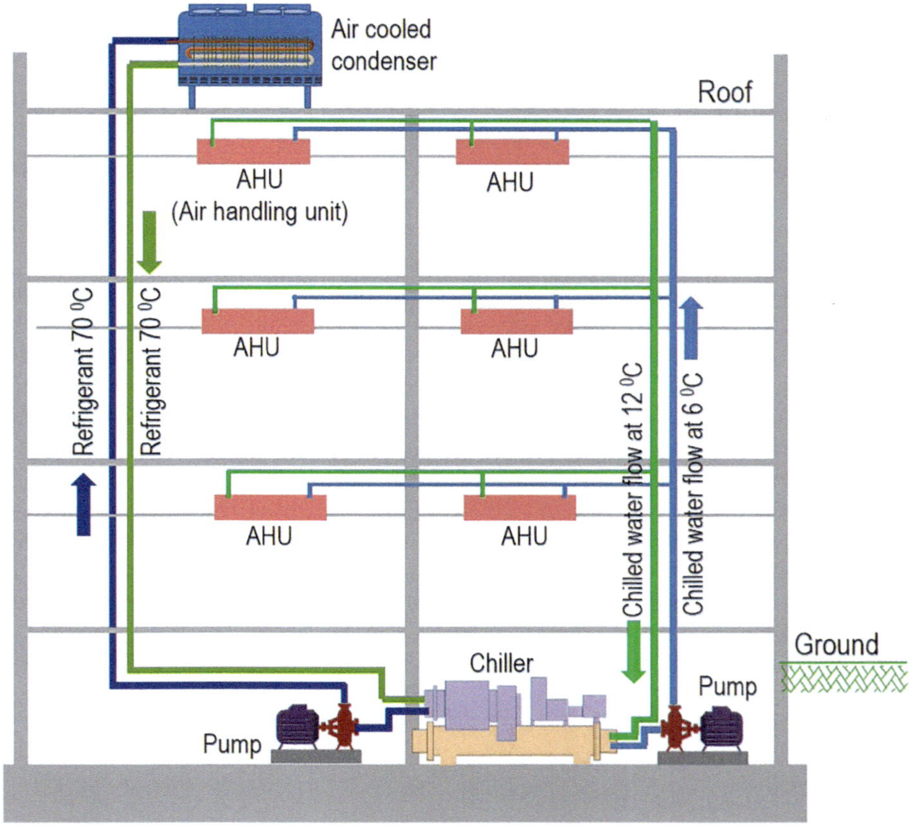

Figure 2.3 Chilled water supply using an air-cooled chiller for air conditioning.

Closed chilled water supply systems may include some additives like corrosion inhibitors, biocides, and antifreeze.

Advantages of closed systems:

1 These systems are energy-efficient and environmentally friendly.
2 These systems provide better control of temperatures and require small makeup water, which greatly simplify control of potential waterside problems. and most importantly, make systems less susceptible to biological growth from slime and algae deposits than open systems.
3 A closed system greatly reduces corrosion problems, as the recirculating water is not continuously saturated with oxygen, as in an open system. The only points of possible oxygen entry are at the surface of the surge tank or the hot well, the circulating pump packing, and the makeup water. For the small amount of makeup water required, proper treatment eliminates corrosion-related problems easily.

Components of a chilled water supply system

The chilled water supply system for air conditioning consists of the following major components:

1 Chiller
2 Pump

3 Piping
4 Controlling devices and appurtenances

Chiller

A chiller is a combination of multiple devices incorporated to lower the ambient temperature of water. A thermodynamic process occurs in the components of the chiller to generate chilled water, absorbing heat from the ambient temperature of the water to be chilled and ultimately releasing the absorbed heat to the atmosphere.

A chiller is primarily composed of a condenser and an evaporator, or heat exchanger. The excess heat of water to be chilled is passed through an evaporator, in which the flowing refrigerant takes this excess heat to the compressor to compress the heated refrigerant tightly together and then transported into a condenser. The condenser then picks up that heat and transfers it from the refrigerant and into the water, which circulates separately through a finned coil loop around the condenser. The refrigerant and water are always kept separated by a metal wall to transfer heat between them.

The condenser water is pumped up to the cooling tower, usually located on the roof, and it rejects that absorbed heat from the condenser water into the ambient air; thus, the condenser water cools down, warming the surrounding air. This cooled condenser water is then returned to a much cooler temperature and given the ability to absorb more heat from the condenser again.

A chiller consists of the following four basic components, as shown in Figure 2.4:

1 Compressor
2 Condenser
3 Expansion valve
4 Evaporator

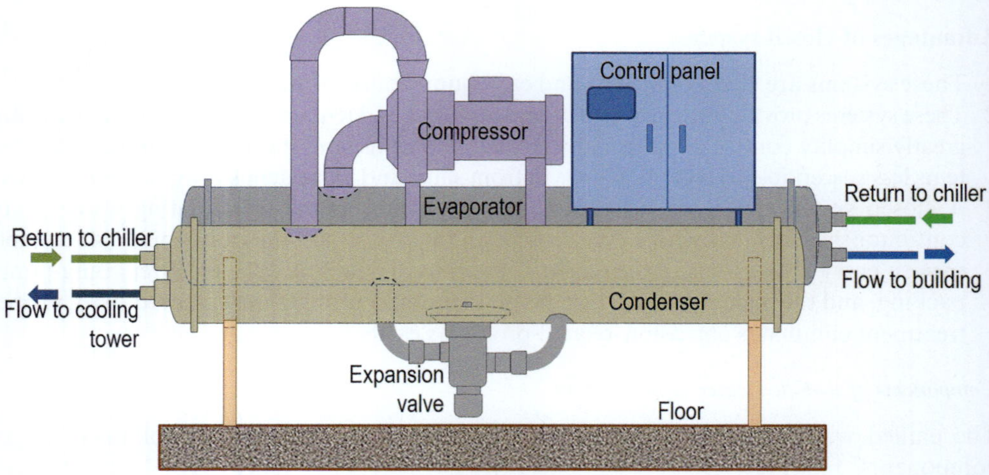

Figure 2.4 A chiller unit.

Types of chillers

There are various types of chillers, based on the condenser cooling. Two of the most commonly used types of chillers, based on the condenser used, are listed below. They primarily differ in the process of rejecting heat into the environment.

1 Air-cooled chiller
2 Water-cooled chiller

Air-cooled chiller

Air-cooled chillers absorb heat from the process water to chill it, and the absorbed heat is then transferred to the air around them, as shown in Figure 2.5. Air-cooled chillers are mostly placed outside of a building, so they easily remove heat from the chilled water by exhausting the heat directly into the surrounding air. As air is a very vital part, the installation of air-cooled chillers should be in such a position that no obstruction occurs to airflow on both the air discharge and air intake sides.

Water-cooled chiller

Water-cooled chillers use water from an external cooling tower to reject heat from the refrigerant in the condenser as the condenser water flows through the chiller and picks up heat. Heated water is then cooled in a cooling tower by releasing heat into the atmosphere, as shown in Figure 2.6.

The cooling tower is located outside, on the roof of the building, to release heat from the condenser water by evaporating some of the condenser water into the atmosphere. Due to the evaporation of some of the condenser water, heat is removed from the condenser water, which is then cooled down; the cooled condenser water then returns back to the chiller. This process is then repeated all over again.

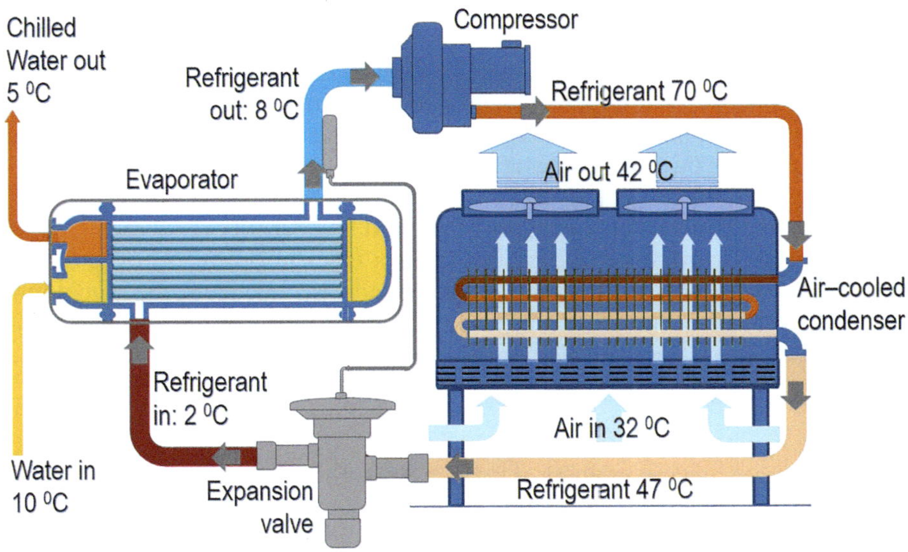

Figure 2.5 Functional diagram of an air-cooled chiller.

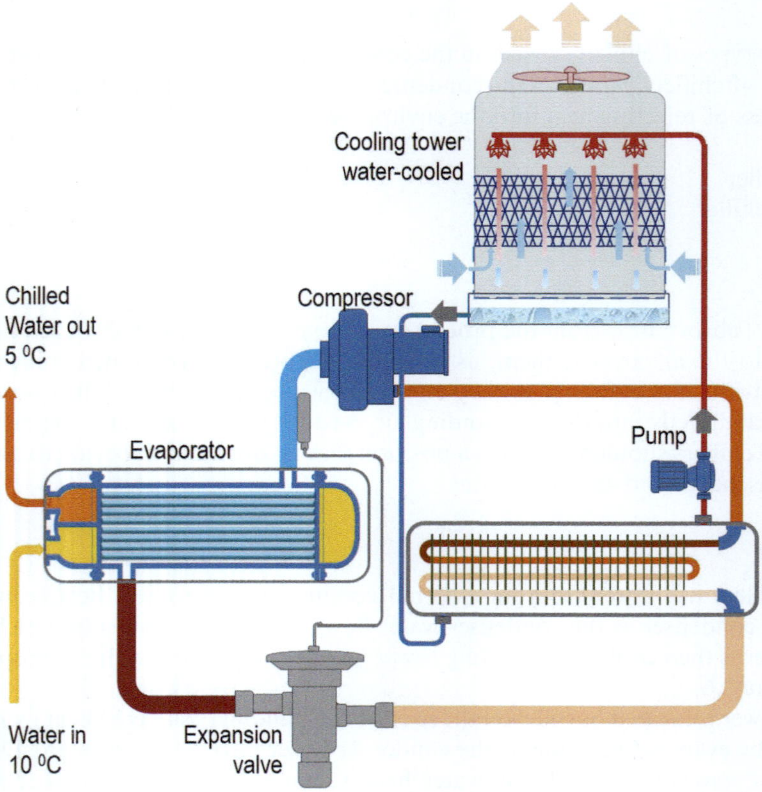

Figure 2.6 Functional diagram of a water-cooled chiller.

Comparisons between air-cooled and water-cooled chillers are as follows:

1 Air-cooled chillers require less maintenance than water-cooled units.
2 Air-cooled chillers do not require a cooling tower or a condenser water pump.
3 These chillers consume approximately 10 percent more power than a water-cooled unit due to wet surfaces transferring heat better than dry surfaces [1].

Where good water is plentiful, easily available, and inexpensive, a water-cooled system, often involving a cooling tower, becomes the most efficient option. On the other hand, where water is unavailable or expensive, the water supply is inadequate, the water is unclean, and the average relative humidity is high during the hot season, air-cooled systems are preferred.

Condenser

The condenser is a device used in a water chiller to condense a refrigerant—a heat absorber—from a gaseous state to a liquid state by cooling it. Relatively, hot and high-pressure refrigerant vapor or gas enters the condenser, where gaseous refrigerant transforms into liquid refrigerant that exits the condenser.

A condenser is basically a heat exchanger that is generally classified by the kind of refrigerant or liquid used to cause the cooling. It can be housed inside the chiller or can be located outside; in both cases, it rejects the heat into the air after absorbing heat from the refrigerant.

The condenser is placed between the compressor and the expansion valve. Hot refrigerant vapor enters the condenser at high pressure, and liquid refrigerant exits the condenser, absorbing heat from the surroundings. The absorbed heat from inside the building is rejected outside via the cooling tower. Condensers are designed to cause minimum heat transfer loss, thereby maximizing efficiency. Typically, on the inner wall of the condenser, lime, rust, and scale accumulate on the waterside of the exchanger, resulting in a loss of efficiency in heat transfer, which may lead to damage to the condenser.

Two types of condensers are in use:

1 Air-cooled condenser
2 Water-cooled condenser

Air-cooled condenser

An air-cooled condenser is basically a finned tube type that acts as a "radiator", as shown in Figure 2.7. It uses ambient air to cool and condense the hot refrigerant gas. It uses a motorized blower to force air across a grid of refrigerant lines. Blowing air over the condenser coil removes heat from the refrigerant.

Unless they are specially designed for high-ambient conditions, air-cooled condensers require ambient temperatures of 35°C or below to operate effectively. An air-cooled condenser consumes more energy than a water-cooled one.

Water-cooled condenser

A water-cooled condenser performs two steps to complete the heat transfer. First, heat moves from refrigerant vapor into the condenser water. Then, the warm condenser water is pumped to a cooling tower, where the process heat is ultimately discharged into the atmosphere.

Refrigerant

A refrigerant is a chemical compound typically found in either a liquid or gaseous state that can readily absorb heat from the environment. So, these are used in water chillers. There are various types of refrigerants, such as water, chlorinated fluorocarbons (CFCs), hydro-chlorofluorocarbons (HCFCs), Freon's R-11, R-12, R-21, R-22, R-502, etc.

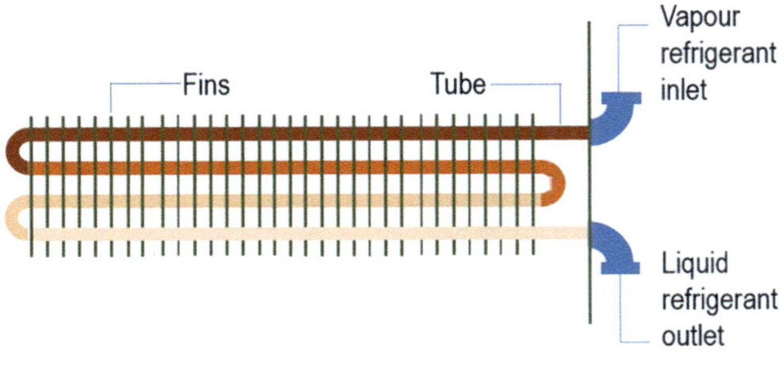

Figure 2.7 A finned-tube condenser.

Refrigerant shall be chosen based on the required cooling temperature; its appropriation is also governed by its ease of maintenance, physical space requirement, etc. Furthermore, ozone depletion potential (ODP) and global warming potential (GWP) of the refrigerant need to be considered for the compliance of environmental protection regulations.

All the refrigerants work on the basic principle of compression and phase-change from a liquid to a gas and back to a liquid phase. This process of changing from a gas to a liquid and back to a gas phase again by absorbing heat and then cooling the refrigerant is called the refrigeration cycle. To achieve these phenomena, refrigerant circulates through the evaporator, compressor, condenser, and expansion devices included in the chiller.

Evaporator

The evaporator is basically a heat exchanger in a chiller, but it is more specific than just a common heat exchanger. The evaporator absorbs heat energy from the heat captured from either air or water and transfers it to a refrigerant, causing the refrigerant to boil and change into a vapor. As the refrigerant evaporates, changing from a low-pressure liquid phase to a vapor phase, the temperature of the process coolant reduces.

There are three main types of evaporators: coil type, shell and tube type, and plate type. The types are chosen based on the following aspects:

1 Heat transfer efficiency
2 Requirements for water quality
3 Convenience of maintenance and
4 Space occupation

Generally, shell and tube evaporators are used in chilled water applications, as illustrated in Figure 2.8. When the cool liquid refrigerant flows through the tubes encased in a shell, the water to be chilled fills the shell, flowing around the tubes. As heat is transferred from the filled water to the refrigerant gas, which then boils inside the tubes, the resulting vapor is drawn into the compressor.

Relatively tempered water will enter the evaporator shell at one end, and chilled water will leave through the opposite end.

The refrigeration cycle starts with a low-pressure liquid-gas mix entering the evaporator. In the evaporator, which is basically a heat exchanger, heat is absorbed from the water to be chilled, which transforms it from a low-pressure liquid to a low-pressure gas. The

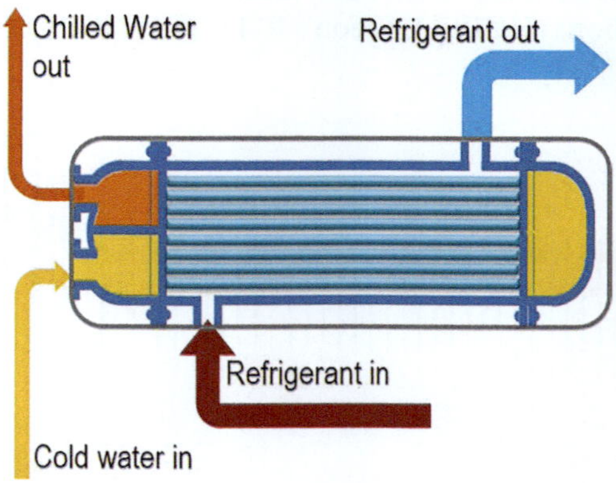

Figure 2.8 An evaporator.

low-pressure gas is then flown to the compressor, where it is compressed into high-pressure gas. The high-pressure gas enters the condenser, where ambient air or condenser water releases heat to cool it down in a high-pressure liquid. The high-pressure liquid travels to the expansion valve, which controls how much liquid refrigerant is to be entered into the evaporator, thereby beginning the refrigeration cycle again.

Every chiller has to maintain a minimum permissible evaporator flow rate by maintaining a low water velocity limit. To ensure minimum flow through the chiller, a bypass with a modulating control valve is employed.

Compressor

A compressor provides power to the refrigerating cycle that increases the pressure of a refrigerant gas by reducing its volume.

A compressor performs multiple functions. It removes refrigerant from the evaporator and ensures that the pressure in the evaporator remains low enough to absorb heat at the correct rate. It also raises the pressure of outgoing refrigerant vapor so that its temperature can be maintained high enough to release heat when it enters the condenser.

Compressors are classified by their type of performance, as mentioned below:

1 Centrifugal
2 Variable speed
3 Scroll or reciprocating
4 Rotary screw

Centrifugal compressor

The centrifugal compressor incorporates dynamic technology that causes an increase in velocity as a means to increase pressure for compression of the refrigerant gas. The refrigerant flows in through the suction end, hits the center of the impeller, where it is centrifuged by the vanes when rotated, and imparts an angular velocity onto the particles of the refrigerant. This angular velocity makes the refrigerant particles fly out at high velocity in all directions and collect in the volute (bowl), where kinetic energy transforms into an increase in pressure in the fluid flow to pass it down into the condenser. This centrifugal compressor shown in Figure 2.9 is a very common type of compressor set up for a central plant in large buildings. A compressor may be multistage, having multiple impellers in series.

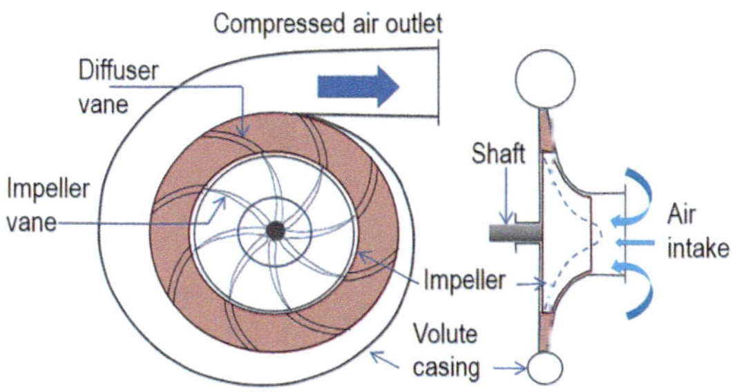

Figure 2.9 A centrifugal compressor.

The capacity of a centrifugal chiller can be greatly compromised in low-load conditions. At low-load conditions, the flow of refrigerant may not be sufficient to maintain flow through the compressor, causing a stall or a surge if flow is momentarily reversed until pressure is sufficient for flow to resume. However, this flow issue can be mitigated with variable-speed drives. Due to their wide range of capacity and potential applications, the compressors of the centrifugal chiller can be run by multiple types of drivers, such as electric motors, steam turbines, reciprocating engines, etc.

Variable speed compressor

Because a chiller is likely to handle a range of heat loads and cooling temperatures, it is to be sized for the highest heat load and lowest temperatures needed to be handled. For long-term operating costs, a chiller with a variable-speed compressor for maximum energy efficiency in partial-load situations is to be considered.

Scroll compressor

Scroll compressors use two inter-leaving scrolls shown in Figure 2.10, which are used for compressing or pressurizing the refrigerant gas in the compressor. Generally, one scroll remains attached to the compressor body and the other scroll orbits eccentrically without having rotation. Orbital motion is provided by an eccentric shaft. The orbiting motion generates numerous gas pockets traveling between the two scrolls. On the outer portion of the scroll, the pocket draws in gas, and then moves it to the center of the scroll, where it is discharged. As the gas moves in the smaller inner pocket, the temperature and pressure increase to the desired discharge pressure.

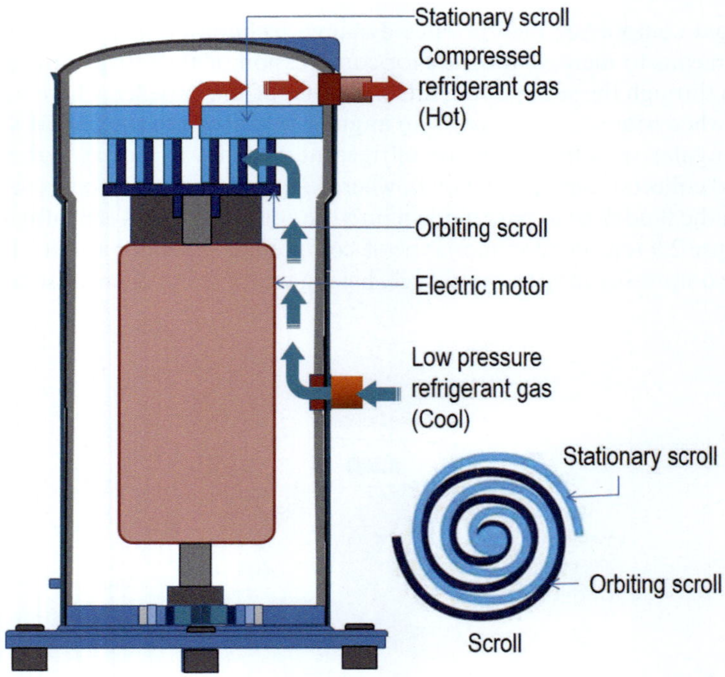

Figure 2.10 A scroll compressor. Courtesy: www.airchange.com.au [2]

Water chillers with Hermetic scroll-type compressors are suitable for industrial cooling or commercial applications.

Screw compressor

A screw compressor is a positive displacement machine that uses a helical rotary screw-like element in a close-tolerance housing. When this is rotated by a driver, the screw-like grooves gather refrigerant vapor through the intake port, trap it in the pockets between the grooves and compressor housing, and force it to the discharge port along the meshing point path.

Screw compressors are extensively used in large air conditioning and industrial refrigeration applications. Two types of screw compressors are in use: single-screw and twin-screw compressors.

Single-screw compressor: The single-screw compressor consists of two gate rotors and a main rotor with matching teeth. The main rotor is driven by a prime mover, typically an electric motor, which meshes with the gate rotors to produce the compression process. The rotation of the rotors raises the trapped gas pressure to the discharge pressure.

Twin-screw compressor: A twin-screw compressor, shown in Figure 2.11, consists of accurately matched rotors: one male and one female rotor that mesh closely when rotating within a close-tolerance common housing. One rotor is driven while the other turns in a counter-rotating motion. As the two mesh, refrigerant flows in the axial direction. This compressor is very common in commercial and industrial applications.

The mating rotors rotate between very close tolerances, so they require cooling and lubrication, which is done by forcing oil into the screw-rotors. The oil also acts as a seal for the clearances. The oil is entrained by the flowing refrigerant gas, leaves the screw rotor, and is ultimately recovered by an oil separator for reuse after cooling and filtering. Both the refrigerant and oil flow into the oil separator together. Incorporating centrifugal force, the oil is separated from the gas by pushing it to the wall of the separator. Refrigerant gas is ejected through the discharge port, while oil drops down along the wall to an oil tank below. The oil tank is on the high-pressure side of the system, which helps circulate the oil. The compressive action of the screw itself provides the necessary pressure differential.

Based on the compressor type used in the chiller, chillers are also named accordingly, such as scroll chillers, etc.

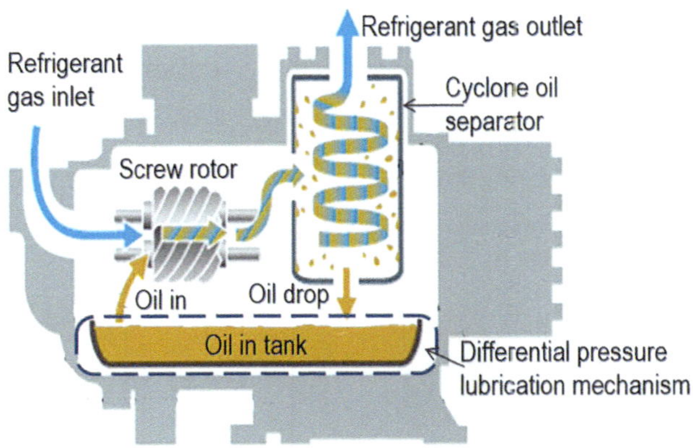

Figure 2.11 A twin-screw compressor. Courtesy: PT Metropolitan Bayutama [3].

Expansion valve

An expansion valve is a device in the chiller that controls the refrigerant flow released into the evaporator to regulate the vapor leaving the evaporator and change the flow based on the cooling load. It helps facilitate the change of liquid refrigerant with high pressure in the condenser to gaseous refrigerant with lower pressure in the evaporator.

Types of expansion valves: There are four types of expansion valves that are generally used. The expansion valves are named below:

1 Automatic expansion valve
2 Thermostatic expansion valve
3 Capillary tube
4 Float type expansion valve

Automatic expansion valve

An automatic expansion valve is used to regulate the flow of refrigerant in the pipe line from the condenser to the evaporator, which is performed by a pressure-actuated diaphragm. The diaphragm, being flexible, actuates the poppet valve. An increasing pressure in the sensing bulb presses down the poppet to open the valve further. There is an adjustable spring providing a closing force on the valve, which is used to control the superheat.

It helps maintain a constant pressure in the evaporator. This valve is not efficient when the load fluctuates; hence, this type is not suitable for use in air conditioning, as the load fluctuates a lot during its operation.

Thermostatic expansion valve

These valves, also known as thermal expansion valve shown in Figure 2.12, use a valve mechanism to control the flow of liquid refrigerant into the evaporator coil. The flow is controlled by the pressure in the evaporator. There is a sensing bulb clamped firmly to the coil to ensure proper sensing, as shown in Figure 2.13 which detects the temperature of the evaporator. When the temperature of the evaporator increases due to the demand for cooling, the pressure in the bulb increases; hence pushing the spring to open the valve. Reversely, when the temperature of the evaporator reduces due to a lack of demand for cooling, the pressure in the bulb will drop, causing the spring to close the valve.

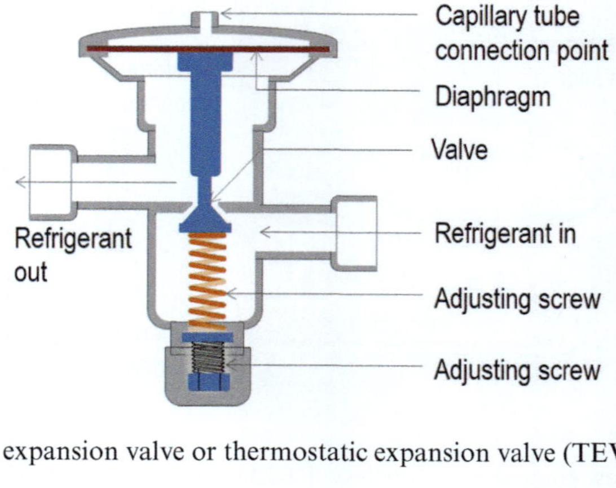

Figure 2.12 A thermal expansion valve or thermostatic expansion valve (TEV or TXV).

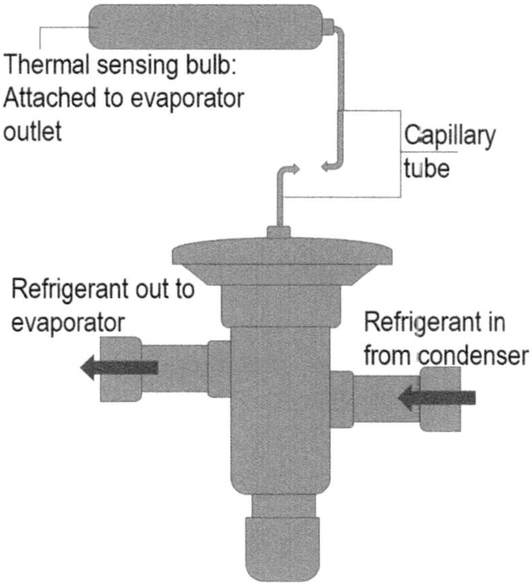

Figure 2.13 A thermal expansion valve with a sensing bulb.

This type of expansion valve operates satisfactorily even when the load fluctuates and, hence, is suitable for use in air conditioning systems. The valve provides a higher flow rate as the evaporator warms; conversely, when it cools, it causes a reduction in the flow rate.

Capillary tube expansion valve

A capillary tube having a small internal diameter, which could be coiled for part of its length, is installed in the suction line. A filter-drier is sometimes fitted before the tube to remove dirt or moisture from the refrigerant.

Float type expansion valve

These valves are seldom used today. It is used to maintain liquid refrigerant level in the evaporator and to control refrigerant flow from condenser to evaporator.

Capacity range of chillers

The capacity of a water chiller is typically measured in tons or kilowatts (kW). The capacity range of water-cooled chillers can vary greatly, from as low as 5 tons or 17 kW to as high as 1500 tons or 5300 kW, or even higher, whereas air-cooled chillers come in sizes ranging from 7.5 to 500 tons [4].

The capacity of scroll chillers nominally ranges from 35 to 1050 kW (10 to 300 tons). To provide a larger capacity, a scroll chiller will typically have multiple blocks of smaller compressors (e.g., 70 kW each) [5].

Screw chillers are available with capacities ranging from 105 to 2813 kW (30 to 800 tons). Screw chillers typically maintain design efficiency (kW/ton) at partial loads. The screw compressor can vary capacity in response to changing loads, from 20 percent to 100 percent of capacity [5].

Centrifugal chillers incorporate dynamic technology that uses a change in velocity as a means to increase pressure for compression of the refrigerant gas. Centrifugal chillers are available in capacities ranging from 100 to 1000 tons [5].

Cooling tower

A cooling tower is basically a heat rejection machine. Heat rejection is done by bringing air and water into contact to cool a coolant, mostly water, and release excess heat into the atmosphere.

There are two methods by which heat is removed from water: Loss of latent heat through evaporative cooling and cooling through sensible heat transfer. Sensible heat transfer in a cooling tower is either by conduction, convection, or a combination of both. Heat transfer methods differentiate cooling towers into two types of designs, as mentioned below:

1 Open-circuit or wet cooling towers and
2 Closed circuit or fluid cooling towers

Open-circuit or wet cooling tower

Open-circuit or wet cooling towers operate mainly on the principle of evaporative cooling. When the cooling water comes into contact with the air, a small portion evaporates, thus lowering its temperature and through partial heat transfer between air and water. So, this process is also known as "evaporative cooling".

The main advantage of wet cooling towers is their efficient cooling. It is possible to cool water below the ambient temperature or the dry-bulb temperature by using a wet cooling tower.

In the open-circuit cooling tower, warm water is sprayed through the nozzles into the air stream, as shown in Figure 2.14, to maximize contact with the air and increase the cooling effect. There is a large cooling fan located on the top of the tower, which blows air through the unit and thus helps the cooling process. The cool ambient air enters and comes into direct contact with the sprayed condenser water (in an open cooling tower). This process

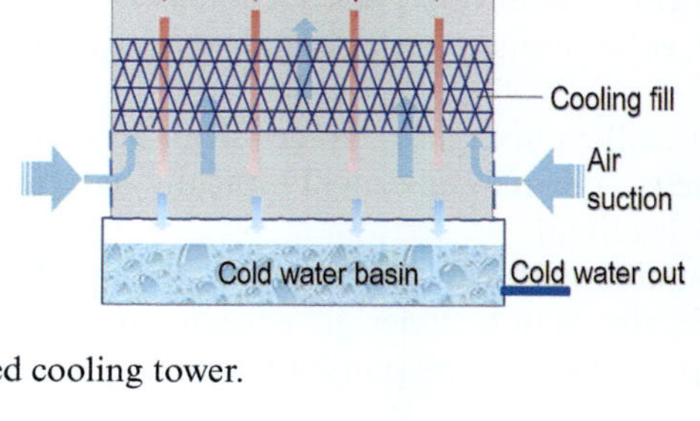

Figure 2.14 A water-cooled cooling tower.

allows the heat of the condenser water to transfer into the air, and this air is then blown out into the atmosphere. There is a drift eliminator that removes the tiny droplets generated during the evaporation of the airflow in the tower. An open-circuit system is preferred due to its lower initial cost.

Closed-circuit or fluid cooling tower

In this system, the higher-temperature water returned by consumers passes through tubes or coils where sensible heat transfer occurs. Water is sprayed outside of these tubes like what is done in water cooling towers. Thus, the condenser water is isolated by the tubes or coils in the cooling tower and cannot come into direct contact with the atmosphere. In the sprayed water, both the latent and sensible heat are removed by coming into contact with the air stream. The main advantage of this system is that consumers use water that is free from contamination. Figure 2.15 represents a closed-circuit cooling tower.

A closed-circuit cooling tower is preferred when the quality of the condenser water is to be highly maintained.

Cooling tower size

The appropriate size of a cooling tower depends on four factors. Among those, the size varies directly with the heat load and inversely with range, approach, and wet bulb temperature.

Range is the temperature difference between the incoming hot and outgoing cold water at the tower. By increasing the range, the capital cost and energy cost of the tower can be reduced. Approach is the temperature difference between what is being produced and the "power source" that creates the product. In a cooling tower, the "product" is the cold water leaving the tower, and the ambient wet bulb is the driving force that creates the cold water. If a cooling tower produces 29.44°C of cold water when the ambient wet bulb is 25.55°C, then the cooling tower approach is −13.88°C.

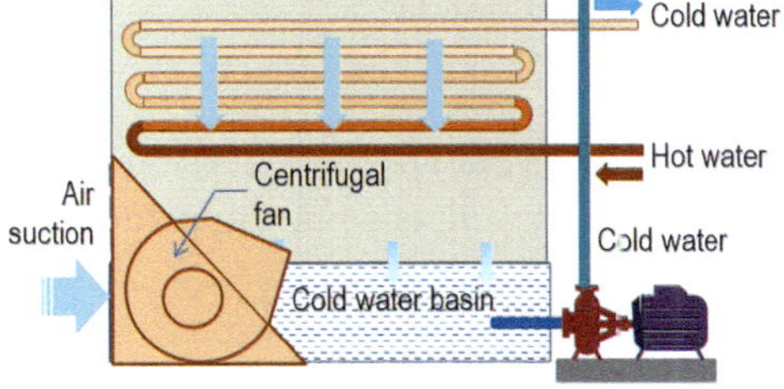

Figure 2.15 A closed-circuit cooling tower.

Wet bulb temperature: Ambient wet bulb temperature is a condition of air measured by a device called a psychrometer. A psychrometer has a thin film of water on the bulb of the thermometer that is twirled in the air. The thermometer will show a reduced temperature after about a minute. The low point where additional twirling does not reduce the temperature is termed the wet bulb temperature. The measured wet bulb temperature is a function of the relative humidity and ambient temperature of the surrounding air. The wet bulb temperature indicates how much water vapor the atmosphere can hold under current atmospheric conditions. A lower wet bulb temperature indicates the air is drier and can hold more water vapor than it can at a higher wet bulb temperature.

$$\text{Approach} = \text{Cold water temperature} - \text{wet bulb temperature}$$

$$= 35^{\circ}\text{C} - 25^{\circ}\text{C} = 10^{\circ}\text{C}$$

$$\text{Range} = \text{Difference between hot and cold water temperature}$$

$$= 42^{\circ}\text{C} - 35^{\circ}\text{C} = 7^{\circ}\text{C}$$

$$\begin{aligned}\text{Cooling tower efficiency} &= \text{Range} / (\text{range} + \text{approach}) \times 100 \\ &= 7^{\circ}\text{C} / (7^{\circ}\text{C} + 10^{\circ}\text{C}) \times 100 \\ &= 41.17\%\end{aligned}$$

A cooling tower with a larger diameter will consume much less energy, so it would be more efficient [6].

The thumb rule is each ton of a water-Cooled Chiller requires 1 ton of cooling tower.

Cooling tower materials

Historically, it was known that cooling towers were constructed with wood. Sometimes the cold-water basin was made of concrete. Today, a variety of materials are used to construct cooling towers. But materials are chosen with respect to enhancing corrosion resistance, reducing maintenance, and promoting reliability and long service life. In these perspectives, galvanized steel, various grades of stainless steel, fiber-glass, concrete, and plastic are widely used in cooling tower construction. HDPE plastic is becoming more popular for tower construction. The advantage of HDPE cooling towers is that they are highly corrosion-resistant; have a low risk of damage to surfaces, etc.

Valves and appurtenances

In chilled water supply systems, various types of valves are used to control its flow and manage the temperature variation, as well as some appurtenances installed in the system, as mentioned below:

1 Three-way valve to direct or bypass the flow
2 Balancing valve to maintain the required flow
3 Gate valves for isolation
4 Butterfly valves for isolating and throttling

5 Globe valves for throttling the flow
6 Strainers for filtration
7 Flexible bellows for arresting vibration
8 Check or non-return valve to prevent backflow
9 Pressure and temperature gauges to monitor pressure and temperature
10 Chilled water buffer tank (CBT) to provide the required water

Three-way valve

This valve has three ports: two acts as inlets and one as an outlet, but is operated by one actuator or handle. Three-way valve assemblies can be controlled by pneumatic, hydraulic, and electric actuators. These valves can be used to either divert the flow of fluids, or to mix fluids entering through two inlets and then delivering mixed fluids through a single outlet.

In Figure 2.16, three situations of a three-way valve have been shown. When the valve is in its lowest position, it closes the inlet B; thus, fluid flows enter through inlet A and exit through outlet AB. Again, when the valve is in its highest position, it closes the inlet A; thus, fluid-flows enter through inlet B and exit through outlet AB. In these two conditions, the valve acts as a diverting valve. But when the valve is at its midpoint, it opens both inlet A and inlet B; thus, fluid-flows enter through inlet A and inlet B, mix inside the valve, and finally outgo through the outlet AB. In this condition, the valve is acting as a mixing valve.

Three-way valves are more cost-effective and space-saving compared to utilizing multiple two-way valves for the same purposes. These valves are also used as bypass valves in primary and secondary loop applications.

In a chilled water supply system, a three-way valve is virtually used to save energy; three-way valves are used to switch the flow of coolant to the equipment from chillers, depending on the ambient temperature; thereby, by directing the coolant from the free-cooler to the equipment, low ambient temperatures are used for cooling coolant, resulting in energy savings.

Balancing valve

A balancing valve, also known as a double regulating valve, is mainly used to regulate and balance the flow to achieve consistent water flow through a section of the chilled water supply system, by adjusting the flow through it. When a flow rate fluctuates, following a

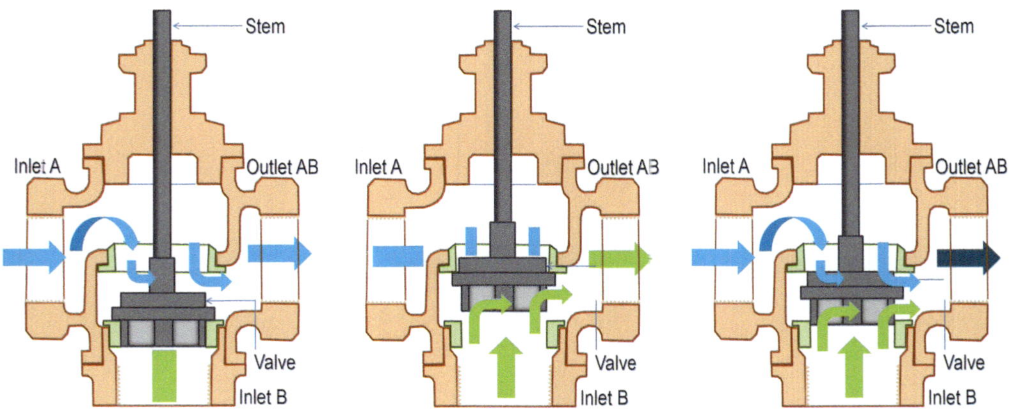

Figure 2.16 Three-way valves in three operating conditions.

balancing valve, the flow rate will remain consistent. Thus, it helps maintain a balanced flow rate and pressure distribution in a chilled water supply system, so that a branch or section within the system can maintain the required flow rate, to maintain the desired temperature.

The valve has two ports: the inlet and the outlet ports, with an adjustable orifice or flow-restricting element, as shown in Figure 2.17. When fluid flows through the valve, a pressure drop occurs across the orifice due to the flow restriction. The pressure drop thus occurs, creating a pressure differential between the inner and outer ports of the valve.

Balancing valves can be operated manually or automatically. Manual valves typically have a handle or a screw mechanism for adjustment, while automated valves are supported by actuators that respond to signals from control systems to maintain the desired flow rate automatically.

Double regulating valve (DRV): DRV valves are also one kind of balancing valve but perform two major functions. Firstly, they balance the flow rate by adjusting the valve's position; secondly, they measure the flow rate and pressure with a flow meter and pressure gauge installed on it.

A balancing valve is installed on the return pipe of the chiller, and the flow is controlled and balanced to the desired flow by adjusting the valve spindle. Earlier, the pressure drop across the coil was measured. To achieve the required flow, the valve spindle is adjusted accordingly.

There are four types of commonly used balancing valves, as mentioned below:

1 Static balancing valves: These valves are the most common type and are designed to balance the flow in hydronic-tempered water flow systems. They adjust the flow statically by setting the fixed position of the valve in the system balancing process.
2 Dynamic balancing valves: Also known as automatic or self-acting balancing valves, adjust the flow rate automatically in response to changes in flow rate in the system. So, these valves are installed to maintain constant flow in the pipe, regardless of pressure fluctuations in the system.
3 Thermostatic balancing valves: These valves are designed to balance the flow rate based on temperature requirements.
4 Differential pressure control valves (DPCV): DPCV's main purpose is to maintain a constant differential pressure across a specific part of the system. There is a built-in differential pressure regulator that automatically changes pressure to stabilize flow within the components and meet the required temperature demand.

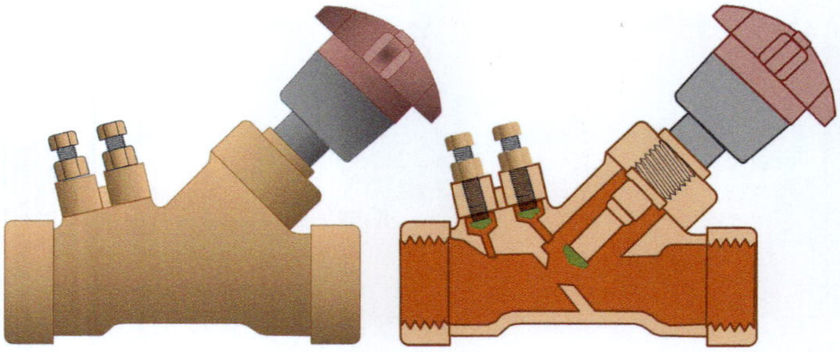

Figure 2.17 A static balancing valve.

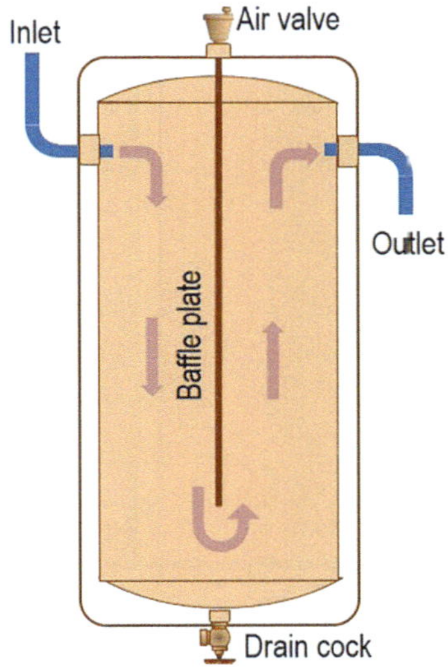

Figure 2.18 A buffer tank with its accessories.

Buffer tank

The buffer tank, shown in Figure 2.18, is primarily installed in a closed chilled water supply system, to provide the extra water volume required in the system, including in the chiller with inadequate water volume capacity, with respect to the chiller capacity. The buffer tank is a pressurized tank with at least 50 mm of insulation to prevent temperature gain. Buffer tanks are composed of an inner tank made of stainless steel and an external cabinet made of galvanized steel sheet. The buffer tank is provided with an internal baffle plate. This ensures efficient water circulation and full use of the tank water. The tank is installed on the suction side of the circulation pump.

In an inappropriate-sized low-capacity chilled water supply systems, insufficient water flow for the chiller will cause poor temperature control, erratic system operation, and excessive compressor and chiller cycling. So, in such cases, a buffer tank can provide the extra water flow needed for the system. If there is an adequate volume of water in the system, no extra storage in a buffer tank is required. In cases of insufficient system water flow, the source temperature will be reached quickly, and the compressor will shut off. Chiller compressors, in general, can start three times in an hour. If the compressor is off and there is a demand for chilled water, the demand cannot be met because the compressor cannot start as frequently as when required. The added water capacity provided by the buffer tank thus reduces the cycling of the compressor unit, which improves temperature control and provides more consistent system operation.

It is recommended to maintain a minimum "loop volume" in dedicated chilled water distribution systems, which is a minimum of 11.36 liters of water in the chilled water circuit per ton of cooling capacity the loop delivers. In applications where more precise control of chilled water temperature is required, a maximum of about 37.85 liters of water volume per ton of cooling capacity is preferred. If the distribution system cannot provide the required volume, buffer tank is to be installed in the way as shown in Figure 2.19, to bring the circuit volume up to the required value.

The flow through the buffer tank changes direction depending on the flow rate of the distribution system versus the flow rate through the chillers. In either case, a buffer tank helps provide the required amount of water to stabilize chiller operation.

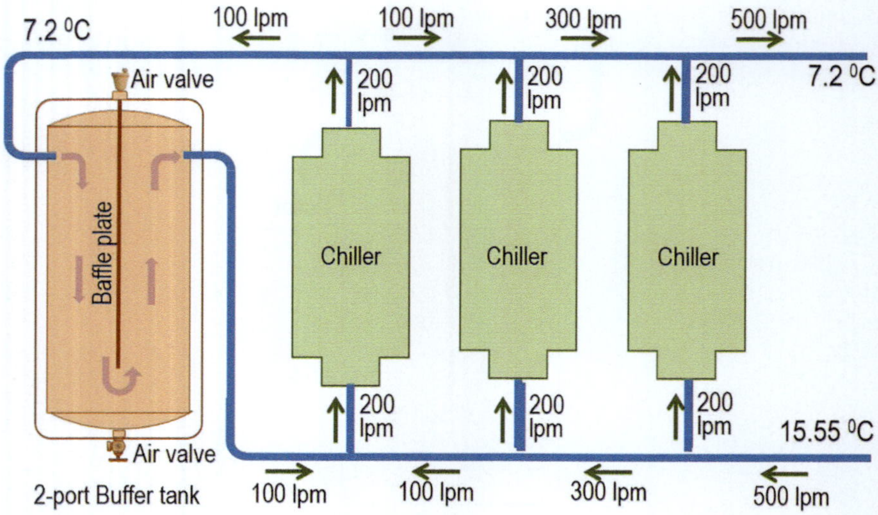

Figure 2.19 Buffer tank connection with chillers.

Pumps for chilled water supply

In a chilled water supply system, both chilled water circulation and condenser water circulation systems use pumps, which are virtually centrifugal pumps. The pumps may be inline or end suction, horizontal split casing type centrifugal pumps, and they may also be constant or variable speed types. The pump is coupled with electric motors to perform as one unit. Chilled water pumps should have low-temperature-resistant seals. In some cases, two or more pumps installed in parallel to run simultaneously and as standbys.

Chilled water circulation pump

A pump is used to circulate chilled water in a closed system. This pump pushes the water through the chiller and around the building. The chilled water pump delivers chilled water and returns the relatively higher-tempered chilled water from the air handling units and fan coil units back to the chiller in an air conditioning system and from the drinking fountains in the chilled water supply system for drinking purposes. The chiller then cools the relatively higher-tempered chilled water, and the chilled water of the desired temperature is then supplied back to the points needed.

The pump must be sized properly to circulate the correct amount of flow at the correct pressure, in order to maintain the desired temperature in the supply system and achieve the cooling intended for.

The typical motor speeds of these pumps are between 1780- and 3600-rpm. 3600-rpm pumps are smaller and lower in cost, while 1780-rpm pumps typically have lower required NPSH and higher efficiency. 1780-rpm pumps should not be used with 3600-rpm chillers, although this is less of a concern with variable-speed operation.

To select the right size of pump for a chilled water supply, the flow rate in liters per minute (lpm) and the total dynamic head in meters (m) are to be determined.

Flow rate: The flow rate for the chilled water pump is typically determined by the maximum flow of the chiller or through the maximum flow rate required by the fan coils, air handling units, or drinking fountains.

Total dynamic head: the total dynamic head of a chilled water supply pump is the total pressure drop in the most hydraulically remote run from the chilled water pump to the chilled water return.

Condenser water circulation pump

The condenser water circulation pump is installed to circulate condenser water, either in a closed or open-circuit system. The condenser water pump delivers relatively higher-temperature chilled water from the chiller to the cooling tower; the cooling tower then cools the condenser water, and the cooled condenser water is supplied back to the chiller.

The condenser water pump must be sized correctly to circulate the correct amount of flow at the correct pressure in order to achieve proper chilling of the water. If there is insufficient flow and pressure, then there will not be enough chilling. If there is too much pressure and flow, then the electrical system will be overburdened, and if there is no speed control, then electricity may be wasted due to excess pumping. The elevation difference between the pump suction and the water level in the cooling tower is the static head of the pump. The dynamic head is to be calculated to determine the total head of the pump, considering the pressure drops in all the pipes, fittings, valves, appurtenances, appliances, etc.

Chilled water distribution piping

In chilled water supply systems, chilled water distribution systems are divided into three categories:

1 Constant-rate chilled water distribution system
2 Primary (constant) and secondary (variable) chilled water distribution system
3 Variable primary flow chilled water distribution system

Constant-rate chilled water distribution system

A constant-flow system is the simplest chilled water distribution scheme. In this system, a set of constant-speed pumps distributes a fixed amount of chilled water at all times, and the temperature varies to meet the load. The system uses 3-way control valves at air handling units that allow some water to bypass the cooling coil during part-load conditions. At low loads, the chilled water flow through the cooling coil is restricted (in response to supply air temperatures to the space), but the total quantity returned to the chiller remains constant. Figure 2.20 illustrates the schematic diagram of the constant-flow rate primary system.

In constant-flow systems, chilled water flow rate (lpm) is considered for peak design conditions, and the difference between the chiller supply and return temperature (ΔT) varies in response to the load (kW).

Advantages

1 The lowest installation cost
2 Lesser foot print than primary-secondary system
3 Easy to commission, control, and operate

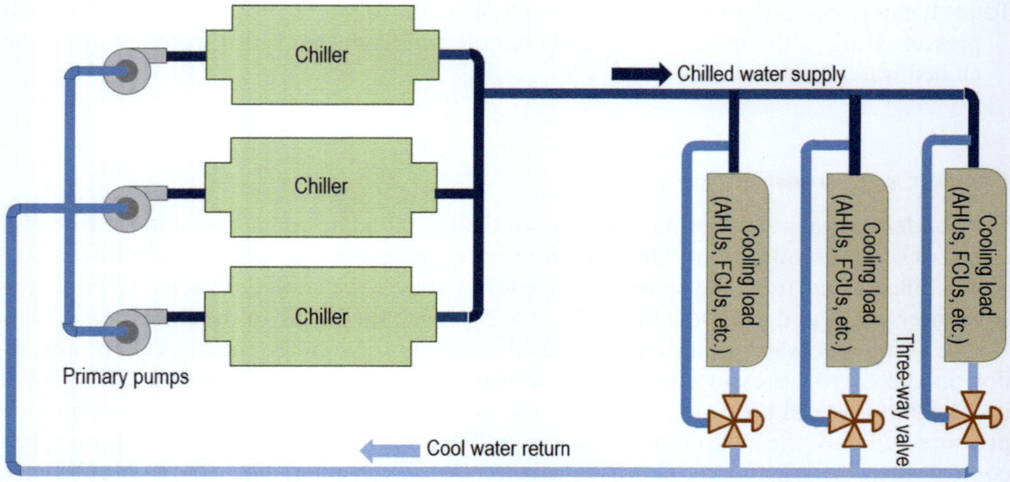

Figure 2.20 A schematic diagram of the constant-flow rate primary chilled water distribution system.

Disadvantages

Highest energy cost as the system runs at full capacity, even at low loads.

Primary-secondary chilled water distribution system

According to ASHRAE, energy-efficient design of large buildings requires that all HVAC hydronic systems requiring a total pump system power exceeding 10 hp be capable of flowing at 50 percent of the design value or less. This standard highly recommends the use of primary-secondary pumping systems.

A primary-secondary pumping system for chilled water supply has two distinct circuits or loops: a primary circuit or loop where chilled water is produced, and its principal components are the chiller and pumps. The pumps of primary circuits have constant volume at low heads, which is desired to provide a constant flow through the evaporator. These are usually placed in tandem with a chiller, though they can be arranged in a header. The secondary circuit, or loop, distributes the chilled water to the terminal units. The components of the secondary circuits are pumps, terminal units like fan coils, air handling units, control valves, etc., as shown in Figure 2.21. The secondary pumps can be constant speed or variable speed and are sized to cause the desired flow rate and head required to overcome the pressure drop of the secondary circuit only. A neutral bridge consists of two tees that are typically located at the suction header of the secondary pumps and at the suction header of the primary pumps, is connected by a de-coupling pipe. This decoupling pipe separates the primary and secondary circuits, or loops. This common pipe is designed for negligible pressure drop at desired flow, which is considered the heart of primary-secondary pumping, allowing the two pumps to operate independently.

Advantages

1 Easy to commission and control
2 Constant flow through the evaporator
3 Easier trouble-shooting
4 Separating loops lowers total pump energy
5 Lower plant energy (can sequence chillers and ancillary equipment)
6 Versatile: multi-circuit capability

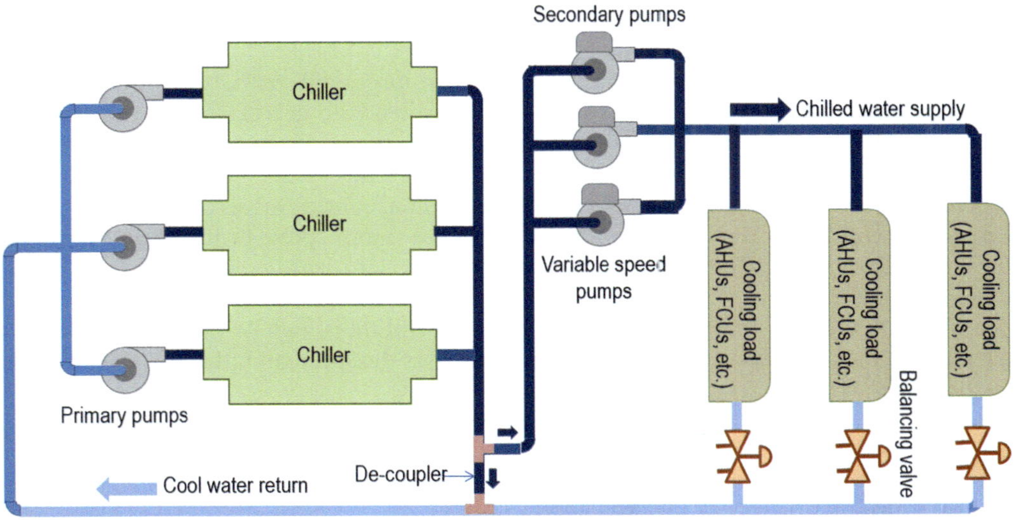

Figure 2.21 A schematic diagram of the primary-secondary chilled water distribution system.

Disadvantages

1 Higher operating and energy costs
2 Highest installation cost for secondary pumps, piping, etc.
3 Potential for higher plant energy loss because of low ΔT syndrome
4 Larger foot print

Variable primary flow chilled water distribution system

In a variable primary flow (VPF) distribution system, chilled water flow is allowed to vary throughout the distribution piping. The primary pumps are variable-speed pumps, as shown in Figure 2.22; so, they provide variable flow to supply system demand to the extent tolerated by the chillers. A bypass line is installed with a closed control valve that opens

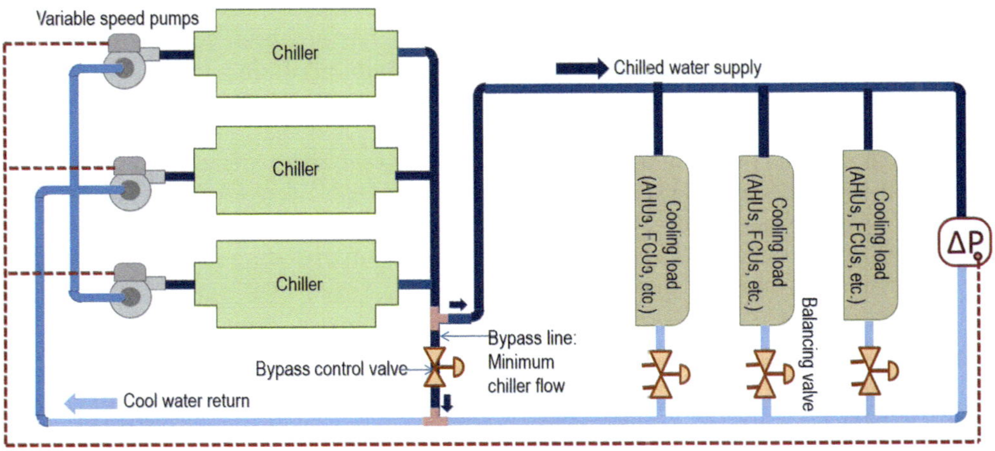

Figure 2.22 A schematic diagram of the variable primary flow chilled water distribution system.

only to maintain minimum flow through the chillers. The bypass is a smaller pipe sized for the minimum flow of the largest chiller.

The variable-speed pumps operate to maintain a target differential pressure (ΔP) at a specific point in the system. This pressure difference tends to decrease when the terminal units (air handling units or fan coils) two-way control valves open in response to increasing loads. To restore the differential pressure (ΔP) across the system, the pump controller increases the pump speed. Conversely, when the terminal control valves close in response to decreasing coil loads, the pump controller slows the pump speed to maintain the target differential pressure (ΔP). Meanwhile, the plant controller turns the chillers on and off to match cooling capacity with system load. The pumps are normally sized to deliver 120 percent of the design flow. If the return water temperature is high or the differential temperature (ΔT) is low, the variable speed pump increases the speed and, thereby, the flow rate to the chillers, thus increasing the load on the chillers.

Advantages

1 Lower installation, operating, and energy costs
2 Smaller foot print
3 It is not prone to low ΔT syndrome caused by the lowering of the return chilled water temperature due to inefficient heat transfer
4 Best chilled water pump energy consumption
5 Lower pump head

Disadvantages

1 Requires a more complex and finely calibrated control system
2 Coordinated control of chillers, isolation valves, and pumps is required
3 Comparatively longer commissioning times to tune the system
4 Need skilled manpower to operate and maintain

Venting chilled water supply piping

In chilled water supply piping, proper ventilation is to be provided; otherwise, various problems will be created, particularly in its flow.

Prevention of the flow of water leaves the heat exchanger dry, for which it no longer cools. This is especially common after a new machine has just been commissioned, and additionally, the chilled water contains a considerable amount of air.

Air-vent valves are mechanical valves that purge air out of the chilled water piping system automatically. They are installed on the top of vertical pipes and apex points, if any, or in the middle of a horizontal pipe, as shown in Figure 2.23. Air-vent valves are installed upright. An isolation valve should be installed before the air-vent valve to enable its replacement when needed.

Air-vent valves vents automatically with float principle; generally, found of cast iron body having nonferrous internal parts; allowable maximum operating pressure is 1034 kPa and 38°C maximum operating temperature.

Chilled water system for drinking

Like a hot water distribution system, chilled water can also be distributed throughout a building. Mostly for drinking purposes, a chilled water distribution system is developed to supply the drinking fountains installed at different locations in a building. In a central

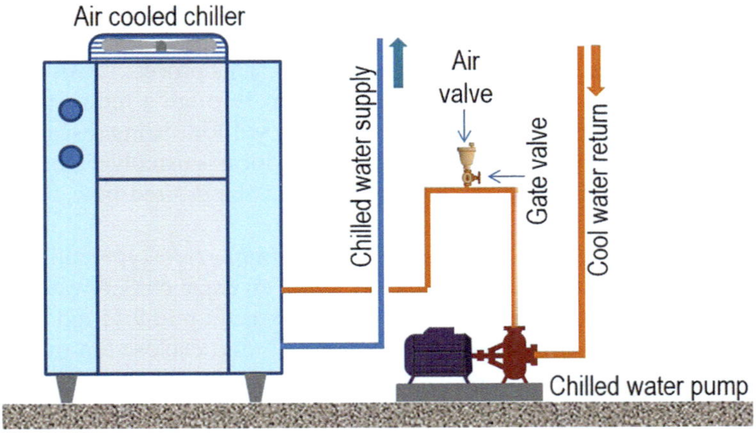

Figure 2.23 Air-cooled chilled water supply systems piping with air vent.

chilled drinking water supply system, water is cooled to 7.22°C, and it is allowed for a 2.8°C temperature increase in the return pipe of the distribution system.

Chilled water is distributed to a building through a piping network system consisting of supply and return piping. The system forms a chilled water loop, which is basically a pair of distribution pipes consisting of a supply pipe and a return pipe. These pipes start from a central chiller plant and end at the points of use through a chilled water dispensers called

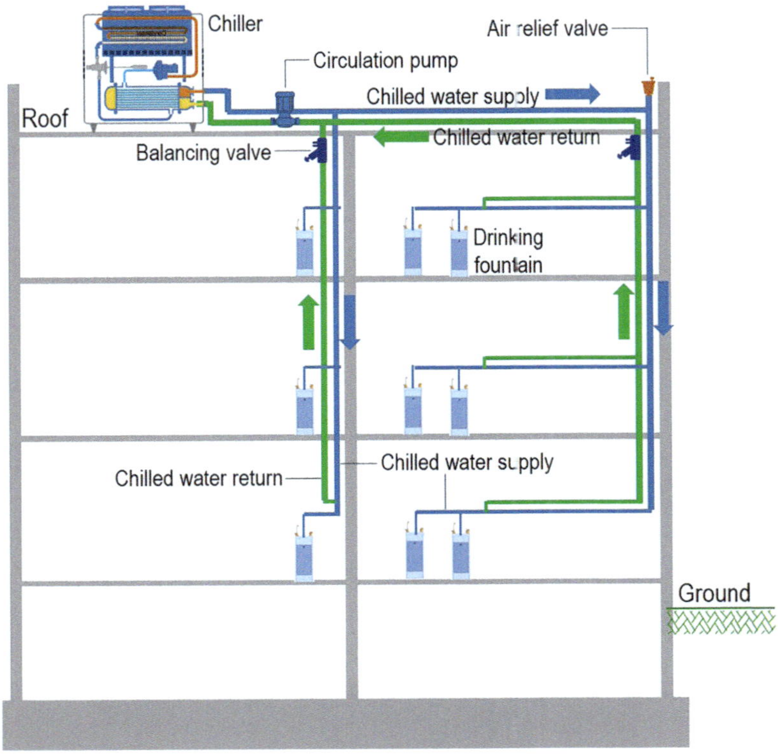

Figure 2.24 A chilled water supply (down-feed) system developed for drinking purposes.

drinking fountains located throughout the building or facility, as shown in Figure 2.24. The water, after being dispensed, returns to the chiller via the piping loop, where it is again cooled before being circulated to the dispensing points. This process forms a "loop", as the chilled water, having gained some temperature, returns through a return pipe back to the chiller, and the chilled water is circulated continuously without being discarded or refilled.

Supply piping includes riser mains and branches. A loop is usually a closed loop. On the return pipe, a balancing valve is installed to maintain the desired flow and, thereby, the temperature within the limit.

A water chiller is connected to a storage tank or immersion-type coil installed in the storage tank. The storage tank must be ideally sized. Average chilled water consumption for a person is considered to be 3.78 lpd, which may exceed up to 9.45 lpd [7].

The chiller used for the drinking fountains may be found enclosed in the cabinet with a dispenser, as shown in Figure 2.25, or remotely located for single or multiple dispensers. A remotely installed chiller for multiple dispensers should have a recirculation system, termed a central chilling system.

Central chilled drinking water distribution system components thereby consist of the following:

1 Chilling unit
2 Distribution piping (supply and return)
3 Drinking fountains (terminal unit)
4 Break tank (make up water storage tank)
5 Pump with controllers
6 Pressure tank (depending upon building height, available pressure, and farthest distance),
7 Valves etc.

Adequately sized storage tank is to be installed to provide sufficient water for fluctuating demand of multiple outlets in the building.

System pressure: Generally, the static pressure in a chilled water supply system is to be maintained at less than 1034 kPa. Chilled water supply pressure typically should not exceed 827 kPa at any point in the system. Differential pressure should not exceed 4 bar, i.e. 400 kPa.

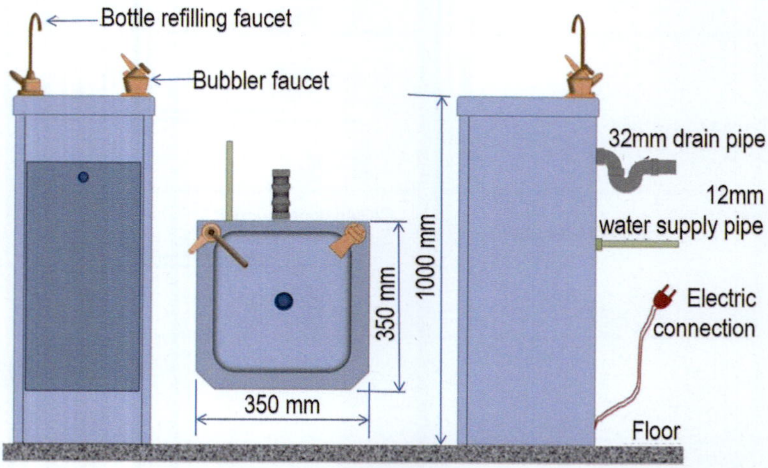

Figure 2.25 A drinking fountain including chiller inside.

Drinking fountain

Drinking fountains are the most common plumbing fixtures installed for facilitating drinking chilled water directly from the unit. These drinking fountains are directly connected to potable chilled water supply pipe and connected to drainage systems for draining out wasted or excess chilled water. Its body is mostly made of steel and stainless steel sheets, designed for exterior installations on the floor, may be semi-recessed or fully-recessed.

There are drinking fountains in which a chiller is made enclosed in the cabinet along with dispenser faucets, as shown in Figure 2.25. In this case only normal water is supplied to the drinking fountain.

Water is supplied to the drinking fountains either from a remote, central chilled water unit or ambient temperature water from the potable cold water system. Water is dispensed under pressure through a bubbler or spout installed on the unit. Wasted water that is not consumed is drained out directly into the floor drain or nearby sanitary drainage system. According to ASPE one drinking fountain is to be provided for every 75 people, at every floor [7]. In the drinking fountain preferred refrigerant used is HFC-134a instead of CFC gases, which is relatively environment friendly.

Chilled water distribution system for drinking

The chilled water supply system to the drinking fountains can be developed in two ways, as mentioned below:

1 Up-feed or pressurized distribution system
2 Down-feed or gravitational distribution system

A down-feed or gravitational chilled water distribution system for drinking purposes has been illustrated in Figure 2.24, in which the chiller has been placed on the roof of the building. In up-feed or pressurized chilled water distribution system for drinking purposes the chiller is generally placed in lowest level floor of the building.

Chilled water supply piping material

Pipes of various materials are used in a chiller. Of them, ASTM A53 Steel of schedule 40 and 80, ASTM B88 Copper Types K, L, and M, ASTM D2241 PVC (SDR 26), and ASTM F2389 Polypropylene (DR 9) are popularly used.

In a chilled water supply system, copper pipe is preferred, but brass or ductile iron pipe is also chosen when metal pipe is considered for the durability of the chilled water distribution system. If plastic pipe is considered economical, then CPVC, PP, HDPE, or PEX plastic pipes can be considered for the purpose. Exposed chilled water supply pipes require thermal insulation to prevent the temperature from rising too fast and beyond an acceptable limit.

Design of major components of chilled water supply system

Designing generally means sizing the major components of a chilled water supply system, which includes the following:

1 Sizing a chiller
2 Chilled water distribution pipe sizing
3 Sizing the cooling tower

Sizing a chiller

Sizing a chiller for room cooling:

There is an easy way to determine the size of a water chiller. The following few factors are considered in determining the size:

1 The incoming water temperature (IWT)
2 Chilled water temperature required (CWT)
3 The flow rate

The general formula for sizing a water chiller based on its flow rate is as follows:

$$Q = m \times C \times \Delta T \tag{2.1}$$

where, Q = Heat load in kilowatts (kW; kJ/s)
m = Chilled water mass flow rate in kg/s (flow rate in cum/s); $m = \rho q$

So,

$$Q = \rho q \times C \times \Delta T$$

where,

ρ = density of water, kg/cum
q = chilled water flow rate, cum/s
C = specific heat of the (fluid) water in kJ per kg per Kelvin (kJ/kg/K).
Water has a specific heat capacity of 4182 J/kg°C or 4.2 kJ/kg-K, or 1 Cal/gram°C and the specific heat of air at constant pressure is 1.005 kJ/kg-K or 0.24 Cal/gram°C
ΔT = temperature difference between inlet and outlet, in degrees centigrade

Example:

Let us consider that the standard chilled water temperature is supplied at 6.7°C and returned at 12.2°C. Assuming that the cooling load is 2200 kW, the required chilled water flow rate can be calculated as follows:

The chilled water density is 999.78 kg/cum, and the specific heat capacity is 4.19 kJ/kg/K. So,

$$2200 = 997 \times q \times 4.187 \times (12.2 - 6.7)$$
$$q = 2200 / 22959.41$$
$$= 0.0958\,\text{cum/s} = 95.8;\ \text{say, } 96\ \text{liters per second (lps)}.$$

Sizing a buffer tank

The formula for a chilled water buffer tank with two ports is as follows:

$$\text{Required System Capacity}(l) = 4 \times \text{Chiller kW}(l)$$

$$\text{Required Buffer tank volume}(l) = \text{Required System Capacity}(l) - \text{Total System Volume}(l)$$

If the chiller is 4-port, it's suggested to go with 1l (l) per kW of chiller capacity for calculating the required system capacity [8].

Example:

A 1000 kW chiller will require a system capacity of 4000 liters. If the total system volume (TSV) is only 3500 liters,

then the required buffer tank volume = 4000 − 3500 = 500 l.

So, a 500 liter buffer tank will need to be installed.

Cooling (refrigerant) load

The refrigeration capacity required to remove the total heat gain load from the environment, the calculated total cooling load, is divided by 16, since the device is intended to operate for 14–16 hours a day, say 16 hours a day.

So, the refrigeration capacity that the chiller will need = 155.47/16 = 9.71 kW.

Determining chilled water flow rate

If 1000 kg/hr of air is to be cooled from 45°C to 10°C in the heat exchanger with chilled water, then the chiller capacity required would be as follows:

The chilling duty (Qa) of the heat exchanger.

$$\begin{aligned} Qa &= \text{Air flow rate} \times Cp \text{ of air} \times \text{temperature}(\text{air in} - \text{air out}) \\ &= 1000 \times 0.24 \times (45 - 10) \\ &= 8400 \text{ kcal / hr} \\ &= 8400 \times 0.00033 = 2.78 \text{ TR} (1 \text{ kcal / h} = 0.0003306934 \text{ ton refrigeration}) \end{aligned}$$

For estimation, it is considered that a typical AHU requires a chilled water flow rate of 2.4 gallons per ton, i.e., 2.6 l/kW. But the value shall be varied according to the temperature changes (ΔT) of AHU using the Equation (2.1).

In reality, all air handling unit loads do not operate at their peak capacity simultaneously across different zones or areas within a building; rather, generally, they operate at different loads concurrently. So, it is wise to reduce the total demand for chilled water flow by multiplying its peak capacity by a factor called the diversity factor to meet the varying cooling demand.

A diversity factor represents the ratio of the maximum simultaneous load to the sum of individual peak loads occurring at different times within a system or across different zones. Rightly considering diversity factor helps in the selection of an efficient chiller that meets the varying cooling demands of different zones or equipment without oversizing or under-sizing the chiller capacity. A diversity factor of maximum 0.8 (80 percent) [9] can be considered for large establishments.

Sizing chilled water distribution pipes

Generally, a pipe for chilled water supply is sized for the desired total flow rate and available or allowable pressure drop in the pipe.

Generally, the branch piping of the chilled water supply system should be designed with a velocity of 0.9–1.8 mps, depending on the actual length. According to ASHRAE, the golden rules are that pipe size should be 50 mm and smaller, based on the maximum 1.22 mps velocity. Size 63 mm pipe and larger based on a maximum pressure drop of 0.762–1.22 of pipe [10]. For 63 mm piping and larger, the maximum velocity of chilled water flow shall be 2.4 mps [11]. To ease the determination of friction loss for fittings, valves, etc., assume the design pipe length is 1.5–2.0 times the actual length to account for the losses. After the pipe diameter is selected, evaluate the influence of each fitting, valve, etc. for confirmation.

After determining the chilled water flow requirement for a particular pipe serving the loads, the required diameter can be determined from the values furnished in Table 2.1 for the mains and Table 2.2 for the lateral branch pipes.

Table 2.1 Main chilled water distribution pipes: maximum allowable flow, velocity, and pressure drop [12]

Pipe size (mm)	*Flow (lps)*	*Velocity (mps)*	*Pressure drop (Pa/100 m)*
75	4	0.9	22712
100	9	1.1	21827
150	24	1.3	18877
200	41	1.3	13371
250	69	1.3	11700
300	114	1.4	12093
350	139	1.6	11208
400	208	1.8	11405
450	284	1.9	11503
500	379	2.0	11503
600	694	2.4	11700
750	1199	2.6	10913

Table 2.2 Lateral chilled water distribution pipes: maximum allowable flow, velocity, and pressure drop [12]

Pipe size (mm)	*Flow (lps)*	*Velocity (mps)*	*Pressure drop (Pa/100 m)*
75	6	1.3	43949
100	13	1.5	42179
150	32	1.7	31364
200	57	1.8	24383
250	95	1.9	20745
300	151	2.1	20647
350	196	2.2	20647
400	284	2.4	20549
450	379	2.5	19565

Sizing a cooling tower

The simplest way to determine the cooling capacity of the cooling tower (CT) is to use the basic Equation (2.1).

$$Q = mC\Delta T;$$

where m = water flow rate (cum/h);

C = water-specific heat (1 kcal/kg/°C) and
ΔT = water temperature difference (°C).

By putting the values in the equation, the end result will come in kcal/h, which can be converted to tons (refrigeration) by dividing by 3023.9, as 1TR ton refrigeration is equal to ~3023.9 kcal/h.

Also, the value in kcal/h can be converted to kW by multiplying by 0.001162, as 1 kcal/h is equal to 0.001162 kW.

Operation and maintenance

The chilled water supply system involves more equipment in comparison to other water supply systems. So, it needs more monitoring and maintenance operations to achieve consistent throughput. Maintenance operation depends on the type of machinery or equipment controlling devices like valves and accessories like sensors, etc., their size, and their climatic condition. Some of the monitoring and maintenance activities to be performed are listed below:

1 Rigorous cleaning of the cooling towers is to be ensured using a cleaning chemical that circulates and flushes out the system. Pressure washing and vacuuming can also be carried out as cleaning processes.
2 Spinning elements like fans, motors, and belts need to be inspected, cleaned, and adjusted when necessary.
3 Electrical components like capacitors and wiring should be inspected and replaced when necessary.
4 Strainers and nozzles shall be checked on a regular basis and cleaned.
5 Water treatment options like water modeling, green chemical usage, filtration, and softening processes shall be monitored and refreshed periodically, at least every month.
6 Best practices shall be carried out primarily for chilled water flow balancing throughout the system, for which the following routine measures are to be taken:

   a Ensure proper distribution of chilled water flow in the system by monitoring, regulating, and maintaining the hydronic balancing valves in the system. Pressure and temperature shall have to be maintained automatically for proper distribution of chilled water.
   b Consider the use of pressure-independent control valves, which automatically adjust flow rates to maintain set temperatures, enhancing system efficiency.
   c Check the condition of the insulation of chilled water pipes to minimize heat gain, prevent condensation, and maintain the desired water temperature.
   d Engage skilled manpower to ensure that all components are operated, monitored, and maintained as intended.

References

[1] www.waterchillers.com "Air cooled chillers" Retrieved from https://waterchillers.com/air-cooled-chillers/#:~:text=Air%20cooled%20chillers%20absorb%20heat,discharges%20is%20not%20a%20factor on 21 August 2024.
[2] www.airchange.com.au "The scroll compressor" Retrieved from https://www.youtube.com/@airchange255/featured on 21 April, 2025.
[3] PT Metropolitan Bayutama "Screw compressor" Retrieved from https://www.bayutama.co.id/index.php/screwcompressor/ on 21 August, 2024.
[4] Reynold India Private Limited (2023) "Water-cooled chiller capacity range" Retrieved from https://www.reynoldindia.com/blog/water-cooled-chiller-capacity-range/#:~:text=The%20capacity%20of%20a%20water,5300%20kW%20or%20even%20higher on 24 March 2024.
[5] Consulting-Specifying Engineer (2019) "Chiller selection: Much more than capacity" by Stacey Ayuthia, Retrieved from https://www.csemag.com/articles/chiller-selection-much-more-than-capacity/#:~:text=The%20capacity%20of%20scroll%20chillers,(e.g.%2020%20tons%20each) on 24 March 2024.
[6] Delta Cooling Towers, Inc. (2016) "Determining proper cooling tower size" Retrieved from https://deltacooling.com/resources/news/determining-proper-cooling-tower-size on 26 March 2024.
[7] Mohammed Abdul Muqeem Siddique (2021) "Designing central drinking chilled water system (for offices, institutional buildings (jails), manufacturing units, housing camps, gym etc)" Retrieved from https://www.linkedin.com/pulse/designing-central-drinking-chilled-water-system-offices-siddique/ on 30 March, 2024.
[8] Masterflow Solutions (2024) "How to size a chilled water buffer tank correctly" Retrieved from https://masterflow.net.au/how-to-size-a-chilled-water-buffer-tank-correctly/ on 23 August, 2024.
[9] Syed Shaaz Hussain (2023) "Understanding diversity factor in air cooled chiller selection" Retrieved from https://www.linkedin.com/pulse/understanding-diversity-factor-air-cooled-chiller-syed-shaaz-hussain-vromf/ on 23 August, 2024.
[10] ASHRAE (1997) *ASHRAE Fundamentals Handbook* Chapter 33 Retrieved from https://takyifwasalama.com/wp-content/uploads/2020/07/ASHRAE-Pipe-Sizing.pdf on 31 March 2024.
[11] Yale University (2009) "Steam and chilled water utilities design guidelines" Retrieved from https://facilities.yale.edu/sites/default/files/files/Steam%20and%20chill%20water%20design%20guide.pdf on 31 March 2024.
[12] The Engineering ToolBox (2014) "Cooling water - pipe lines" Retrieved from https://www.engineeringtoolbox.com/cooling-water-piping-systems-d_1884.html on 01 April 2024.

3 Dual water supply system

Introduction

Mostly, buildings are provided with potable water supply system. It is not necessary that a building should have only potable water supply systems for all types of water needs, as ensuring the quality of potable might be costly, due to its limited availability at present and increasing high demand as well. Buildings need water for various purposes, and providing potable water is not only a loss of money but also logically unethical and in many areas might be illegal from a regulatory point of view. In this context, a separate water supply system is developed, generally in conjunction with the potable water supply system in a building. Commonly, a non-potable water supply system is developed side by side with a potable water supply system in a building, which is generally termed a dual water supply system.

The source of non-potable water may be of a different nature, but its quality must be of acceptable quality as assigned by the concerned regulatory authority.

Water supply authorities may develop dual water supply mains consisting of two separate mains: one conveying drinking (potable) water and the other conveying appropriately treated non-drinking (non-potable) water. Individual buildings may also develop a dual water supply system. The development of a dual water supply system needs additional precautionary measures to be undertaken for the safety of the system and the users as well. In this chapter, the development of dual water supply systems for particular purposes in a building is discussed.

Potable and non-potable water

From a human perspective, water can be broadly termed "potable" or "non-potable". "Potable" water represents the quality of water that is suitable for human consumption directly. On the other hand, "non-potable" water has all other qualities of water that are not suitable for direct human consumption, such as for drinking and culinary purposes, but depending on its level of quality, it may be used for many other purposes.

Non-potable water may be obtained from various sources, as discussed below:

1. Untreated or partially treated surface water
2. Harvested rainwater that has not been filtered or conditioned
3. Recycled or reclaimed wastewater which is the water that is used more than once, either treated or untreated, before discharging into the drainage or natural water course.

DOI: 10.1201/9781003724056-3

Typically, "recycled water" is used to refer to the reuse of treated wastewater for various purposes, to minimize the use of costly and limited potable water.

Non-potable water, even though it looks apparently clear, may contain microorganisms, including viruses, bacteria, such as Salmonella, Campylobacter, and E. coli, etc., and gastro-intestinal parasites, such as Giardia, Hydatid, or Cryptosporidium. These pathogens can cause severe gastroenteritis, such as vomiting or diarrhea. Infants, the elderly, and people with suppressed immune systems are at greater risk for these health problems.

Quality parameters of potable and non-potable water

The quality of potable water shall be according to the parameters mentioned in Table 3.1.

On the contrary, in buildings, the recommended quality of non-potable water or rainwater should comply with the parameters provided in Table 3.2.

Table 3.1 Drinking water quality standards as recommended by the FEPA and WHO [1]

Sl. no.	*Parameter**	*FEPA[a] standards*	*WHO[b] standards*
1	pH	6–9	6.5–9.2
2	Total hardness	–	300
3	Total dissolved solid	2000	500
4	Electrical conductivity	–	300[c]
5	Total coliform count (100 ml)	0	0
6	Sulfate	20	200
7	Sodium	–	200
8	Ammonium	0.01	1.5
9	Zinc	5.0	5.0
10	Iron	0.05	0.3
11	Lead	0.01	0.05
12	Cadmium	0.05	0.01

* All values are in mg/l, except pH, EC (μS/cm), and total coliform count (CFU/ml).
a FEPA (1991).
b WHO (1997).
c WHO (2003).

Table 3.2 Non-potable (rain) water quality standards as prescribed by the uniform plumbing code [2]

Sl.no.	*Parameter*	*Value*
1	pH	6.0–7.0
2	BOD	Not greater than 10 mg/l
3	NTU	Not greater than 2
4	Fecal coliform	No detectable fecal coli in 100 ml
5	Sodium	No detectable sodium in 100 ml
6	Chlorine	No detectable chlorine in 100 ml
7	Enteroviruses	No detectable enteroviruses in 100 ml

Non-potable water use

Non-potable water, depending on its quality, can be used for various purposes other than directly affecting human health. Some areas and ways of using non-potable water from various sources may be as follows:

1. Stored rainwater can be used for fire protection, cooling towers and boiler feed, makeup water, car washing, toilet flushing, clothes washing, pet bathing, etc.
2. Fountains, water falls, and ornamental ponds
3. Swimming pool backwash water can be used for toilet flushing, building washing, floor or road washing
4. Construction projects such as concrete mixing and dust control
5. Environmental enhancements such as fountains and landscape ponds
6. Agricultural wastewater can be used for landscape irrigation, golf courses, sports fields, and crop irrigation
7. Wastewater from a sewage treatment plant can be used for irrigation
8. Quarry water for dust suppression
9. In cooling towers

Recycled water may be prohibited for the following purposes:

1. Drinking
2. Culinary or other kitchen purposes
3. Baths, showers, hand basins, or personal washing
4. Swimming pools
5. Water contact recreation (playing under a water fountain)

Dual water supply system

Dual water supply systems in a building involve two separate distribution systems that supply potable water through one distribution network and non-potable water through another. Both networks work independently within the building, with pipes running mostly parallel to each other and maintaining a distance of separation of not less than 100 mm. and at the left side [3].

The potable-water supply system, also known as the primary supply system, operates like any normal potable-water supply and distribution system and requires an independent potable water source, storage, pumping, and piping system. On the other hand, the non-potable water supply system, also known as the secondary supply system, distributes water from a non-potable water source. The non-potable water system includes separate storage, pumping, and distribution piping, which may or may not have a treatment system.

In this system, corrosion-resistant materials must be used for the corrosive non-potable water to minimize maintenance problems. Distribution pipes used for conveying non-potable water are either tinted or painted purple to distinguish them from potable water conveying pipes.

Urban dual-water supply system

In an urban water supply system, there might be a dual water supply system, in which one pipe network supplies potable water and the other pipe network that runs parallel to the potable water supply network conveys non-potable water, as shown in Figure 3.1.

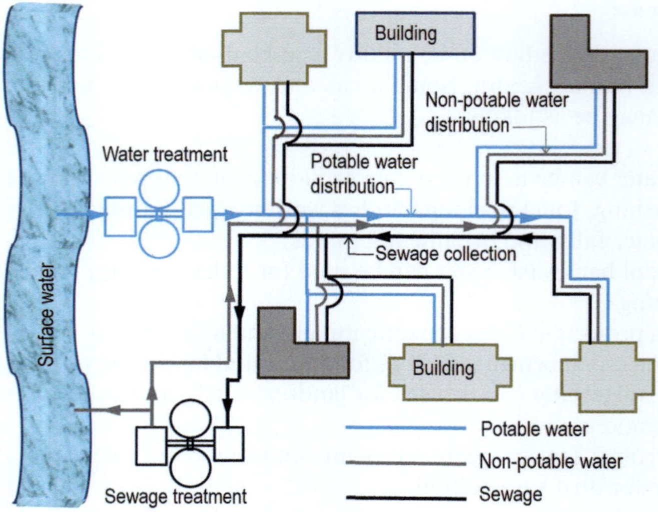

Figure 3.1 Schematic diagram of an urban dual water supply system.

A dual water supply system in a building

In a building, sometimes a dual water supply system is developed, particularly when a dual water supply system is developed in the concerned area. In other areas, one may develop a dual water supply system in a building as a concern for water conservation or as a measure of water crisis management that may prevail in the area. In a dual water supply system in a building, one water-distributing pipe network supplies potable water and the other runs mostly parallel to the potable water supply network, which conveys non-potable water as shown in Figure 3.2.

In some cases, non-potable water supply is to be supported by potable water supply to supply potable water into the non-potable supply network, where there is possibility of shortage in non-potable water at source and during maintenance of non-potable water source components. To address these issues, potable water is connected to the non-potable water pipe network, but to prevent contamination of potable water, it is supported by backflow preventers. Both the distribution system should follow either down-feed or up-feed system.

Advantages and disadvantages of a dual water supply system

A dual water supply system, though developed for managing certain limitations, has some advantages and disadvantages as well, as discussed below.

Advantage

A dual water supply system ensures the use of lesser-quality waters for non-potable purposes, which reduces the use of limited or very costly freshwater resources for such purposes.

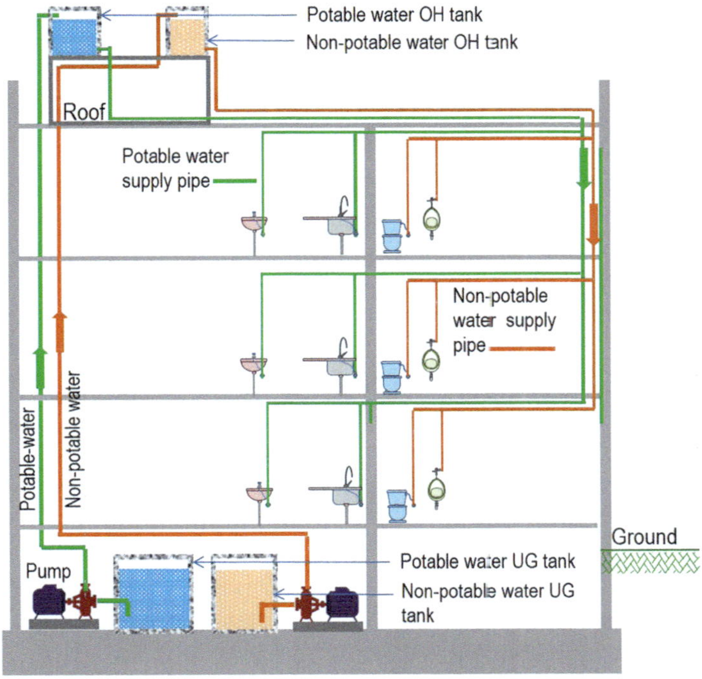

Figure 3.2 Down-feed dual water supply system in a building.

Disadvantages

In a dual water supply system, two distribution systems are to be installed, resulting in high capital, operating, and maintenance costs. In addition, there are risks of polluting the surrounding area when leaks in non-potable systems occur. Furthermore, there are risks of cross-connections between potable and non-potable systems. There is also a risk of misuse of potable water and accidental use of non-potable water while intending to use potable water, especially by uncommon users. Above all, cultural and psychological barriers might be potential hindrances to its development and use.

Guidelines for a dual water supply system

Dual water supply systems must follow some additional guidelines than those for normal potable water supply systems. The guidelines are as follows:

1. Water supply mains conveying non-potable water as part of a dual water supply system shall be constructed from pipes that shall be colored purple.
2. Non-potable water meter: The recycled water meter shall be purple and shall be installed beside the potable water meter with at least 100 mm separation distance between the two meters. Like the potable water meter, the non-potable water meter shall be installed above the ground. The non-potable water meter coupling threads are different from those on the potable water meter to prevent interchange of the meters.

Table 3.3 Safe distance of installing non-potable storage tank from some elements [4]

Element	*Minimum horizontal distance from storage tank, in meter*
Lot line adjoining private lots	1.50
Sewage systems	1.50
Septic tanks	1.50
Water wells	15.00
Streams and lakes	15.00
Water service	1.50
Public water main	3.00

For SI: 1 foot = 300 mm.

3 Potable water piping must not be installed within 100 mm above ground or 300 mm below the ground of any parallel non-potable water plumbing.
4 Fixtures and fittings above and below ground must be clearly and permanently marked and labeled with "Non-potable water; – DO NOT DRINK".

Storage of non-potable water

A non-potable storage tank shall be located at a safe distance from some elements that might be affected due to some misappropriation of the tank. Storage tanks shall be installed with a minimum horizontal distance between various elements as furnished in Table 3.3.

If rainwater is stored as non-potable water, then it may be stored in a single reservoir with two compartments, one for potable water and the other for rainwater storage, being separated completely by a partition wall [5]. The separation wall shall have no opening in the portion above the topmost storage water level.

Supply of potable and non-potable water in a building

The non-potable water supply system must be designed to maintain appropriate water pressure and flow and, most importantly, to avoid contamination by non-potable water.

The dual distribution systems involving potable and non-potable water are developed by drawing these two different waters from two different sources and then supplying them through two separate distribution networks. The two systems work independently of each other within the same service areas in a building. Both systems include pumping and distribution pipes, including valves and other accessories.

In a dual water distribution system, there is every possibility of back flow through cross-connections between potable and non-potable water and thereby cross-contamination of the potable water supply, particularly when provision is made for supplying potable water through a non-potable pipe network due to its likelihood of exhaustion anyhow.

Back flow

Backflow in a water supply pipe may occur simply because of the loss of pressure causing the flow in that water supply pipe. Backflow, or the reversal of flow from its normal flow direction in a pipe, is usually caused either by the development of backpressure or back-siphonage.

The backpressure in a supply pipe might be caused by various reasons, such as a difference in elevation, a pump, a steam boiler, or other means. Backflow due to backpressure may be created due to the installation of a booster pump, heating boiler, etc. on water supply piping.

Back-siphonage

Back-siphonage may occur when the water pressure within the distribution piping falls below that of the faucet it is supplying. Back-siphonage is a backflow caused by a negative pressure, vacuum, or partial vacuum in the supply piping, as illustrated in Figure 3.3. Back-siphonage occurs when the pressure inside a pipe is reduced below the atmospheric pressure. Due to the effect of back-siphonage, water flows in the back direction of the sub-atmospheric pressure zone.

For example, suppose an automatic lawn sprinkler system is spraying a lawn when, all of a sudden, a back-siphonage occurs due to a fire truck pumping water or water rushing from a broken water main. The resulting backflow from the lawn sprinkler system will flow into the plumbing system and then into the water distribution system. As the water backflows, it can suck contamination such as insects, pesticides, herbicides, fungicides, fertilizer, worms, and other contaminants, into the lines through the sprinkler heads. Once in the distribution lines, the contamination could go anywhere in the public water system.

Backflow hazard rating

Back flow and cross connections are rated at three degrees of hazard in connection with the water supply system. The hazards ratings are as follows:

1 High hazard: In a water supply system, there is any condition, device, or practice that has the potential of contaminating the water to such a degree that it will cause the death of the consumer.
2 Medium hazard: In a water supply system, there is any condition, device, or practice that has the potential of contaminating the water to such a degree that it will endanger health of the consumer.
3 Low hazard: In a water supply system, there is any condition, device, or practice that has the potential of contaminating the water to such a degree that it will constitute a nuisance but does not endanger health or cause injury.

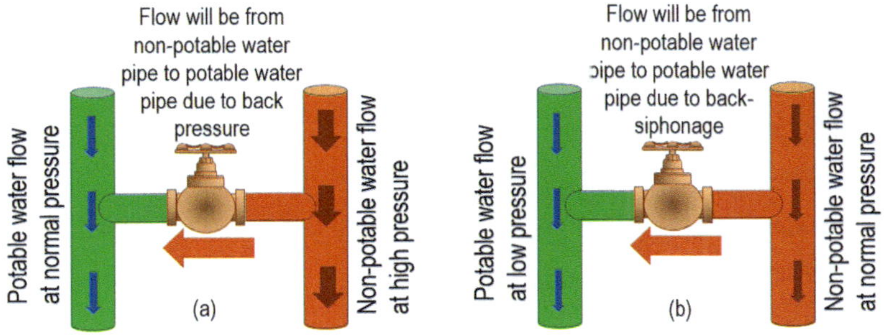

Figure 3.3 Example of back-pressure and back-siphonage.

Backflow preventer

Backflow preventers are controlling devices installed in the piping system to ensure that water is flowing in one direction in the pipe that cannot flow in the opposite direction anyway. Mostly, these kinds of devices are used to protect potable water supply piping against contamination from various kinds of cross-connections made with non-potable piping, thereby keeping potable water safe and clean.

The principal method of backflow prevention suggested by plumbing codes is by keeping an air gap at the supply to a fixture. When an air gap cannot be provided, a backflow preventer shall be used.

Types of backflow prevention assemblies

There are several types of backflow preventers available to choose from. The primary factor in choosing the befitting one is the potential for cross-contamination between the potable and non-potable water supplies.

Among various types of backflow preventer assemblies, the following backflow prevention assemblies are popular in use:

1 Reduced pressure principle assembly (RP)
2 Double check valve assembly (DC)

Reduced pressure principle assembly (RP)

A reduced-pressure (RP) principle backflow prevention assembly consists of two independently acting, internally loaded check valves, a differential pressure relief valve, four properly located test cocks, and two tightly closing shutoff valves at two ends, as shown in Figure 3.4. This assembly provides the highest level of back flow protection, so it is used where a high hazard level condition exists.

In this backflow protection device, there are two independently-acting, spring-loaded check valves that are separated by a differential pressure relief valve, which determines if the supply pressure is greater. The check valves will remain closed; if the difference in

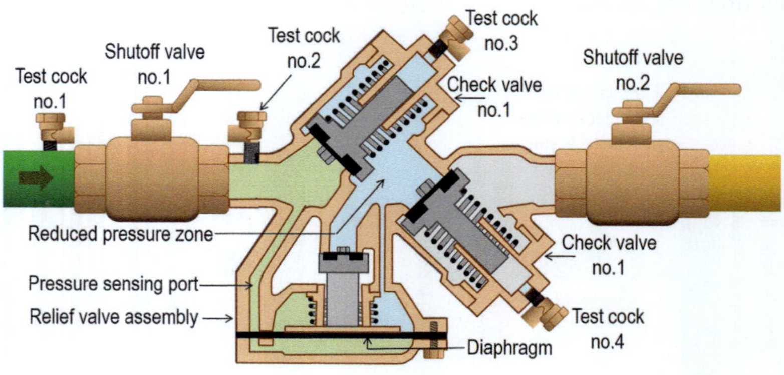

Figure 3.4 A cut section of the reduced pressure principle assembly.

pressure between the supply and zone is reduced, the relief valve will open. During normal operation, the pressure between the two valves is maintained at least 13.8 kPa lower than the supply pressure. The relief valve maintains a zone of relative differential pressure between the two check valves, unless one or both of the check valves malfunction, mostly by leaking. If either check valve leaks, the difference in pressure between the supply and zone is reduced, and the pressure relief valve will open, discharging water out of the system to maintain the pressure differential between the pressure of the fluid in the zone and downstream pipe. This device provides the maximum protection of any valve and can be used in high-hazard situations.

An RP assembly should be properly installed, for which some guidelines are to be followed, as mentioned below:

1 It should be installed at a minimum of 300 mm over a floor and away from a wall.
2 If the assembly is installed higher than 1.5 m over the ground, a permanent platform is required for access and maintenance.
3 Provision for adequate drainage is required, as there is a potential for discharge from the unit.
4 It should be protected from freezing.

Double check valve assembly (DC)

The double check valve assembly (DCVA) is essentially two single check valves assembled within one body and furnished with four test cocks and two shut-off valves at two ends, as shown in Figure 3.5. These valves are designed for use in non-health hazard cross-connections and continuous pressure applications subject to backpressure or back-siphonage incidents, such as lawn sprinklers, fire sprinkler lines, etc., but are not allowed for irrigation systems.

A DC assembly should also be properly installed, for which some guidelines are to be followed, as mentioned below:

1 It should be installed at a minimum of 300 mm over a floor and away from a wall.
2 If the assembly is installed higher than 1.5 m over the ground, a permanent platform is required for access and maintenance.

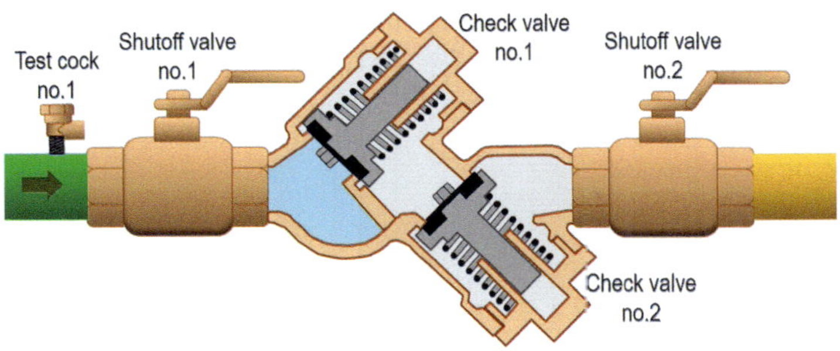

Figure 3.5 A double check (DC) valve assembly.

Separation of potable and non-potable water main

In installing potable and non-potable water pipes side by side, the guidelines given below must be followed.

1 Potable and non-potable mains shall be separated by at least 1.5 m at outside walls.
2 The bottom of potable mains is at least 305 mm higher in elevation than the top of non-potable mains and a sanitary or combined sewer.
3 Non-potable mains are constructed with potable water class pipe and pressure tested for water tightness.
4 Non-potable water piping shall be separated from the building drain or sewer underground by 1.5 m of undisturbed or compacted earth.
5 Non-potable water piping shall not be located in, under, or above any cesspools, septic tanks, drain fields, or seepage pits, etc.

Switching over

In a dual water supply system, switching over from non-potable water supply to potable water supply is sometimes required when non-potable water supply is ceased due to some unavoidable condition like an emptying in the source or a fault in the supply piping, etc. To overcome such a situation, the system is equipped with automatic switching over devices. The switching device used floats to automatically detect when sufficient potable water was available for use in the building. When non-potable water levels were low, the device automatically switched back to potable water, ensuring an uninterrupted supply of water to the building water supply system.

Cross-connection between potable and non-potable water piping

A cross connection is an additional or potential connection between any part of a potable water supply system and a non-potable water supply system in a manner that, under undesired circumstances, would allow potable water to enter the non-potable water system. A cross-connection is any connection or arrangement, physical or otherwise, between any potable water supply systems, either directly or indirectly connected to a water main, fixture, storage tank, receptacle, equipment, or device through which there may be possibilities of entering any non-potable, contaminated, polluted, used, or unclean water, or any other substance, into any part of such potable water system under any condition.

Another way to define a cross connection is any actual or potential connection between the potable water works and any source of pollution, contamination, or other material or substance that could change the quality of water in a potable water supply.

Cross connection may be of two types as follows:

1 Temporary cross-connection
2 Permanent cross-connection

Temporary cross-connection

A temporary or potential cross-connection is a connection for which something must be done to complete the connection.

A hose connected to a hose bib may be a cause of contamination when the end of the hose could be connected to or placed into a non-potable water; e.g. water in a swimming pool, lawn irrigation etc. Therefore, in these cases the plumbing codes require backflow prevention device to be installed on the hose bib.

Temporary cross-connection between the potable water system and the non-potable water system may be permitted when the non-potable water system is supplied from the drinking water system. Temporary cross-connections shall be subject to Water Agency approval. Temporary cross-connections shall be provided at locations and in accordance with the requirements of the Water Agency. It is the responsibility of the Water Agency to ensure the management of temporary cross-connections and their removal when non-potable water becomes available.

Permanent cross-connection

A permanent or an actual cross-connection is the connection that exists at all times.

Any permanent cross-connection between a potable and non-potable water supply has a high hazard rating. There shall be no permanent cross connections between the potable water and non-potable water systems within the network downstream of storages without the permission of the concerned Water Agency.

If potable water is to be filled in a non-potable water reservoir, then the inlet of potable water shall be at least 300 mm above the highest water level in the reservoir.

All backflow prevention devices must be commissioned and tested by a licensed plumber with backflow prevention accreditation.

Avoiding cross-connection

Where the non-potable water supply needs to be supplemented by water from the potable water supply, this shall be provided through a fail-safe backflow prevention device with built-in independently acting non-return valves, preferably by installing reduced pressure zone device (RPZD). In some instances, the Water Agency may require the installation of two RPZDs in parallel to provide alternative supply. Other approved backflow prevention devices like double check valve, as shown in Figure 3.6, can be installed in case of low hazards. Only backflow prevention devices are to be installed on any cross-connection between the potable water and non-potable water systems within the Water Agency's area of responsibility. In any event, the device needs to be located above ground level and in a secure surrounding compound, but not in a cabinet without drainage or ventilation.

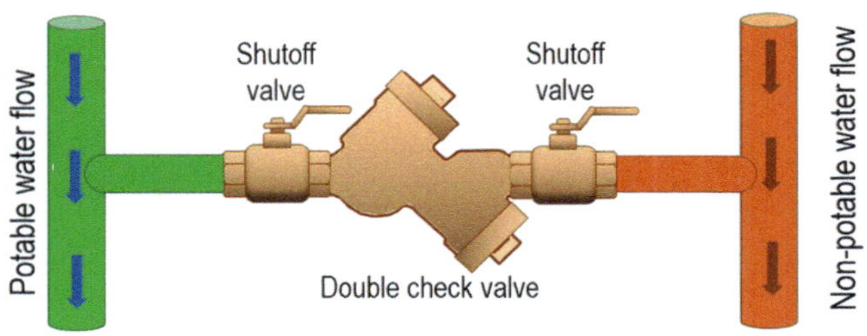

Figure 3.6 A cross-connection supported by double check valve.

References

[1] Adeolu O. Aderemi, Ada V. Oriaku, Gbenga A. Adewumi and Adebayo A. Otitoloju (2011) "Assessment of groundwater contamination by leachate near a municipal solid waste landfill" Retrieved from https://www.researchgate.net/publication/255178792_Assessment_of_Groundwater_Contamination_by_Leachate_near_a_Municipal_Solid_Waste_Landfill/figures?lo=1 on 4 April, 2024.

[2] Uniform Plumbing Code (2015) *Non-potable Water Systems* Chapter 13 Retrieved from https://up.codes/viewer/texas/ipc-2015/chapter/13/nonpotable-water-systems#13 on 4 April, 2024.

[3] www.southeastwater.com.au "South East Water technical guidance – Recycled water installation information" Retrieved from https://southeastwater.com.au/content/dam/building-and-development/South%20East%20Water%20technical%20guidance%20Recycled%20water%20installations.pdf on 29 August, 2024.

[4] Uniform Plumbing Code (2021) *Water Supply and Distribution* Chapter 29 Retrieved from https://up.codes/viewer/virginia/irc-2021/chapter/29/water-supply-and-distribution#29 on 29 August, 2024.

[5] S. Azizul Haq (2017) *Harvesting Rainwater from Buildings*. Switzerland: Springer international Publishing.

4 Fire standpipe system

Introduction

All buildings are more or less prone to getting caught by fire; the consequence may be catastrophic in terms of life or property loss. The building structure itself, finishing, decorative materials, and its contents may also be susceptible to fire. In a building, there might be the occurrence of unintended fire; generally, uncared intended fire, due to overheating of weak or overloaded electrical elements, and in addition, some uncared-for human activities with fire might become the cause of fire in a building. Therefore, buildings need to be developed to ensure an acceptable level of fire safety: minimize the risks of unwanted occurrence of fire and its rapid spreading; maximize the scope of evacuating because when there is an unwanted fire all occupants are supposed to leave their places to a safe refuge area; and install a proper fire suppression system to contain heat and smoke caused by fire.

When an unwanted fire occurs or is identified, it is to be doused without delay. Fire in a building occurs for many reasons, which mostly escalate due to the burning of flammable hydrocarbon materials present in the building. Such a fire can be easily doused by pouring or spraying water thereon. For large fires, water is thrown from a safe distance; so, a special water supply system is to be developed, commonly known as a fire hydrant system, to help spray sufficient water on fire for a certain time and from a distant location. In this chapter, one of the fire hydrant systems, known as stand pipe system, is discussed, which is developed in buildings for the purpose of dousing fire safely and effectively.

Fire

Fire is a state, process, or instance of combustion—a series of very rapid chemical reactions in which fuel or other material is ignited after reaching a level of heat in the presence of oxygen, giving off light, heat, and flame. So, fuel, heat, and oxygen are the principal elements that cause a fire, which are termed a triangle of fire as represented in Figure 4.1.

For a combustion reaction to occur, a fuel source is required to be heated to a certain level called the ignition point of that fuel in the presence of oxygen. Fuels are materials such as wood, coal, gas, or oil that are burned to produce heat or power. Surrounding air is the source of oxygen, as it contains about 21 percent oxygen. Heat is a thermal energy that is caused by the violent movement of tiny particles like atoms, molecules, or ions in solid, liquid, and gaseous matter. Heat is transferred from one object to another due to the difference in temperature between the two objects. Flames of fire are the visible manifestation or outcome of the combustion reaction. As a fire grows, it spreads; the rate of spreading is subject to the fuel load, primarily the volume available for spreading and the passage for spreading through.

DOI: 10.1201/9781003724056-4

Figure 4.1 Triangle of fire.

Causes of a fire in a building

There might be innumerable causes of minor to major fire incidences in a building. Of them, there are some common causes of fire occurrences, as listed below:

1 Electrical short circuits
2 Heated electrical items
3 Oven fire
4 Open flames
5 Burning cigarettes
6 Lightening, etc.

Whatever the cause of initiating the occurrence of a fire, it grows big due to the burning of some fast-ignitable fuels like paper, fabric, plastic, or wood materials present around, and the fire not being doused at its initial minor or small stage. The fire then might become unmanageable by some individuals' initiatives to suppress the fire, which then starts growing and spreading.

Fire suppressing

When the fire is too small, it can be suppressed or doused by an individual who might cause the fire accidentally or detect the fire. When the fire is comparatively small and remains contained, generating no toxic smoke, then a trained individual or a small group of trained individuals, present on the concerned floor of the building with safe exits, can fight the fire to suppress it with the help of an available portable fire extinguisher or by drawing water from a nearby source or a fire hydrant. It is to be noted that water is a very common agent to extinguish a fire. Water extinguishes a fire by cooling the fuel and the surrounding environment, which removes heat because of its ability to absorb massive amounts of heat as it converts to steam vapor.

Fire hydrant

A fire hydrant, also termed a fireplug or fire-cock (archaic), is basically a point of connection on a standpipe by which one, firefighters in particular, can tap into a water supply at high pressure to draw water for mainly suppressing a fire nearby. It is one of the most important building services and a major component of an active fire protection system in a building.

The buildings or places in the following configuration should have indoor fire hydrant systems:

1 Plants and warehouses with a building footprint greater than 300 sqm
2 High-rise public buildings and residential buildings with a building height than 21 m
3 Building height greater than 15 m or volume greater than 10,000 cum of office, academic buildings, and other single, multistory buildings
4 A compartment having a volume greater than 5,000 cum, like in stations, terminals, airports, exhibition halls, store houses, manufacturing buildings, libraries, gymnasium buildings, etc.
5 Theaters or cinema halls with more than 800 seats and auditoriums with more than 1200 seats

Advantages of a fire hydrant system

1 A fire hydrant system is an effective means of suppressing both the initial and final stages of fire in a building.
2 The system provides the opportunity to fight the fire from a safe distance due to its large piping system.
3 As it covers each point of the building, it has more penetration capability as compared to other fire protection systems. So, high accessibility is its main advantage.
4 Hydrant valves can be installed at any strategic location throughout the building, leaving no room or area unprotected within the building and its premises.

Disadvantages

1 Water flow through a hydrant might be so strong that its operation requires a minimum of two trained personnel to manage the thrust exerted by the hose.
2 Hose reels might be damaged during operation or while in the cabinet.
3 Fire hydrant water can cause damage to sensitive elements or equipment during operation.
4 The pumps may malfunction, and water may be exhausted during operation.

Fire hydrant types

For the protection of a building from fire, the following types of fire hydrant systems can be developed:

1 Fire standpipe system
2 Fire sprinkler system (discussed in chapter 5)
3 Combined standpipe and sprinkler system

Fire standpipe system

A fire standpipe system is an assemblage of a series of components that together provide a source of water to assist fire fighters in suppressing fire. The components of a fire standpipe system are as follows:

1 Water reservoir
2 Fire pumps

3 Stand pipes and branch pipes
4 Panels
5 Hoses
6 Nozzles
7 Fire brigade connections
8 Landing valves
9 Appurtenances

A schematic diagram of a fire hydrant in a multistory building has been shown in Figure 4.2 to show its various components.

Types of fire standpipe systems

There are basically two types of standpipe systems that ensure how the water is delivered through the outlets, to suppress a fire:

1 Wet standpipe system
2 Dry standpipe system

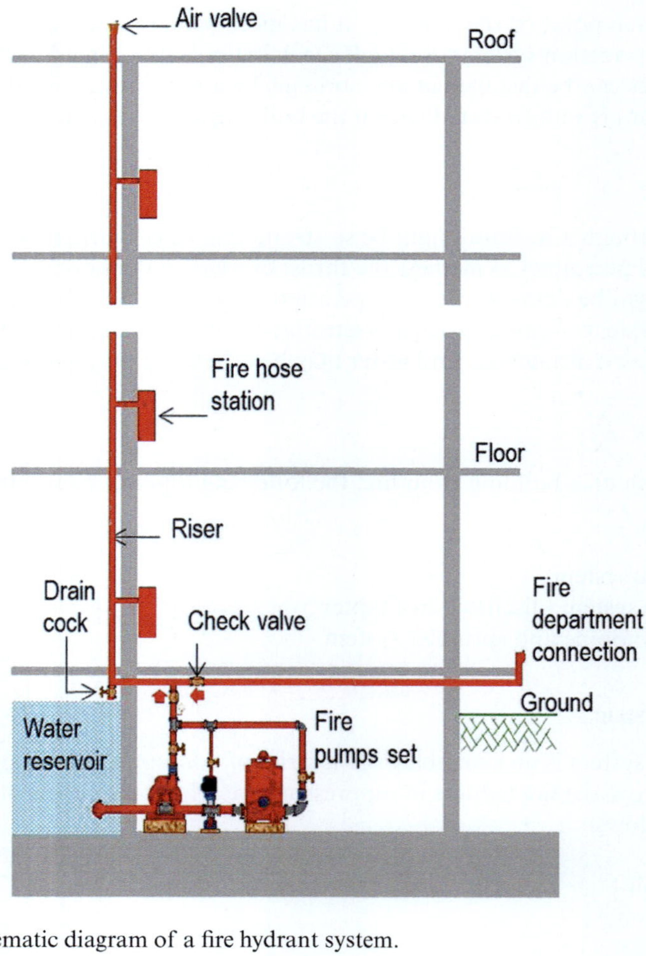

Figure 4.2 A schematic diagram of a fire hydrant system.

Wet standpipe system

In the wet stand pipe system, water in the pipes is kept filled and pressurized at all times. When a hose nozzle is opened, water flows through it. This system is generally preferred in high-rise buildings where it would be challenging for the firefighters to pump water at high pressure to deliver water to the high-level floors of the building. Wet standpipe systems are designed in two ways, as follows:

1 Automatic wet standpipe systems
2 Manual wet standpipe systems

Wet standpipe systems are less preferred for areas where very low atmospheric temperature will cause water to freeze and thereby damage the system.

Automatic wet standpipe systems: In an automatic wet pipe system, water instantly flows from the nozzle as the system remains filled with water under pressure. When water flows through a nozzle, pressure drops in the system, which activates the pumps to cause a supply of water and develops pressure in the system to make it continuously operable during a firefighting.

Manual wet standpipe systems: In this wet system, water is kept filled in the pipes at all times, but not under pressure, to fight fires by simply opening the landing valve and the nozzle. This system is mainly for the use of the fire department, which uses a pumper to cause an external water supply to enter the system through the fire department connection (FDC) installed outside the building and connecting with the system. These systems are suited for high-rise buildings where water is not going to freeze. The biggest advantage of having water in the pipes at all times is that it helps detect leaks before an emergency.

Dry standpipe system

As opposed to a wet standpipe system, in a dry standpipe system, the standpipe riser is kept devoid of water to avoid freezing at low temperatures; instead, it is filled with air under pressure and supplied with water on demand when a landing valve and nozzle are opened, as the system is connected to a water reservoir through pumps. Dry standpipe systems are designed in three ways, as follows:

1 Automatic dry standpipe systems
2 Semiautomatic dry standpipe systems
3 Manual dry standpipe systems

Automatic dry standpipe systems: In this system, when the landing valve is opened, it releases the pressurized air. A pressure drop in the air trips a differential dry-pipe or deluge (pre-action, discussed in chapter 5) valve installed downstream of the standpipe riser, causing water to gush into the riser. As the water is under pressure in the riser piping downstream of the deluge valve loss pressure, the fire pump starts running to deliver water at the desired pressure. This system is best suited for freezing environments below 4°C as well as high-rise parking garages.

Semiautomatic dry standpipe systems: This dry standpipe system has mildly pressurized air in most of the piping that runs from the deluge (pre-action) valve to the hose outlets, but water remains pressurized in the piping downstream of the deluge valve. To operate the system, firefighters connect hoses to the landing valve and send a message to activate the manual deluge valve to allow contained water to flow through the stand-piping system.

Manual dry standpipe systems: This dry pipe system is essentially an empty pipe system, which is generally used by the fire department to get connected through the FDC or by starting the pump to supply water. These systems are obviously less expensive, but they are considered less reliable than other types because it is hard to spot flaws or leakages in the piping due to having no water filled in.

Water reservoir

For firefighting, a comparatively large reservoir is required to store water, which must ensure a fast-flowing volume of water when necessary. Most commonly, a water reservoir is to be incorporated where an adequate supply of water from the main is not available or reliable.

The important factors to be considered, with respect to the water reservoir for firefighting systems, are its size and location. Determining the size of the water reservoir for firefighting depends upon various considerations, as mentioned below:

a Fire flow rate, which is to be determined.
b Duration for firefighting, which is to be estimated

Fire flow

The water flow rate, i.e., the fire flow rate, is primarily based on the type and the maximum floor area of the building considered for firefighting. According to the National Fire Protection Agency (NFPA) of the USA, fire flow is defined as the rate of water flow under a residual pressure of 137.9 kPa for a specified duration that is necessary to control a major fire in a specific structure or building. The minimum fire flow rate mentioned in NFPA 1142 has been tabulated in Table 4.1, which is applicable for areas where there is no water or inadequate water for firefighting.

The volume of water to be stored for firefighting is the fire flow required multiplied by the recommended duration of fire flow as mentioned in Table 4.2.

Table 4.1 Water delivery rate for firefighting in a building [1]

Sl.no	*Total water supply required (liter)*	*Water delivery rate (lpm)*
1	<56,780	950
2	56785 – 85,170	1900
3	85,175 – 113,560	2850
4	>113,560	3800

Table 4.2 Required fire flow and flow duration [2]

Fire flow area (Square meter)					*Fire-flow (lpm)*[b]	*Flow duration (hours)*
Type IA and IB[a]	*Type IIA and IIIA*[a]	*Type IV and VA*[a]	*Type IIB and IIIB*[a]	*Type VB*[a]		
0–2109	0–1180	0–762	0–548	0–334	6000	2
2110–2806	1181–1579	763–1013	549–734	335–446	7000	
2807–3595	1580–2025	1014–1198	735–910	447–576	8000	
3596–4487	2026–2248	1199–1617	911–1171	577–715	9000	
4488–5481	2249–3084	1618–1979	1172–1431	716–874	10000	
5482–6587	3085–3688	1980–2369	1432–1709	875–1050	11000	
6588–7776	3689–4376	2370–2796	1710–2026	1051–1245	12000	3
7777–9077	4377–5100	2797–3270	2027–2406	1246–1449	13000	
9078–10470	5101–5890	3271–3772	2407–2722	1450–1672	14000	
10471–11957	5891–6726	3773–4311	2723–3112	1673–1914	15000	
11958–13555	6727–7627	4312–4877	3112–3521	1915–2165	16000	4
13556–15255	7628–8584	4878–5491	3522–3967	2166–2443	17000	
15256–17038	8585–9578	5492–6132	3968–4431	2444–2722	18000	
17039–18924	9579–10647	6133–6810	4432–4924	2723–3029	19000	
18925–20922	10648–11771	6811–7534	4925–5444	3030–3345	20000	
20923–23012	11772–12951	7535–8287	5445–6076	3346–3679	21000	
23013–25195	12952–14177	8288–9077	6077–6559	3680–4032	22000	
25196–27490	14178–15468	9078–9894	6560–7154	4033–4404	23000	
27491–Greater	15469–Greater	9895–10758	7155–7776	4405–4785	24000	
–	–	10759–11659	7777–8417	4786–5175	25000	
–	–	11660–12588	8418–9095	5176–5593	26000	
–	–	12589–13545	9096–9922	5594–6020	27000	
–	–	13546–14558	9923–10517	6021–6466	28000	
–	–	14559–15598	10518–11269	6467–6931	29000	
–	–	15599–16667	11270–12040	6932–7414	30000	
–	–	16668–17782	12041–12848	7415–7906	31000	
–	–	17783–Greater	12849–Greater	7907–Greater	32000	

Note:

a Types of construction are based on International Building Code
- Types IA and IB are considered noncombustible and fire resistive.
- Types IIA and IIB are also noncombustible, but have little to no fire resistance.
- Types IIIA and IIIB are considered combustible.
- Type IV is a combustible construction type, but limited to heavy timber construction.
- Types VA and VB are combustible construction, but slightly less fire-resistive than the Type III construction types.

b Measured at 138 kPa residual pressure.

Fire pump

Fire pumps are generally a set of pumps that are one of the essential components of water-based fire protection or fire hydrant systems, generally installed in high-rise buildings and structures. Fire pumps work on any one of the two working principles: centrifugal and positive displacement. Positive displacement-type fire pumps are highly preferred for non-water firefighting systems; particularly those using foam concentrate or water mist.

Fire pumps used in buildings are basically centrifugal pumps with a specialty that is used to boost the water pressure to cause the required flow of water in a piping system, and to throw or spray water on a fire to suppress it. A fire pump is necessary when the fire hydrant system is fed from a non-pressurized water tank when the other sources of water for the purpose have inadequate pressure and firefighting equipment cannot reach the critical areas of a building, higher floor levels in particular. Majority of the fire pumps are driven by both electric and diesel motors.

Types of fire pump

Various types of fire pumps are available, mostly based on their special features and performance. The selection and installation of the right type and size of pump help avoid excessive costs and excessive pressure developments that might damage the system and put business assets and lives at risk. Centrifugal fire pumps of various types, generally used in fire hydrant systems are as follows:

End suction pump

In an end suction pump, the discharge outlet is made perpendicular to the suction inlet, as shown in Figure 4.3. These pumps are more compact and require less floor space for installation. Typically, the largest pump of this type has a limited flow capacity of approximately 5,678 lpm.

Horizontal split-case pump

These pumps are comparatively reliable and manufactured in a wide range of flow and head capacities. For being horizontally split, periodical maintenance work can be done easily for having relatively easy access. However, these pumps typically need comparatively more space than all other types of fire pumps. A horizontal split-case driven by a diesel motor has been shown in Figure 4.4.

Vertical turbine pump

This type of pump can be started under negative suction pressure. So, water can be easily lifted from any below-grade water source, like an underground water tank or surface water bodies like ponds, lakes, rivers, etc.

Multistage multiport pump

These pumps have multiple impellers in series connected by a horizontal shaft in a single casing. The pump casing having a single discharge port is termed simply a multistage pump,

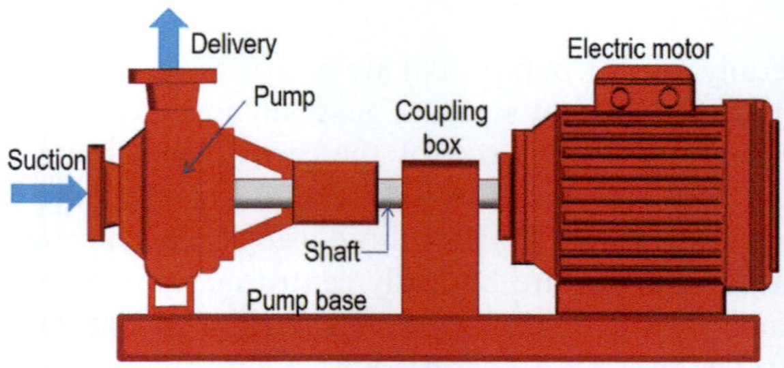

Figure 4.3 Electric motor-driven end suction fire pump (side view).

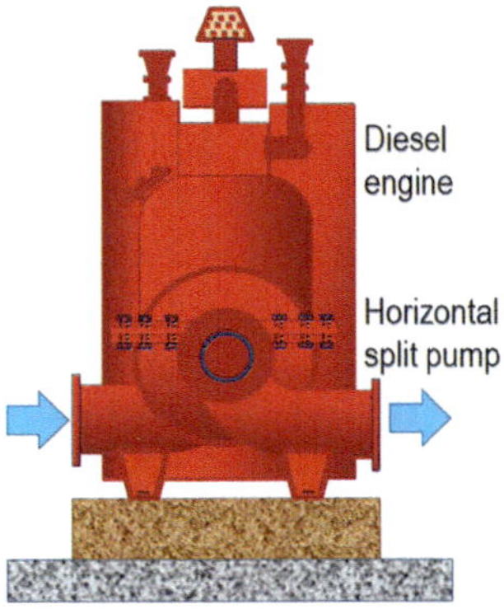

Figure 4.4 Diesel engine-driven horizontal split fire pump (front view).

and the pump having multiple ports or discharge outlets discharging at different pressures—each port delivering at increased pressure from the consecutive series impellers—is termed a multistage-multiport pump.

A multistage-multiport pump could be efficiently used in a high-rise building with multiple zones with different pressure requirements. Each pressure zone is served by a particular port of the multistage-multiport pump. The port delivering at the lowest pressure serves the lowest pressure zone, while the highest pressure port delivering the highest pressure serves the highest pressure zone, and so on.

Advantages of a multistage-multiport pump

1 Requirement of fewer drivers
2 Less pipe work and therefore requirement of fewer valves, joints, etc.
3 No requirement for water storage tanks in upper pressure zones
4 Lower floor space, structural loads, and associated costs
5 Lower energy consumption and pollution

Inline pump

These pumps are vertical shaft types, with the driver directly above the pump. In these pumps, the suction inlet and the discharge outlet are on approximately the same plane. These pumps are installed on pipes, so they occupy no floor space. But pump maintenance and repair can be difficult because the motor must be lifted off and removed to gain access to the pump, making it one of the more expensive pumps from the repair and maintenance point of view. Inline pumps with a maximum flow capacity of 5,678 lpm are made and can only be driven by electric motors, which limits their diversified applications.

For exceptionally tall buildings, more than one fire pump may be required to create the pressure required for the topmost floor. In such cases, the installation of inline pumps in series would be preferable. NFPA 20 recommends a maximum of three pumps to operate in series [3].

Fire-pump specialty

Though fire pumps are basically centrifugal pumps, for their special use for accidentally occurring fire, they need some specialty in their construction, as mentioned below:

1 Since the fire pumps are rarely in operation except for testing, internal components should be made of non-corrosive materials like stainless steel, bronze, etc. to avoid the pump clogging or seizing.
2 Fire pumps have low NPSH curves to avoid cavitation, causing decreasing flow and pressure. So, the pump selection must be based on the required NPSH value.
3 Maximum water flow is ensured mainly to avoid cavitation of the fire pump and additionally to ensure enough power has been reserved [4].
4 Fire pumps cannot operate in parallel because the discharge check valve is forced closed when the pressure on the outlet side of the valve is higher than that on the inlet side. For this reason, multiple fire pumps cannot be added in parallel to boost the pressure or flow in a fire hydrant system [3].
5 Self-primed pumps should be avoided as there might be several causes of failure in the operation of a self-primed pump due to malfunctions that hinder self-priming.
6 Fire pumps are to be supported by adequate and alternate power reserves.
7 Fire pumps shall be easily maintainable. The primary cause of fire pump engine failures is inadequate maintenance.
8 The fire pump should be supported by a drain cock to drain all the water after an operation.
9 Most fire pumps are painted red.

Suction pipe requirement

The suction pipe by which the pump sucks water from the bottom of a reservoir shall be as small, straight, and joint-less as possible. If bending is unavoidable, then the bend shall be at a distance of 10–12 times the diameter of the suction pipe. The purpose of bending a suction pipe after such a long distance of pipe length is to prevent turbulent flow from entering the fire pump and impellers and to avoid forming air pockets that can cause wear and cavitation over time.

Anti-vortex plate, also known as vortex inhibitor, as shown in Figure 4.5, is fitted at the inlet of the suction pipe in the tank. According to NFPA 22, where a reservoir is used to

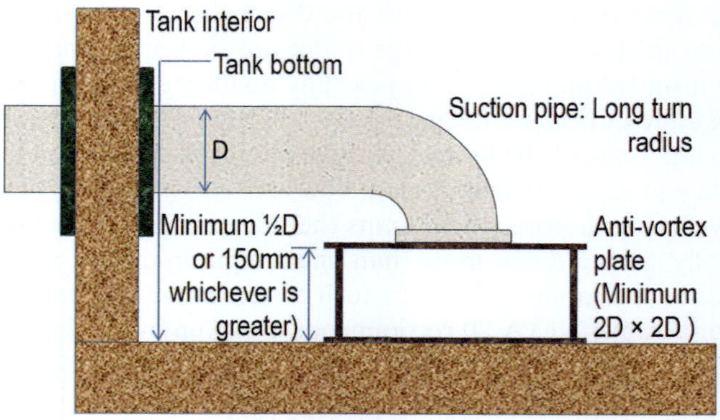

Figure 4.5 Anti-vortex plate assemblies.

draw water for a fire pump, the suction inlet of the pump shall be equipped with an assembly that controls vortex flow to reduce vortices and dead water area, minimize the air entrainment into the pump, and reduce the flow, with a view to ensuring the pump and the tank are working at their maximum efficiency.

Fire pump driver

Pump drivers, or motors, are basically energy conversion machines that convert electrical or chemical energy to mechanical energy. Fire pumps are mostly driven by electric motors or, alternatively, by diesel engines in the absence of electricity. To ensure uninterrupted power supply, both electrical motor and diesel engine-driven fire pumping systems are adopted, as there is every possibility of disruption in electric supply during a fire. All precautions must be taken to ensure that electricity or enough diesel fuel is available to operate the driver in case of a fire. Exclusively, diesel engines are generally used as drivers for fire pumps in areas where there is no or an unwanted risk of an irregular, unreliable, or unsuitable electrical power source.

Electric motor for a fire pump

The electric motor used to drive a pump is energized by electrical power, primarily to spin a rotor inside, which spins a long shaft connected to the impeller of the pump to rotate the impeller, creating pressure for pumping.

Electrical motor drives are cost-effective where electricity is easily and sufficiently available. Electric motors are also easy to operate. Due to having few moving parts, very little maintenance is needed. But electric motors are considered as reliable as the electric supply is reliable.

Diesel engine for a fire pump

In many cases, a diesel engine-driven fire pump is installed as a redundant fire pump along with the electric-driven fire pump. A diesel motor comes mounted on the same skid as the pump, as shown in Figure 4.6.A diesel generator is basically a machine that converts the chemical energy of diesel fuel to mechanical energy to produce electricity through the

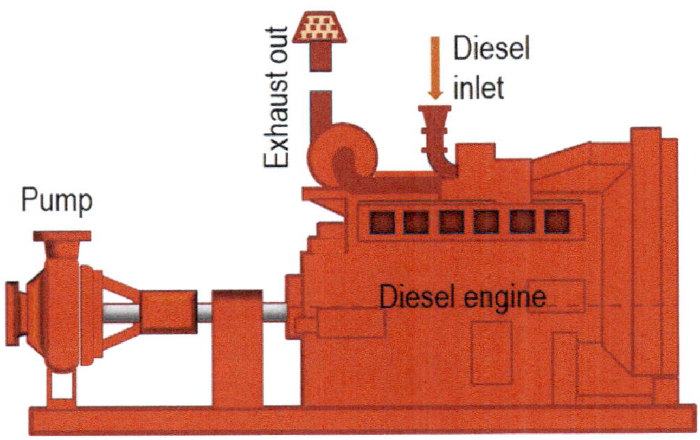

Figure 4.6 Fire pump driven by a diesel engine (side view).

combustion of fuel. These machines are therefore a combination of a diesel engine and an electric generator to generate electricity.

The diesel motor is assembled with a pump on the same pump skid, mounted on a common base element. Diesel engines use the power generated by the combustion of diesel fuel to turn a shaft that spins the impeller. Unlike electrical motors, diesel-driven pumps require a lot more infrastructure and maintenance, such as on-site diesel storage fuel tanks, batteries for engine startup, combustion air for the engine, increased ventilation, an engine exhaust system in the room, and a governor system to control the power. The governor system is important because of the risk of the diesel engine producing too much power and spinning the impeller too fast.

The level of reliability of diesel engine driving systems significantly decreases as the system ages. Plastic and rubber components get brittle with age. Wiring harnesses and connectors get compromised through vibration and chafing. Fuel, oil, and coolants all deteriorate with age and, if not replaced regularly with the correct fluids, can cause clogging in injectors and cooling systems. Diesel engines generate noise and exhaust emissions during operation. The combustion of diesel fuel releases gases polluting the indoor and outdoor environment.

The most common diesel engine voltage employed is 12 volts (V) direct current (DC) for engines rated less than 300 kilowatts (kW), or approximately 400 brake horsepower (BHP).

Jockey pump

A jockey pump, also known as a pressure-maintenance or make-up pump, is the smallest pump among the pumps of a multiple-pump set, mostly used in a fire hydrant system in a building with a view to maintaining set pressure in the system. The fire hydrant system is supposed to remain at its set desired pressure, but for various reasons, leakages in particular, pressure might drop in the system, which is undesirable. When there is a pressure drop in the hydrant, it is sensed by the automatic controller, which will cause the jockey pump to start first among others in the pump set. Basically, Jockey pumps are nothing but multistage centrifugal-type pumps. The jockey pumps are mostly vertical-type multistage centrifugal pumps, as shown in Figure 4.7.

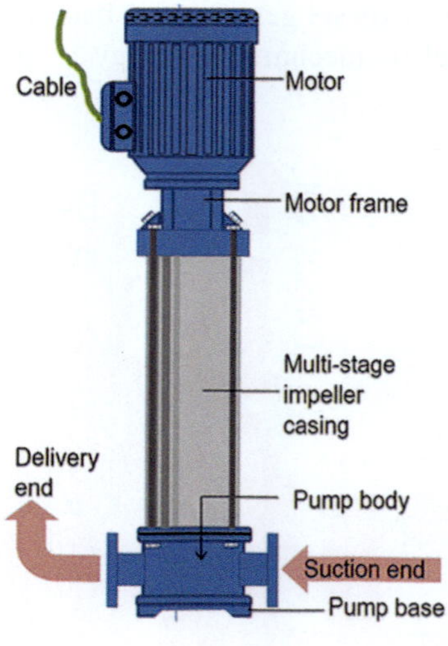

Figure 4.7 A jockey pump.

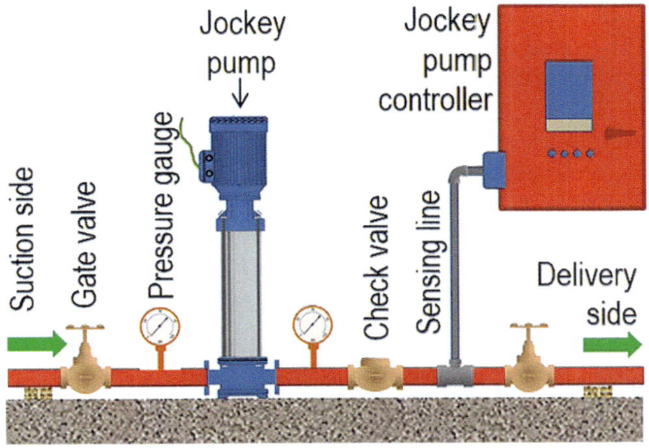

Figure 4.8 Jockey pump connections.

In a fire hydrant without a jockey pump, from time to time, due to water leaks or any other causes, there may be an unwanted drop in pressure level in the hydrant, causing the fire pump to start, though actually it was not needed as there was no incidence of fire. Such undesirable start and stopping of the fire pump for such unwanted conditions will cause a shortening of the fire pump's life expectancy. To overcome such occurrences, a jockey pump is included in the fire pump set to run it instantly when pressure in the hydrant system drops, thus maintaining relatively constant desired pressure on the fire hydrant system.

The suction pipe of the Jockey Pump shall be connected upstream of the fire pump suction outside screw and yoke (OS&Y) valves and downstream of the backflow preventer check valve, and its discharge line shall connect downstream of the fire pump discharge valve as shown in Figure 4.8. To assure that both the fire and the jockey pump run when they are supposed to, and the jockey pump won't interfere with the fire pump run, the sensing system of the fire pump set controller should be set up in the following arrangement:

1 The jockey pump should stop at a pressure level equal to the pump churn pressure—the pressure at zero flow plus the minimum static supply pressure.
2 The jockey pump start running pressure level should be at least 68 kPa (0.68 bar) less than the jockey pump stop pressure level.
3 The fire pump should start running at a pressure level of 34 kPa (0.34 bar) less than the jockey pump's starting pressure level. When multiple fire pumps are installed in series, the successive pump run pressure level difference shall be 68 kPa (0.68 bar).

According to the NFPA, the jockey pump is sized for a flow rate less than the leakage rate in order to ensure a normal system pressure drop. The jockey pump is generally sized considering its flow rate is sufficient to make up the allowable leakage rate within 10 minutes or 3.8 lpm, whichever is larger. There is another consideration for the flow rate of a jockey pump, which is considered to be 1 percent of the flow rate of the fire pumps. Some old code suggested having a minimum of 3 percent and a maximum of 10 percent of the main pumping flow requirement at the same pressure, subject to a minimum discharge of 180 lpm [5] and maximum of 450 lpm. According to NFPA 20, the head of the jockey pump shall be considered to have 68 kPa–138 kPa higher pressure than the main fire pump [6].

Installation of an oversized jockey pump will create water-hammering effects in the hydrant piping. This is due to the pump quickly coming on and going off. The pump will also cause short cycling with a quick supply of water, which can damage the pump and shorten its life.

The fire water supply system is designed for a minimum residual pressure of 7.0 kg/sqcm at the hydraulically remotest hydrant point, ensuring the designed flow rate at that point [7]. If the designed pressure of 21 kg/sqcm is desired to be maintained in the fire hydrant piping at all- times, then the operational sequence of the jockey pump may be set as follows: auto start at 20 kg/sqcm and auto stop at 21 kg/sqcm, the difference shall be 5 percent. It is good practice to have two jockey pumps: one as the lead pump and the other as standby, both operating in auto mode.

Fire pump controller

A fire pump controller is a device primarily to start and stop a fire pump as and when required, and secondarily to monitor the conditions in an ongoing manner that could hinder or prevent the smooth operation of the pumps. Fire pump controllers are designed to operate and control a fire pump set, including either electric motor-driven fire pumps or diesel engine-driven fire pumps, or both, along with a jockey pump. Figure 4.9 shows a fire pump's set, including one electric motor-driven fire pump, one jockey pump, and a diesel engine-driven fire pump.

Fire pump controllers can be manually operated to start fire pumps, either by activating a button switch or a lever installed on the controller cover or door. But most commonly, fire pump controllers are set up to operate automatically, either by a drop in system pressure or via a signal from a remote device, such as a flow switch or deluge valve. A fire pump controller can be designed to run continuously until manually shut off, or a timer can be used to automatically stop the fire pump after a set amount of time has passed, assuming that by this time the pressure in the system has reached the desired level and stabilized.

Bourdon-type pressure switches with mercury are used to detect pressure drops and initiate a signal to start and stop the fire pump controllers. Now, electronic pressure

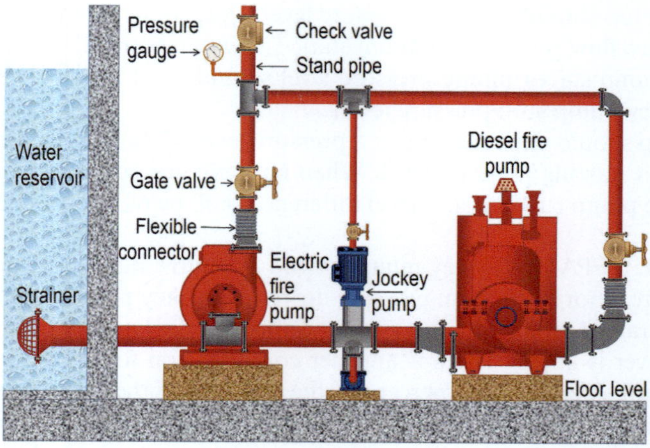

Figure 4.9 Fire pumping including main electric pump, diesel pump and jockey pump and piping connections.

transducers, which allow for real-time electronic communication with modern computer screen interfaces, or "operator interfaces." There is a need to set an "ON" setting and an "OFF" setting for when the fire pump is needed to turn on and when it is needed to turn off. A pressure in the system is the most reliable sign that a hydrant in the building has been activated, and the fire pump is set to turn on once this happens automatically.

Once started, a fire pump controller can be set to automatically reset or (more commonly) require a manual stop, whereby the pump operator has to manually push a button on the controller to stop it. For safety reasons, qualified and trained personnel should operate and service these controllers.

As fire pump applications require an alternative backup power source, there is a need for a transfer switch; generally, it is standard to switch over a diesel engine. An automated Programmable Logical Controller (PLC) system is used in the control panel, which enables transferring from running a Jockey Pump to starting a main electric pump or starting a diesel engine instead of an electric pump when there is no electricity.

Stand pipe

A standpipe in a fire hydrant system in a building is the main rigid water supply piping, which is installed mostly in a vertical position with or without offsetting, to which a branch pipe is made to connect to a fire hose for the manual application of water to the fire. A standpipe system virtually serves as a conduit to convey water under pressure from a water reservoir to hose stations at one or multiple locations within a building for firefighting purposes.

Types of standpipe systems

According to NFPA 14-2007, there are various types of standpipe systems, as mentioned below, that can be developed in a building for firefighting purposes.

1 Automatic standpipe system: This is a standpipe system that is connected to a water reservoir or supply system capable of supplying the water demand at all times and requires no action other than opening a landing valve only to get water into the hose for firefighting.
2 Wet standpipe system: This is a standpipe system having piping containing water at all times under a desired pressure and being connected to a permanent water supply or storage.
3 Dry standpipe: This is a standpipe system designed to have piping remaining devoid of water. In a non-automatic system, supply of water is made when the system is being utilized. It can be an automatic system as well.
4 Manual standpipe: This is a manual standpipe system that is operated exclusively being connected with its fire department connection (FDC) to supply water for the system demand.
5 Combination standpipe and sprinkler system: This is a system where the fire protection piping services both 65 mm outlets for the fire department use and outlets for automatic sprinklers (discussed in a Chapter 5).
6 Standpipe multi-zone: A standpipe system that is vertically subdivided, as required by the fire codes, into multi-zones to limit the maximum operating pressure in the system. Standpipe hydrants in each zone shall have their own individual automatic water supply

system. All zones serving occupied floors located higher than 91.4 m shall be provided with primary and auxiliary water supplies.

7 Semi-automatic standpipe system: A semi-automatic standpipe system is designed for situations where there is danger of serious water damage.

Fire hose station

The fire hose station is predominantly the location of a fireplug or cock and associated equipment. Branch pipes of the fire standpipe direct water to the hose stations, where flow is prevented by a closed landing valve that possesses a quick connection point for a hose reel, to get a controlled but steady flow of water jet at high pressure to conveniently throw water streams manually on fire of potential risk, continuously in projectile motion from a safe position. Thus, a hose station provides an accessible, protected, and quick connection point for a fire hose reel.

A fire station is installed in a strategic location with highly visible fire hose reel signage where everyone can easily locate them in a fire emergency. Fire hydrants consisting of standpipes supported by fire hose stations provide some advantages, as discussed below:

Quick suppression: Standpipes with smaller diameter hose systems can be effectively used in the initial stage of fire before the presence of the fire department or before the automatic systems are activated.

One-person operation: This smaller-diameter hose system can be activated and operated by even one person very quickly before developing unmanageable heat, smoke, or flame and preventing spreading of fire to surroundings.

Minimal water consumption: Standpipe hose systems cause minimum water consumption as the required amount of water can be applied directly at the base of the fire locations, whereas indiscriminate spray from automatic systems causes unwanted water loss. Additionally, the hose systems can be turned off immediately after the fire is doused.

Occupancy safety and rescue: Standpipe hose systems allow time for evacuation, protect a means of egress, and defend occupants who cannot evacuate the building.

Fire station coverage

The Australian standard AS2441 defines the concept of system coverage. The length of a fire hose reel or reels shall be and need to be located in a position such that the operator can reach a fire in any part of the building to fight. The maximum length of the hose allowed for this purpose is 36 m with a 4 m spray or throw distance from the nozzle [8]. It is system coverage that will determine if internal hydrants are required.

Fire hose station requirements

The location of the fire hose station shall be fixed following the applicable fire safety codes for buildings in the concerned areas. The following guidelines are found to be common in most of the fire codes, with some deviations:

1 In a fire hydrant system, hose stations are to be provided on each floor of the building.

2 At each hose station, hose connections, valves, fire hose, nozzle, and hose rack shall be housed in an easily operable hose cabinet.

3 A hose cabinet shall be of sufficient size for housing all the accessories and appliances, maintaining sufficient clearance to permit easy handling for operation.
4 The hose cabinet door must not obstruct the required width of a means of egress when fully opened.
5 Fire hose reels shall be clearly marked with white letters on a red sign, placed 2 meters or more above the floor. Operating signage shall be visible within 2 m. of the reel.
6 The hose and the hydrant valve shall be readily accessible and not obstructed by any fixed or removable elements. The height of the reel needs to be between 1.4 and 2.4 m. above the floor.
7 Fire hose reels need to be mounted robustly. The mounting must not only support the weight of the apparatus but also the forces applied by the firefighter pulling on the hose.

Location of the fire station

Fire stations shall be located to satisfy the following requirements:

1 Every indoor portion of the building can be sprayed by a water stream and is within 4 m. of the extended hose end.
2 Fire hose reels must be located within 4 m. of an exit. When a hose reel within 4 m of exits cannot provide system coverage, in such cases, it is permissible to provide an extra fire hose station in the path of travel to an exit.
3 A fire hose reel should not be located such that it is required to take the hose through a fire or smoke door to fight a fire.

Fire station hose cabinet

Fire station hose-reel cabinets are designed to predominantly accommodate fire hose reels fitted with a nozzle. It is also designed to house landing valves and portable fire extinguishers, along with other accessories and equipment for which its size varies. Fire hose reel cabinets may be fixed or swinging, generally installed on a wall.

Fire cabinets can be installed recessed, semi-recessed, and surface mounted on a wall or self-standing, with a single or double door fitted with lock mechanisms of either lock type or key type and a latch mechanism of the turn handle type. Cabinets are generally made from mild or stainless steel, with the sheet having a thickness of 1.0–2.0 mm. Doors are mostly flush, solid-metal types, but glass door types are also made. Glasses may be tempered, wired, or fire-rated for 2 hours.

Common sizes of a fire cabinet are 1000 mm by 1000 mm and 400 mm deep. There are cabinets of height 2700 mm as well. The minimum width and depth of the cabinets are found to be 800 mm and 230 mm, respectively.

Classes of fire hose stations

There are three classes of fire hose stations installed on a standpipe, as mentioned below, depending on the appurtenances they accommodate:

1 Class-I
2 Class-II
3 Class-III

Class-I

In a Class-I standpipe system, a 65 mm hose connection is provided for use primarily by the trained personnel in heavy firefighting operations, if any, or by the fire department as a final response to fire suppression. This class does not have a hose attached. The fire department is to bring hose packs, pumps, and water to the fire spot, connect the hose to the standpipe through a special connection system, and then start their fire suppression operation. These connections must be compatible with the hose coupling utilized by the fire department. Figure 4.10 represents a Class-I fire station.

Class-II

In a Class-II standpipe system, a 40 mm hose reel connection along with the desired length of hose is provided for use primarily by the trained personnel among the occupants of the building or by the fire department as an initial response to fire suppression. The hose pipe is typically kept in a cabinet. Figure 4.11 represents a Class-II fire station.

Class-III

In a Class-III standpipe system, both a 40 mm hose reel connection along with the desired length of hose to supply water for use by trained personnel among the occupants and a 65 mm hose connection to supply a larger volume of water for use by the fire departments and those among the occupants trained in heavy fire streams, if any, are kept in the hose cabinet. Figure 4.12 represents a Class-III fire station.

Landing valve

A landing valve is basically a globe valve, designed to be installed on a wet riser fire hydrant in a building. As the risers are mostly poked through the landing of the stairwell of a building and the valve is operated standing on the landing, the valve is named so.

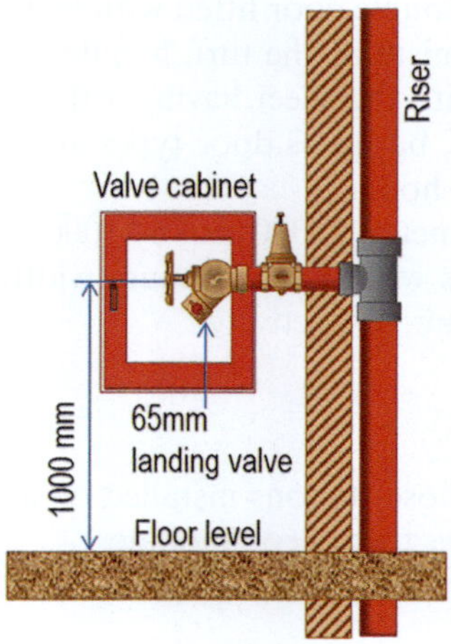

Figure 4.10 Installation of a fire hose connection for a Class-I fire station.

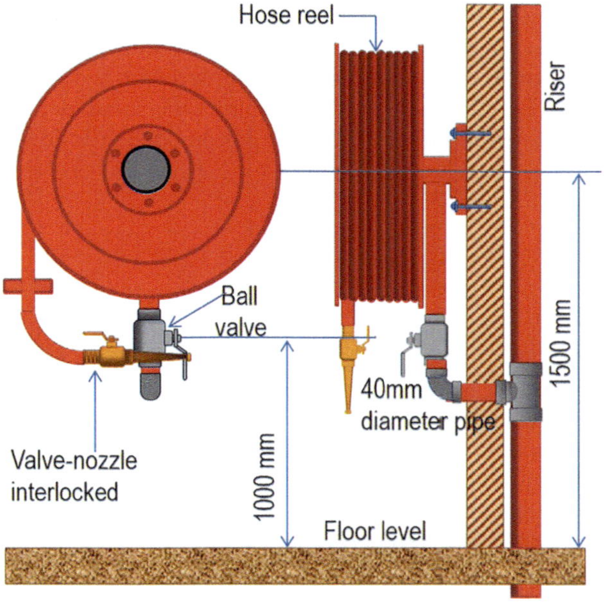

Figure 4.11 Installation of a fire hose reel for a Class-II fire hose station.

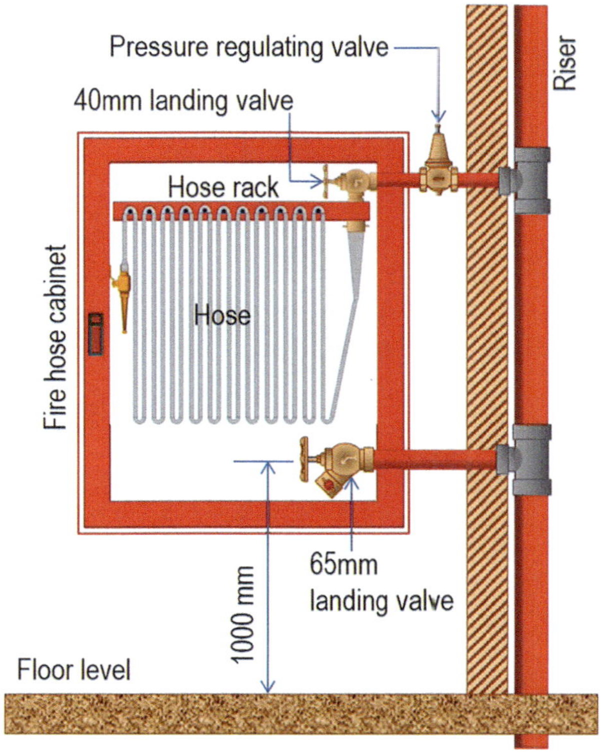

Figure 4.12 Installation of a Class-III fire hose station.

Landing valves are installed on a branch pipe on the riser, with a hose connected to the coupling end of the valve. By rotating the handle of the landing valve anti-clockwise, it can simply be activated to cause the flow of water through the hose of the fire hydrant system. Though the basic function of landing valves remains the same, depending upon application, location, and code requirements, different designs of the following types of landing valves are available:

1 Oblique type
2 Right-angle type
3 Horizontal type
4 Bib-nose type
5 Pressure-regulating type

Figure 4.13 shows all types of landing valves used in the fire hydrant system.

Pressure-regulating landing valve: These landing valves are designed to perform as a landing valve with pressure-reducing capability as well. These types of valves are generally designed to provide a range of outlet pressures of 5 bar–8 bar for use at nominal inlet pressures up to 20 bar and a working temperature range of 0°C–38°C.

Most standards allow the use of different materials for the construction of landing valves, which include bronze, aluminum, and stainless steel. The important qualities of valve materials are their mechanical strength, corrosion resistance, and durability.

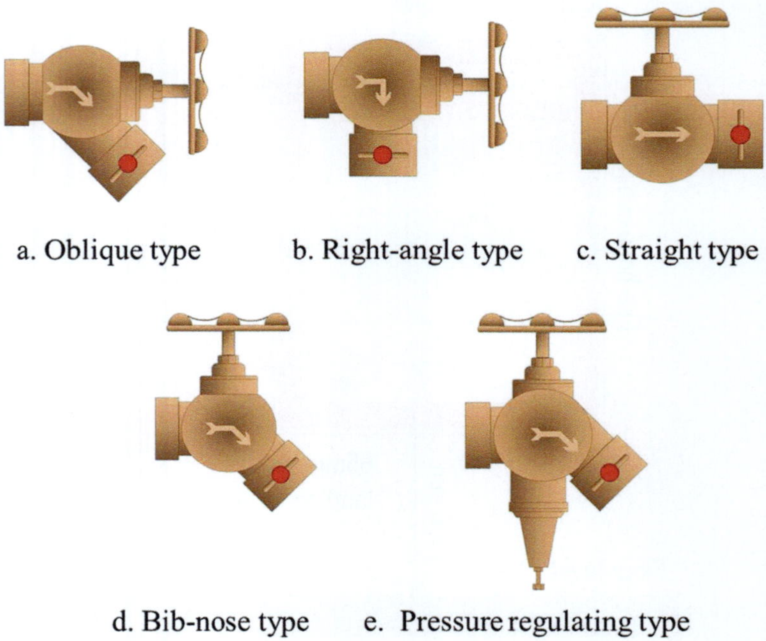

Figure 4.13 Various types of landing valves.

a. Lay flat hose. b. Tube hose c. Reinforced hose

Figure 4.14 Various fire hose pipes.

Fire hose

Hoses are flexible pipes used to convey water. Fire hoses are available in different types for uses in various purposes. In very common uses, there are three types of fire hose, as mentioned below and shown in Figure 4.14:

1 Lay flat hose
2 Tubing hose
3 Reinforced hose

Lay-flat hoses are those that generally lay flat when empty of water. It can be found in sizes from 25 mm up to 63 mm for use in firefighting. It also comes with many different coupling systems. Lay flat hose pipe is generally three-ply with two synthetic rubber layers and textile reinforcement. The working pressure is 12 bar and the burst pressure is 42 bar. The hose reel is hydrostatically tested for 18 bar. For handling fire hose pipe of diameter 63 mm, two trained people are needed to operate it, so it is kept mainly for the use of fire brigade personnel.

Reinforced hoses are made for higher strength for pressure resistance, abrasion resistance, etc. Reinforcement is made from a circular woven jacket made from a special blend of cotton and synthetic yarn with high abrasion and heat resistance. The inner surface is rubber-lined to minimize friction. These are generally coated with UV-resistant epoxy powder. Hoses can be found in lengths of 30 m; their thickness may be about 1.2 mm.

Tube hoses are mostly high-pressure braided rubber hoses; made in different sizes, from 20 mm in diameter to 25 mm, and in length of 30 m. Tube hoses are made from a rubber compound reinforced by synthetic textile yarns to make them strong, abrasion-resistant, and weather-resistant.

Hose reel

The fire hose reel is mainly for spraying water on initial small fires, requiring a smaller water flow, and is supposed to be used by the trained occupants of a building. Hose reels are generally wound in a disc made of 1.2 mm pre-treated electro-galvanized steel sheet painted with electrostatic powder coating. Automatic fire hose reels are fitted with an internal valve installed into the hose reel unit. The valve opens automatically within 3 revolutions of the hose reel while pulled out and automatically closes when the hose is wound back to the reel. Hose reel size may be 19 mm and 25 mm diameter. Hose reels are available in fixed, swinging, recessed, and concealed types with automatic or manual valves.

Hose rack

In a fire station fire hose rack assembly is installed in which a bracket is used for hanging fire hose on pins fixed in the rack as shown in Figure 4.12. After taking out the fire hose and spray nozzle, the automatic discharge mechanism will allow the water to flow through the fire hose. The size of the rack is generally made for 40 mm and 65 mm hose of length 30 m, and it is fixed on right-angle landing valve. The approximate size of a hose rack cabinet is furnished in Table 4.3.

Friction loss of the fire hose

Friction loss in fire hoses plays a significant role in determining the pressure head to be overcome by the fire pump. Friction loss in fire hoses is governed by the following factors:

1 The quality of the hose, particularly its internal lining
2 The length of the hose
3 The diameter of the hose
4 The velocity of flow through the hose

The quality of the hose: The thickness of the inner lining, the age of the hose, and the weave of the jacket all have implications for the quality of the hose, thereby offering friction. Even with the good quality of the inner lining of the hose, some friction will be there; in addition, every little imperfection in the lining will generate friction.

The length of the hose: friction loss increases with the increase in length of the hose. Friction loss for a 30 m length hose is determined, so for actual length, it is calculated as a multiple of that value. For example, if the friction loss in a 40 mm hose is 1034 kPa (of 30 m), then for 6 m lengths, the total friction loss would be 206.8 kPa.

The diameter of the hose: friction loss also increases with the increase in diameter of the hose. So, it is important to increase the diameter of the hose when trying to keep friction loss to a minimum. While the flow is maintained the same, if the hose diameter is doubled, the friction loss in the hose is only 1/32 times the friction loss in the hose of the previous diameter.

Friction loss varies with the square of the velocity. If the velocity of flow is doubled in the hose, the friction loss is quadrupled. If the velocity of flow is quadrupled, the friction loss in the hose will be increased by 16 times.

The simple and easy formula used for determining the friction loss in a fire hose is as below:

$$Fl = C\left(\frac{Q}{100}\right)^2 L \tag{4.1}$$

Table 4.3 The approximate size of a fire hose rack cabinet [9]

Sl. no.	*Hose size (mm)*	*Length (m)*	*Length (mm)*	*Height (mm)*	*Depth (mm)*
1	40	30	600	720	200
2	65	30	760	800	220

Table 4.4 Coefficients of friction loss for various hose sizes [10]

Sl. no.	*Hose size (mm)*	*Coefficient*
1	40	6.0
2	65	0.8
3	125	0.03

where, C = A coefficient, which is a number used for a specific hose size. Table 4.4 shows various hose sizes and their coefficients.

Q = flow, lpm/100
L = length in 30 m

Fire hose nozzle

Fire hose nozzles are one of the most important firefighting gears for firefighters to put out water in various forms on fires to douse it. Fire hose nozzles are handheld elements that are connected to the fire hose end, which provides a water flow stream to extinguish distant fires. This nozzle helps the stream get its shape, attain velocity, and reach the fire. Fire hose nozzles are made of diverse materials like brass, aluminum, and even plastic.

There are three major functions performed by a fire hose nozzle, which are as follows:

1 Provide a distant reach of water by regulating the flow of water through the specially designed outlets of the nozzle
2 Project water spray to the fire spot within range
3 Provide stream mode for long-distance targets and fog mode for short-distance targets

Types of fire hose nozzles

There are five most commonly found hand-line nozzles used by firefighters. The types of nozzles are as follows, and shown in Figure 4.15:

1 Smooth-bore nozzle
2 Constant or fixed orifice nozzle
3 Selectable or adjustable nozzle
4 Automatic or constant pressure nozzle
5 Multipurpose nozzle

Smooth-bore nozzle

The smooth-bore nozzle is the most basic, simple type of nozzle, both in design and function, which produces a solid stream of water with large droplets and can get maximum reach and penetration. The smooth-bore nozzle may be available as a tip or in a break-apart configuration that provides the opportunity for a smooth bore of different tip sizes and also a fog nozzle in one. These nozzles may have a simple ball-valve "shut-off" with an

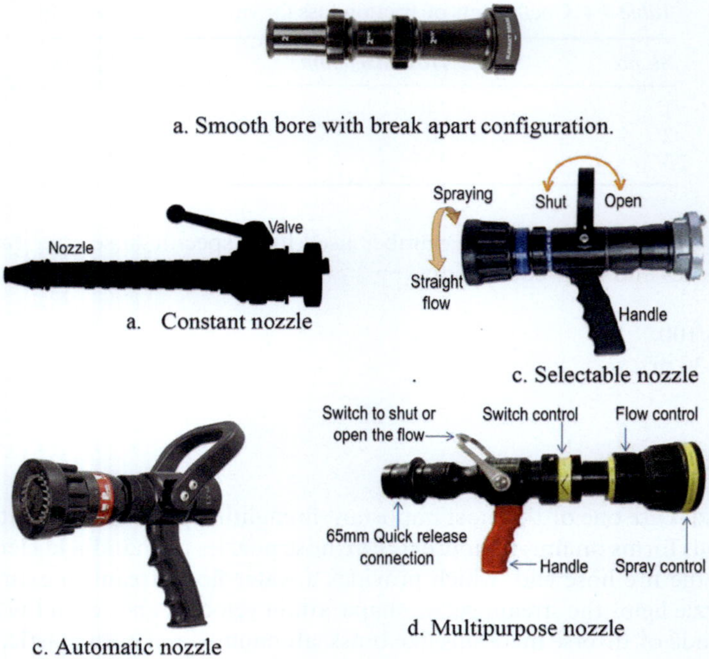

Figure 4.15 Various types of nozzles for firefighting.

attached tip. These nozzles are generally designed to operate at a pressure of 345 kPa to produce a flow rate between 606 and 1003 lpm.

Constant gallonage nozzle

Through this type of nozzle, a fixed flow rate is obtained ranging from 473 to 1136 lpm under an operating pressure range of 345–690 kPa. A mid-range constant gallonage nozzle produces 662 lpm at an operating pressure of 517 kPa. This fixed-flow device saves the fire fighter's precious time by not requiring them to modulate the water flow.

Selectable gallonage nozzle

Various water flows through the nozzle can be selected during operation by the firefighter based on the fire situation. This can be easily and quickly done without turning off the nozzle.

Automatic pressure nozzle

This automatic pressure nozzle maintains varied rates of water flow at a particular pressure to ensure the reach of the fire target, but at the expense of the varied water flow deliverance, as the pressure remains the same but the water flow is regulated automatically inside the nozzle.

Multipurpose nozzle

This is a hybrid nozzle that can be turned to fog mode while in stream mode, so that firefighters can go close to fires while maintaining safety. Its lower pressure can also provide higher water flow that might be needed for the firefighting of high-rise buildings, outdoor fire areas, cooling surrounding the fire areas, etc.

A fog stream consists of tiny water droplets discharged in a uniform direction toward the fire. Fog nozzles are good for auto fires, outside fires, and liquid petroleum fires. The small water droplets of a fog stream absorb heat faster than a solid stream because it has a comparatively larger surface area. In confined areas, fog droplets turn into steam, smothering the fire.

An adequate water stream is needed to have the reach to be able to hit the seat of the fire, for which adequate pressure is to be maintained in the nozzle. To extinguish the fire efficiently, the following nozzle pressures are generally considered:

1 Smooth bore hand-line: 345 kPa
2 Smooth bore master stream: 552 kPa
3 Fog nozzle hand-line: 690 kPa

Flow through a nozzle

The amount of water discharging from a smooth-bore nozzle is determined by its pressure and the inside diameter of the opening. The formula for determining the flow from a smooth-bore nozzle is as follows:

$$Q = k\sqrt{N_P} \tag{4.2}$$

where, Q = Flow rate of a nozzle (lpm);

k = a factor represents the nozzle orifice size for a particular flow and pressure. The k-factor of different nozzle orifice can be found from the Table 4.5.
N_P = Nozzle pressure (kPa)

Table 4.5 The k-factor of different nozzle orifice [11]

Sl.no	*Nozzle orifice size (mm)*	***k**-factor (lpm/$\sqrt{bar}$)*
1	9.5	40
2	11.11	57
3	12.7	80
4	13.5	115
5	15.87	160
6	19.05	200

Fire department connection

A fire department connection (FDC) is a means of connection through which firefighters can provide supplemental water into the standpipe or other firefighting system for fire suppression. According to NFPA 14, such FDCs are required on all standpipe systems.

Fire department connections are usually installed at the end of a pipe connected to the fire hydrant standpipe, which is located on the outside of the building to be protected. Through a fire department connection, firefighters supply water to the standpipe system at the desired pressure, drawing water from a pumper fire truck—a fire apparatus with a permanently mounted fire pump of at least 3000 lpm capacity, a water tank of 2000–30,000 l, and a hose body. This allows firefighters to operate fire hoses on the upper floors of the building, above the effective reach of fire department aerial equipment commonly considered to be 23 m.

Single or multiple FDCs of 65 mm diameter each, are installed at a particular location. The number of FDC is determined generally considering one 65 mm inlet for every 1000 lpm of system fire flow demand, which is termed a single FDC. Double-inlet FDCs include two 65 mm inlets, termed "Siamese" connections. So, a triamese FDC has triple-inlets, and will be required for 3000 lpm fire flow demand. Based on number of fire department connections the targeted fire flow and the required stand pipe size have been furnished in Table 4.6. Various FDCs and their stand pipe sizes are shown in Figure 4.16.

Table 4.6 Fire department connection number based on targeted fire flow [12]

Sl.no	*FDC type*	*Inlet number*	*Target fire flow (lpm)*	*Hydrant standpipe size (mm)*
1	Single	1	1000, Maximum 2000	65
2	Siamese	2	2000, Maximum 4000	100
3	Triamese	3	3000, Maximum 4800	150

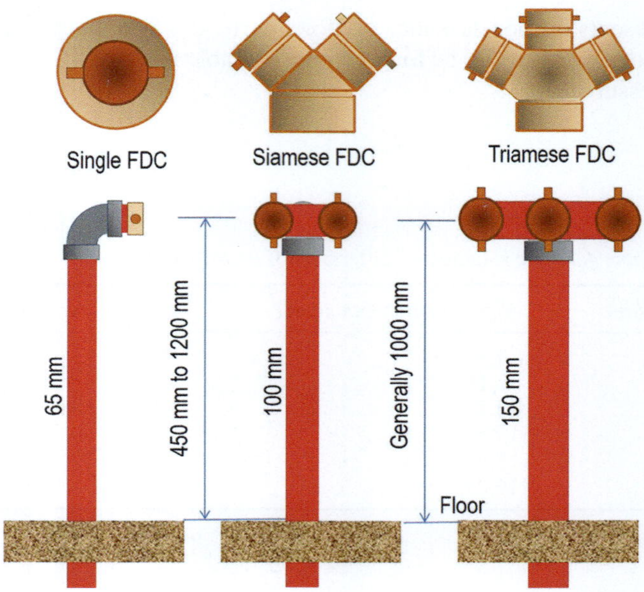

Figure 4.16 Various types of fire departments connections (FDC) and the stand pipe sizes.

Locating the FDCs

In positioning the FDCs following guidelines shall be followed:

1 It should be at a visible and accessible location from the street or nearest fire department access, and preferably on the street side of the building.
2 It should be located within 15.2 m of the street or nearest fire department access.
3 It should not be more than 30.5 m from the nearest fire hydrant, unless approved by the concerned regulatory authority.

All FDCs shall be protected by approved plugs or caps, secured, and easy to remove by the fire department. Its height shall not be less than 450 mm or more than 1200 mm above the grade or floor standing on which the connection would be made.

Quick coupling

Quick-coupling is a jointing device designed to provide a fast, convenient, and reliable way of repeatedly connecting and disconnecting fluid lines, water hose pipe in particular. In quick couplings, there are some elements in common. All have two parts. One is a plug in the male half, and the other is a socket that is in the female half, as shown in Figure 4.17. When connected properly, these two parts are sealed and locked to get joined effectively, to contain internal pressures, and to resist any tensile forces that tend to pull the joint apart. These two parts can easily be disconnected by disengaging the locking mechanism and separating the parts without the help of any tools.

Coupling types: There are a good number of common designs of quick couplings. Here, only the most popular locking mechanisms are mentioned:

1 Ball-lock coupling
2 Roller-lock coupling

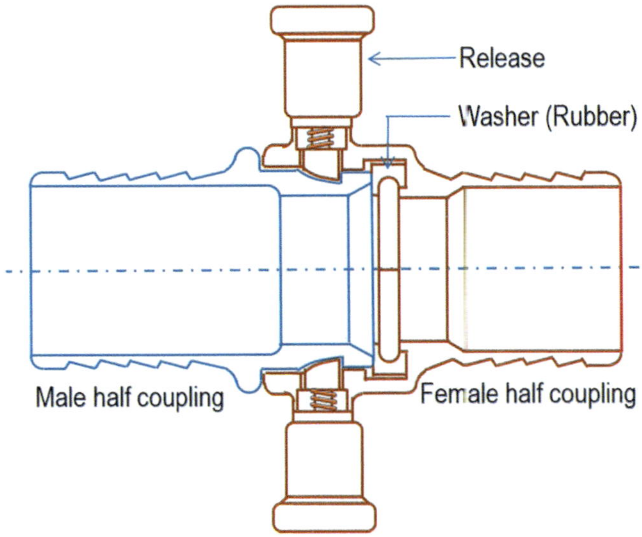

Figure 4.17 A quick coupling (cut section).

3 Pin-lock coupling (claw coupling)
4 Bayonet coupling
5 Ring-lock coupling
6 Cam-lock coupling

Valves and appurtenances

Pressure-reducing valve

In order to keep the maximum allowable working pressure at the inlet of the hose reel, a pressure-reducing valve is used in the hose reel system. Generally, on the branch pipe of the riser stand pipe a pressure-reducing valve is installed, as shown in Figure 4.18, to maintain the desired pressure in the hose and, thereby, in the nozzle. According to NFPA 14, the maximum system pressure in a fire hydrant shall be 24 bar, the static pressure when water is not flowing. In the Class-II system, some code allows a minimum residual pressure of 4.5 bar to be maintained at the remotest 40 mm hose connections. So, hoses near the pumps might have pressure greater than 6.9 bar that is needed to be reduced, which can be done by installing a pressure-reducing valve. According to NFPA14, when the static pressure at any hose connection exceeds 12.1 bar, a pressure-regulating valve shall be installed just behind the outlet of the 65 mm hose connection to limit the pressure. A pressure of 6.9 bar for 40 mm hose connections shall be made available for trained personnel use, and 12.1 bar for higher hose connections to be used for fire department personnel.

Air valve

On a fire hydrant an air valve is installed mostly on the top of the stand pipe, as shown in Figure 4.19, to release accumulated air at the top of the stand pipe. Air pockets may be at each high point of a fully pressurized horizontal branch pipeline or offset of the stand pipe, if any; so, at all such high points, an air valve is to be installed.

Air release valve sizing depends on the volume of air to be released from pipe high points and the diameter of the pipe. In sizing a valve there is a rule of thumb that suggests

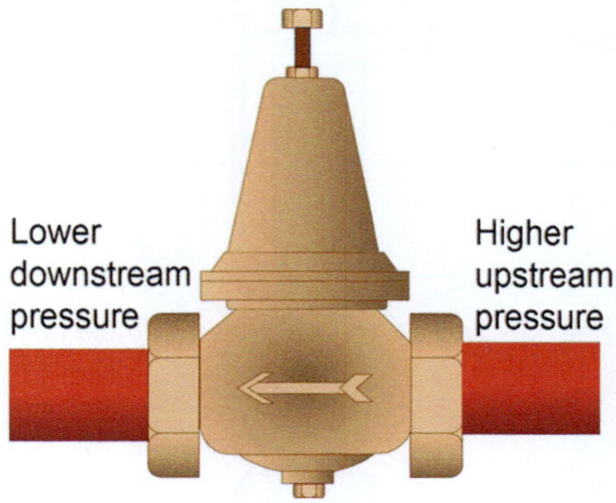

Figure 4.18 A pressure reducing valve on a pipe.

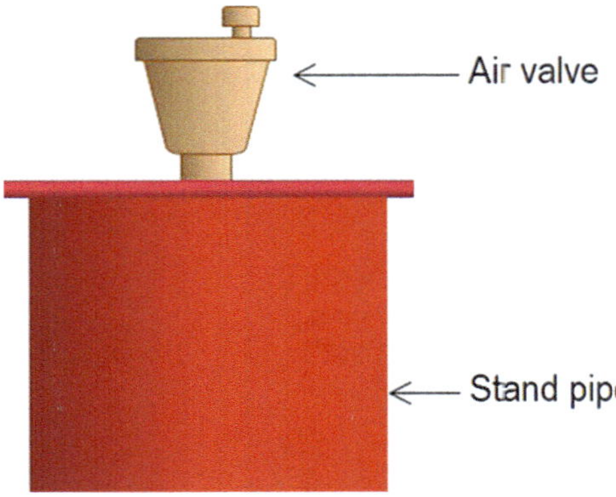

Figure 4.19 Air valve on top of standpipe.

Table 4.7 Sizing of the drain cocks fire hydrant stand pipe [14]

Sl.no	*Standpipe size*	*Drain size*
1	Up to 50 mm	20 mm or larger
2	65 to 90 mm	32 mm or larger
3	100 mm or larger	50 mm only

air release valves be 25 mm per 300 mm of pipe diameter [13]. This means a 100 mm diameter standpipe would have a 8.33 mm diameter air valve.

Drain cock

A drain cock or valve shall be installed at the lowest point of the standpipe piping downstream of the isolation valve. The drainage system shall be arranged to discharge water at an approved location. Sizing of the drain cock shall be as specified in Table 4.7.

Check valve

Fire stand pipe may be provided with water from one of two sources: a water reservoir or through fire department connections, for longer-duration use in a fire emergency. Though multiple potential water sources provide more reliability to fire protection systems, they can also create some problems in the system. Water is supposed to flow in one direction to the standpipe for performance reasons and must not flow into other pipes, causing damage to equipment or appliances installed on it. To ensure water flow in one direction, either from the water reservoir or from the fire department, connections toward the stand pipe check valves are installed both on the pump delivery pipe drawing water from the reservoir and on the pipe connecting the stand pipe and the fire brigade connections, as shown in Figure 4.20.

Check valves are installed to prevent the reverse or back flow of water from occurring. Check valves possess a hinged clapper with a spring, which keeps the valve closed until the water flowing from a source opens it. These clappers only open in one direction: when the

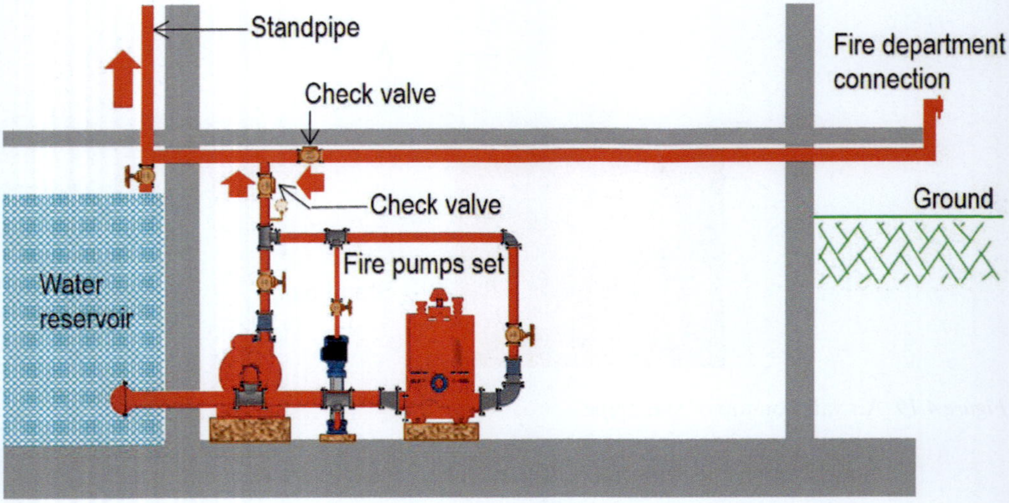

Figure 4.20 Check valve installed on a pump delivery and fire department connection pipe.

flow is reversed or stopped, the spring presses the clapper against a rubber seat, forming a water-tight seal that thus prevents back flow.

Designing fire hydrant elements

The design of major components of a fire hydrant system is an essential job for the engineers concerned. The major elements of the standpipe fire hydrant system designed are as follows:

1 Reservoir
2 Pump
3 Hose length

Sizing a water reservoir

According to NFPA 22, a water storage tank for a fire protection system in a building shall be designed to meet the system demand in terms of required fire flow for the design duration of firefighting.

Fire flow is calculated based on the fire flow area of the building. The flow area is the total floor area of all floor levels of a building, except for Type I (443), Type I (332), and Type II (222); in these cases, the fire flow area would be the area of the largest three successive floors, as shown in Figure 4.21. The fire flow area should be calculated based on the area inside the surrounding exterior walls of each floor and the fire separation walls, if any, used to create separate building occupancy.

A building within 30 m of another building is considered susceptible to fire from an adjacent building and so considered exposed to fire. When the fire exposure distance is 30 m or less from adjacent property, the minimum fire flow furnished in Table 4.8 shall be adhered to. Increases in water storage may be required by the concerned, depending on the area of the exposed structure. When protecting exposures within 30 m or less, the minimum

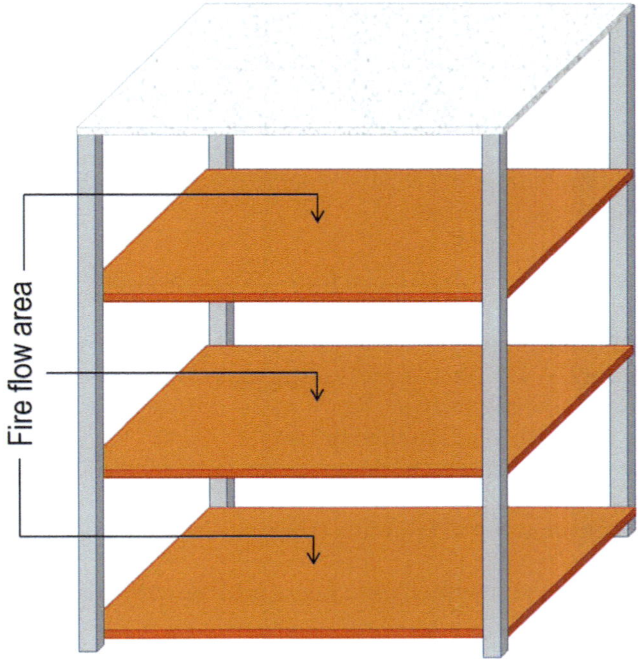

Figure 4.21 Fire flow area of a multistory building.

Table 4.8 Minimum fire flow based on exposure distance [15]

Sl. no	*Exposure distance*	*Minimum fire flow*
1	Over 30 m	1000 lpm
2	10.01–30 m	2000–3000 lpm
3	3–10 m	3000–4000 lpm
4	3 m or less	4000–6000 lpm

If buildings are contiguous, a minimum fire flow of 10,000 lpm will be required.

flow duration shall not be less than 2 hours unless otherwise approved by the concerned fire department.

Water storage requirements range from a minimum of 2,500 liters for a small property (less than 500 sqm) to a maximum of 20,000 liters for rural properties exceeding 1 hectare in size [16].

Sizing a hose reel or pipe

The standard or temporary length available for a fire hose reel of 20 mm or 25 mm is not more than 30 m. But the actual required length for the concerned floor of a building shall be determined by finding the travel length to lay, starting from the hose station to the point about 12 m behind the farthest point of the floor, as the 12 m distance is considered to be the safe distance of water spraying from the nozzle to the fire in an electric line of 440 volts [17].

Sizing fire pumps

For a fire pump set two types of pumps are designed: the main fire pump and the Jockey pump.

Sizing main fire pump

For sizing the main fire pump, the head or engine pressure is to be known, which can be calculated from the following formula:

$$\text{Head}(H) = \text{Static head} + \text{Total friction loss} + \text{Nozzle pressure.}$$

where, static head is the elevation of the water top in the reservoir to the highest or farthest hose station in meters, N_P = Nozzle pressure in m. = 10.2 Nozzle pressure in bar. F_L = Friction loss in risers, hoses, appliances, etc.

Flow (Q) is the fire water flow required in lpm.

Then the brake horse power for the pump is calculated by the formula as follows:

P = motor power consumption in kilowatts (kW)

In order to calculate the pump efficiency in the SI system, the following formula is to be used:

$$\text{Brake horse power} = \frac{Q \times H}{367 \times P} \tag{4.3}$$

where, Q = flow in cubic meters per hour (cum/h)

H = Head or pressure of the pump, in meters (m)

Sizing a jockey pump

NFPA 24 suggests selecting a jockey pump that will make up the allowable water loss rate in 10 minutes or 3.79 lpm, whichever is larger. A rule of thumb for sizing jockey pumps supplying underground piping is to use 1 percent of the fire pump's rated capacity.

For example,

When the fire pump has been designed for a rated flow of 3000 lpm, then the jockey pump flow rate would be 3000 lpm × 0.01 = 30 lpm.

Jockey Pump Head: The pressure development of the jockey pump shall be slightly greater than the pressure of the main fire pump to avoid false starting of the main fire pump. The Jockey pump shall have a higher pressure of at least 0.7 bar or 7 m than that of the main fire pump. So, if the head of the main pump is 100 m, then the head of the jockey pump should be 107 m.

Fire hydrant system maintenance

A fire hydrant system generally remains idle unless any fire event occurs or any fire drill is carried out to train the building occupants on how to respond effectively and efficiently during a fire event. As a result, the major elements or components of a fire hydrant system can seize up from corrosion or silt deposition, resulting in an inoperable condition during times of need. Metals begin to rust; in time, the caps may rust so tight they cannot be removed easily, and valves and joints develop blistering and pitting to the point of becoming unusable. This is often referred to as being "galded." So, this system needs regular monitoring and maintenance work to keep it operable during a fire event. So, this system needs a protocol to follow, as mentioned below:

1 All the nearby places and approaches to the hydrants should be periodically inspected to ensure free and quick accessibility and hydrant indicator plates are in position.
2 The major hydrant components, like inlets, landing valves, drain valves, door hinges, and landing valve boxes, should be inspected every six months.
3 Spindles, glands, and washers of valves shall be checked every six months for any leakages and ensure that they are in operable condition.
4 The cleanliness of the water in the storage tanks is to be ensured.
5 Fire pumps shall be properly maintained. When any part of the outlet on the rising mains is found to be defective but no immediate replacement is available, the whole valve assembly should be removed from the main and replaced with a blanking-off plate or plug so that the system remains operative without disturbing the whole system [18].
6 Regular inspection every 90 days (or as specified by local regulation) should be done to hose reels to ensure that there are no leaks in

    1 Inlet valve
    2 Automatic on-off valve
    3 Glands
    4 Tubing
    5 Shut off nozzle.

7 Annual flushing of the system helps reduce the amount of buildup and bring to attention any repairs that may need to happen to the hydrant in order for it to be an effective system for extinguishing a fire.
8 After a firefighting operation, the fire hose is usually kept hung to be empty and dry because trapped water that remains in the hose for a long time can deteriorate the material and render it unreliable or unusable for further use.
9 The check list is provided below to identify the key concerns of taking immediate measures to ensure the operability of the system:

    i Legibility of the operating instructions
    ii Goodness of warning labels and tags free from damage
    iii Condition of the cabinet, with no cracks or breaks in the glass
    iv Easy operability of the cabinet door
    v Easy operability of the lock of the cabinet
    vi Smoothness of the hose rack swing out, at least 90° from the cabinet
    vii Goodness of safety seals
    viii Visible damage, missing of parts or corrosion to the hose valve
    ix Hose is properly folded or hung in the hose rack
    x Goodness of the hose; whether intact with no breaks, holes, or tears
    xi Presence and goodness of the hose nozzle
    xii Easy accessibility of all the valves, hose nozzles, fire extinguishers, etc.

References

[1] National Fire Protection Association "Water delivery rate in standard on water supplies for suburban and rural fire fighting" NFPA 1142; Fire-Flow Requirements for Buildings.

[2] Uniform Plumbing Code (2021) "Fire-flow requirements for buildings" Retrieved from https://up.codes/viewer/san-antonio/ifc-2021/chapter/B/fire-flow-requirements-for-buildings#B 0n 19 March, 2024.

[3] Tracey Foster (2019) "NFPA 20: fire pump design" Retrieved from https://www.csemag.com/articles/nfpa-20-fire-pump-design/ on 19 February, 2024.

[4] Grundfos Holding "Fire pump" Retrieved from https://www.grundfos.com/solutions/learn/research-and-insights/fire-pumps on 19 February, 2024.

[5] Bureau of Indian Standards (2003) "Installation and maintenance of firefighting pumps — code of practice" Section 6.5. Retrieved from https://www.cracindia.in/admin/uploads/IS-15301.pdf on 19 March, 2024.

[6] Al-Emran Hossain (2023) "Pressure maintenance (Jockey) pumps for fire protection under NFPA 20" Retrieved from https://www.linkedin.com/pulse/al-emran-hossain-uzbic/ on 19 March, 2024.

[7] Safe World HSE (2022) "Capacity of fire jockey pump" Retrieved from https://www.safeworldhse.com/2022/10/what-is-fire-jockey-pump-how-to.html on 21 February, 2024.

[8] Integrated fire Services "Determining the location of fire hose reels" Retrieved from https://integratedfire.com.au/blog-posts/2018/7/12/determining-the-location-of-fire-hose-reels on 22 February, 2024.

[9] Iron Man "Fire hose rack assembly" Retrieved from http://firefightingtool.com/2018/7-4-fire-hose-rack-assembly.html on 18 March 2024.

[10] Lyndon Cousins "Friction loss (Q formula)" Retrieved from https://www.youtube.com/watch?v=Ooi3PARHO98 on 17 March 2024.

[11] AGF Manufacturing, Inc. (2021) "What is K-factor for fire sprinklers?" Retrieved from https://content.agfmfg.com/blog/what-is-k-factor/ on 17 March 2024.

[12] Y. Utiskul and N.P. Wu (2016) *Residential Fire Sprinklers – Water Usage and Water Meter Performance Study*, The Fire Protection Research Foundation, Quincy, MA.

[13] American Water Works Association, AWWA M51 *Air Valves: Air-Release, Air/Vacuum and Combination*, 2nd Ed, 2016, pp. 2–10. Denver, CO: American Water Works Association (AWWA).

[14] National Fire Protection Association (NFPA) "Standard for the installation of standpipe and hose systems" Chapter 7, NFPA 14. Retrieved from https://firesolution.id/fire-solutions/fire-hydrant-system/nfpa-14-standard-for-the-installation-of-standpipe-and-hose-systems-chapter-7 on 25 February, 2024.

[15] San Diego County Fire Authority "Water tank standard for fire protection" in Pioneer Water Tanks Webpage. Retrieved from https://pioneerwatertanksamerica.com/what-are-the-nfpa-22-fire-protection-standards-for-water-tanks/ on 25 February, 2024.

[16] Clark Tanks "Water storage requirements for fire-fighting services" Retrieved from https://clarktanks.com.au/2018/04/26/water-storage-requirements-for-fire-fighting-services/ on 25 February, 2024.

[17] A. M. Buswell and W. V. Upton (1937) *Water Quality for Fire Fighting*. Department of Registration and Education, State Water Survey Division Urbana, Illinois.

[18] Bureau of Indian Standards (1990) "Code of practice for installation and maintenance of internal fire hydrants and hose reels on premises" Retrieved from https://law.resource.org/pub/in/bis/S03/is.3844.1989.html on 16 March 2024.

5 Fire sprinkler system

Introduction

It is universally acknowledged that buildings are prone to fire, mostly originating from their content, predominantly from the activities of their occupants. So, to make a building safe for living in, all possible measures for avoiding the occurrence of fire and minimizing loss by suppressing it at the quickest possible time, various fire suppression measures are employed. In the previous chapter, one of the fire suppression measures using water called the fire standpipe system has been delineated; in this chapter, another fire suppression measure using water will be exclusively discussed, which is known as the automatic fire sprinkler system.

Fire sprinkler system

Water is supposed to be the most widely recognized agent for fire suppression. To spread water automatically and effectively on unwanted fire in a building, an automatic fire sprinkler system is developed. A fire sprinkler system is considered an effective and reliable fire suppression strategy using water, for the buildings. Practically, water droplets of sprinkling water from sprinkler heads help reduce fire byproducts, weigh down smoke particles and prevent speeding of fire, to make the space safer for occupants to breathe and escape therefrom.

In this system, water is caused to fall in a sprinkling pattern on fire, either from the top ceiling or the side walls. To make a sprinkle pattern of water spray, a small device is used. To ensure effective water supply, an acceptable water supply system for sprinkling water is required. This is also a hydrant system as it deals with water supply, alternatively termed a fire sprinkler hydrant system, as at the end, water is discharged in a sprinkling pattern on fire, through sprinkler heads.

The sprinkler fire hydrant system is planned, designed, and developed to provide sufficient flow of water and pressure for a required duration without the need for any human intervention to meet the most hydraulically demanding area for fire suppression in the facility. To ensure that a sprinkler system can adequately protect a property, it must be well maintained as well by skilled manpower familiar with the components, operation, and testing of the systems; absence or weakness in these efforts might appear as potential causes for system failure during need.

It is general and common to realize in the USA that about 5–11 minutes [1] will require a fire department to respond, coming all the way to the spot of a fire to fight, but an uncontrolled fire may grow and spread through the space by this time, causing tremendous smoke

DOI: 10.1201/9781003724056-5

and fire damage, for which an automatic fire sprinkler would be an effective choice as a preventive measure. Although the fire sprinkler system is used generally in factories and large commercial buildings, in present days residential buildings and small buildings are also using this system for being cost-effective. In a building, automatic sprinklers shall be installed in the following conditions:

1 A basement car park or storage, exceeding 200 sqm occupancy area [2]
2 Multilevel basement floors [2]
3 Any room or compartment of a building exceeding 1125 sqm area [2].
4 Departmental stores or shops exceeding 500 sqm covered area [2]
5 All non-domestic floors of mixed occupancy that constitute a hazard and are not provided with staircases independent of the remainder of the buildings
6 Godowns and warehouses as considered necessary
7 Where there are multilevel basement floors

Major components

Automatic fire sprinkler systems consist of various components, major to minor, some being optional, to make the complete system. Following are the major components required:

1 Water reservoir
2 Stand and branch pipe
3 Sprinkler head
4 Fire pump
5 Valve

Minor components are as follows:

1 Alarm
2 Test connection
3 Drain cock

In a fire hydrant sprinkler system, hose stations can also be added as an optional or additional supporting element for firefighting, which is discussed in the preceding chapter; then, the system is termed a combined system. Figure 5.1 shows all the major components of a fire sprinkler system.

Water reservoir

A water reservoir is required for holding sufficient water to ensure adequate water flow for a particular duration and to suppress fire when it occurs in a part of a building. These reservoirs are generally built in the lowest basement floor or underground, occasionally on ground. Water is pumped from the reservoir to generate pressure for the flow of water in the system.

The size of the water tank or reservoir depends primarily on the hazards classification of the building for which the sprinkler system is designed. So, in most cases, the greater the hazard, the larger the volume of water required to suppress the anticipated fire. The desired duration of water application based on the hazard classification causes more water to be

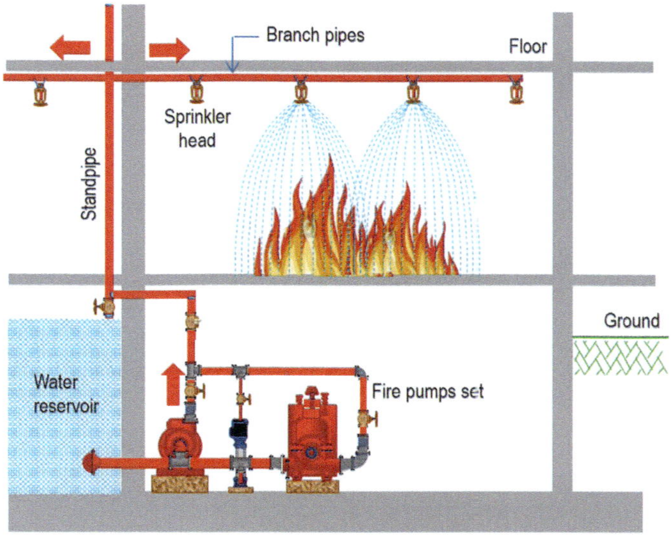

Figure 5.1 Major components of a fire sprinkler system.

reserved for. The factors that need to be considered for determining the amount of stored water for fire protections are as follows:

1 The sprinkler design standard
2 Hazard classification and design density
3 Duration of water supply
4 The capability of water supply mains and the inlet to fill the tank in a reasonable time, according to BS EN 12845 which is 36 hours or having sufficient flow to replenish a tank in 30, 60 or 90 minutes, as necessary [3]

The reservoir tank shall have one inlet. Where the required total water supply is provided by more than one source, multiple sources shall be interconnected so that there is one inlet leading to the reservoir and there should be one branch outlet for the firefighters to utilize the entire required water supply from any one location or connection without interruption of flow.

For enhanced reliability, a very large fire water storage tank of more than 50,000 liters can be split into two half-capacity tanks by providing a separation wall supported by controlling valves in the middle of the tank, as shown in Figure 5.2. A sump can be formed in the base of the tank to allow the pump to draw water to a lower level.

Pressure tank

In some cases, pressurized water tanks, as shown in Figure 5.3, are incorporated to supply water to sprinkler systems. They are often used where public water is unavailable and the demand on the sprinkler system is relatively low. Pressurized tank capacity generally ranges between 19,000 and 75,000 liters.

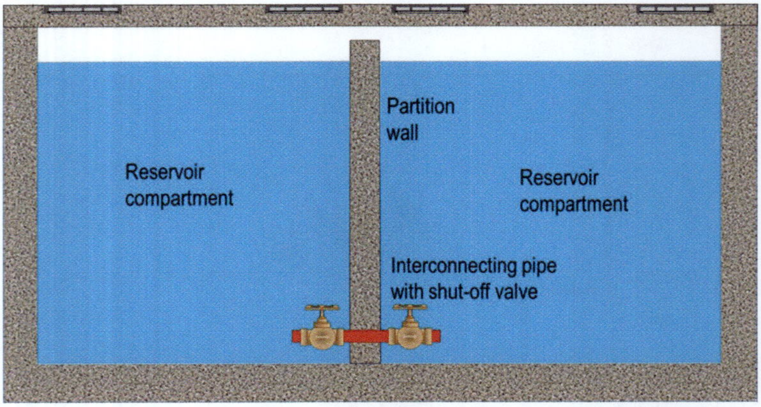

Figure 5.2 Partitioning of a large firewater reservoir.

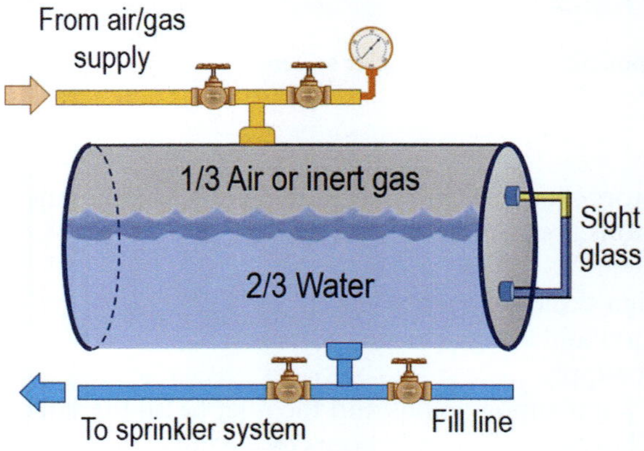

Figure 5.3 Pressurized tanks for fire-sprinkler water supply.

In pressurized tanks, two-thirds of the tank is occupied by water to be used; the rest, one-third, is kept filled by air or inert gas under pressure. The air pressure in the tank is approximately 690 kPa.

As the pressure tanks possess very limited capacity, generally designed to suppress small fires, once the water in the tank is replenished, there will be no more water available to the sprinkler system for additional duration to suppress fire. To ensure its effective performance, it is very important to monitor and maintain water level, water temperature, and air pressure in the tank.

Fire pump for fire sprinkler system

A pump is a vital part of a fire sprinkler system, which draws water from either an underground water tank or other dependable water sources. Being driven either by an electrical or diesel motor, the high pressure created by the pumps distributes water through a riser, and the piping network to spray water through the sprinkler heads, is activated automatically due to the temperature rising up to a certain level caused by an unwanted fire below or to a hose station to through water to a fire. A fire sprinkler pump is also basically a centrifugal pump discussed in the previous chapter.

In an automatic system, fire pump-impellers are kept non-rotating under the system pressure though the power switch is on and start running whenever the activation of a fire sprinkler occurs, causing a drop in the system pressure, which signals the pump to run.

Flow rate

Sprinkler fire pumps are specified for a rated flow provided by the manufacturer of the pump. Fire pumps are rated by the manufacturer, which indicates a rated flow that produces a certain amount of net pressure, typically between 276 and 1724 kPa. The rated net pressure usually only develops at the rated flow of the pump. At flows less than the rated flow, the net pressure is typically higher than the rated net pressure; and at flows greater than the rated flow, the net pressure becomes lower than the rated net pressure.

The sprinkler used for industrial buildings should have a discharge of about 75–150 lpm, and a typical Early Suppression Fast Response (ESFR) sprinkler at a pressure of 340 kPa discharges approximately 380 lpm [4].

The flow rating of fire pumps ranges from 100 to 20,000 lpm, but the most commonly found flow ratings are 1000–4000 lpm. Fire pumps are generally allowed to be run beyond their designed rated flow. According to NFPA 20, fire pumps can be used up to 150 percent of their rated flows. So, a pump rated at 2,000 lpm can be used with fire protection systems having maximum flow demands up to 3,000 lpm, as long as it meets the pressure demands.

All fire pumps are supposed to have on their bodies a nameplate stating the rated flow (lpm), rated pressure (kPa), rated speed (rpm), direction of impeller rotation, and other information specific to the pump.

Occupancy hazard classification

Section 19.3.1.2.4 of *NFPA 13: Standard for the Installation of Sprinkler Systems* categorizes buildings or parts of buildings into six occupancy hazard classifications. These are as follows:

1 Light Hazard
2 Ordinary Hazard (Group 1)
3 Ordinary Hazard (Group 2)
4 Extra Hazard (Group 1)
5 Extra Hazard (Group 2)
6 Special Occupancy Hazard

Design area

Hazard level and floor area are the two major factors required to determine the flow rate. Design area size can vary, but the minimum size for Light and Ordinary Hazard occupancy is 140 sqm, and for Extra-Hazard areas, the minimum size is 230 sqm. Depending on the location and condition, adjustments to the design area's effective size are needed. The conditions that require increases in size are as follows:

1 Sloped ceilings
2 Use of dry pipe or double-interlock pre-action system
3 Non-sprinkled combustible isolated spaces adjacent to the design area

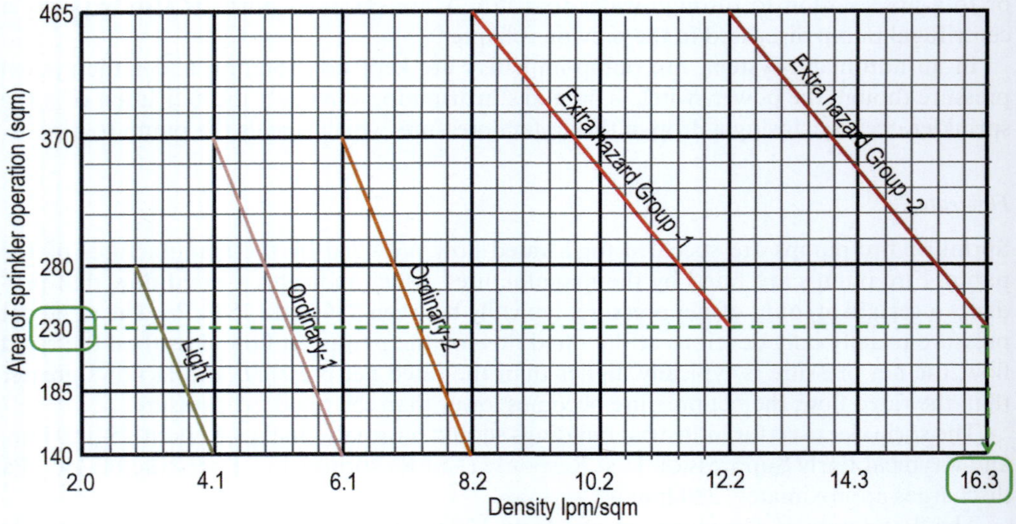

Figure 5.4 Density-per-area curves for sprinkler heads, based on occupancy hazards. SI conversion from source [5].

When specific conditions are met, the use of quick-response sprinklers enables designers to reduce the design area size. About 25 percent reduction is allowed when high-temp sprinkler heads are used in an extra-hazard setting, but the area can go no lower than 186 sqm.

To hydraulically calculate minimum water requirements for sprinkler systems, the required minimum densities of water and remote areas to be served by a sprinkler head are very important to know. NFPA 13 provides density/area curves as shown in Figure 5.4. From the curve, for the appropriate hazard level and for the design area size, the corresponding water density is found. Then, to calculate the required flow, the following formula is used:

$$Q = d \times A \tag{5.1}$$

where,

Q = Minimum required flow lpm
d = Water density lpm/sqm
A = Area sqm.

Required pressure

Once the minimum required flow rate is known, the minimum required pressure can be determined. Pressure and flow rate are both tied to a value called k-factor, which describes how much water flows from a sprinkler head (or a nozzle) at a given pressure. Every fire sprinkler has a specific k-factor. The relationship between pressure and flow rate is:

$$Q = k\sqrt{P} \tag{5.2}$$

Table 5.1 The *k* factor for different sprinkler heads

Nominal k-factor	*k-factor range*	*Nominal orifice size (mm)*	*Thread connection (mm)*
lpm/√bar	*lpm/√bar*		
NFPA 13/UL 199	NFPA 13/UL 199	NFPA 13	NFPA 13/UL 199
40	38–42	10	13
60	57–63	11	13
80	76–84	13	13
115	107–118	14	20 or 13
160	159–166	16	20 or 13
200	195–209	20	20
240	231–254	/	20
320	311–343	/	25
360	349–387	/	25

SI conversion from source [6].

where,

P = pressure
Q = Flow rate
k = A factor

To know the required pressure for the sprinkler in a room, the flow through each sprinkler head in the space is to be known. It is evident that every sprinkler head will have different flow rates at different pressures. But practically, the total required flow is divided by the number of sprinkler heads to be installed to find the required flow from each sprinkler head. Then, a k-factor is selected for the sprinklers, and the pressure is calculated. Table 5.1 provides the k factor for different sprinkler heads.

Head and flow for fire pump

In sizing the fire pump for a sprinkler system, the required pressure head for a particular flow rate for the system is to be determined; likewise, it is determined in a water supply system.

The total head of a fire sprinkler pump is the summation of the static suction head, city water main pressure when connected with, the static delivery head to the highest or farthest sprinkler head, the summation of friction loss in pipes leading to the highest or farthest sprinkler head, and the pressure required for the highest or farthest sprinkler head.

The total dynamic head of the pump varies along its performance curve that is determined by three limiting points: the shut-off, the rating, and the overload. The shut-off point represents the maximum total pressure head when the pump is operating at theoretically zero flow; this is also referred to as the churn pressure. The rating is the range of pressure and flow that the pump should produce when operating at 100 percent capacity. The total head pressure should not be less than 65 percent of the rated total head when the pump is operating at 150 percent of rated flow capacity; this is the overload point. If the system flow exceeds the flow at the overload point, the pump might be subject to possible cavitation and damage.

In determining the flow rate, the most hydraulically demanding area in the building is to be selected. The estimated flow demand for that area is to be determined, as discussed earlier.

Fire pump performance curve

The performance curve made for a particular fire-pump represents its performance in a variety of situations, which is prepared and provided by the manufacturer. These curves show how much pressure a pump produces at a given flow rate, which helps compare performance among multiple pumps to choose the most efficient one.

In the pump performance curve, the head is presented in the x-axis, and corresponding flow values in the y-axis are plotted. The plotted graph of the fire pump is found to be a gradually sloping curve, as shown in Figure 5.5.

For a compliant fire-pump, the churn pressure, the pressure when the pump is in running condition but there is no flowing of water, should be a value below the 140 percent of rated pressure. Rated capacity is the capacity that the pump is designed for to perform at its maximum efficiency.

If the fire-pump draws water from the city water supply system, the system demand must be below the combined supply curve, as discussed in the following example.

Let us consider a fire sprinkler system that has a flow demand of 2000 lpm at a pressure of 40 m head. From the flow test, it is seen that the public water main has a static pressure of 30 m and a residual pressure of 20 m when flow is 1600 lpm, as shown in Figure 5.6. A fire pump rated at 1800 lpm at 40 m is selected for consideration. Figure 5.6 includes a plot of the pump curve with churn pressure being 140 percent of the pump's rated pressure, or 56 m, and 150 percent rated flow, i.e., 2700 lpm at 20 m. When the pump curve and the public water supply main curve are combined, it shows that the system demand is below the combined supply curve. Therefore, the selection of the fire pump can be considered acceptable.

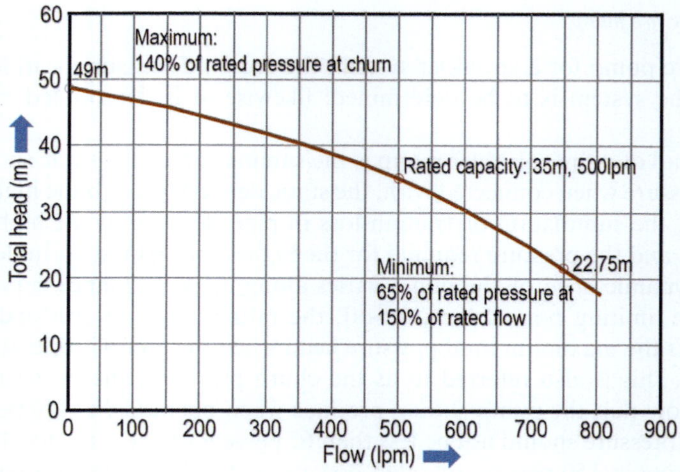

Figure 5.5 Fire pump performance curve.

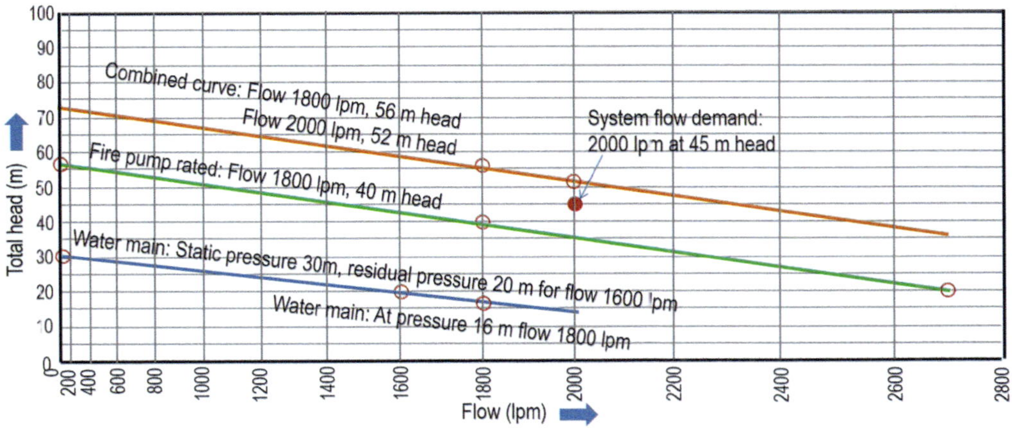

Figure 5.6 The combined supply curve.

Fire pump testing

Both centrifugal and vertical turbine-type fire pumps have the same performance criteria. These types of fire-pumps shall be tested for three specific proficiencies, as mentioned below:

1 Churn pressure not exceeding 140 percent of the rated pressure
2 Rated pressure at rated flow, at 100 percent capacity
3 150 percent rated flow, which represents the pump's flow at 65 percent of the rated pressure

According to NFPA 20 for centrifugal fire-pumps, sprinkler system flow demand should never exceed 150 percent of the rated flow, also called the overload point of the pump.

Excessive pressure development can damage a fire sprinkler system and may lead to system failure. NFPA 13 therefore suggests that the components of above-ground fire sprinkler systems must be rated to at least 1200 kPa [7].

Jockey pump for fire sprinkler system

Jockey pumps, which have already been discussed in Chapter 4, are also used in automatic fire sprinkler systems for the same purposes.

A jockey pump is connected to the fire sprinkler system to maintain desired pressure in the sprinkler pipes. When a fire sprinkler is activated or there are leakages somewhere in the piping and its control devices, there will be a pressure drop in the system, which will be sensed by the jockey pump's automatic controller and will cause the jockey pump to start.

A jockey pump is generally sized for a flow less than the flow to one sprinkler head. According to, if the piping is subject to some leakage, the guideline is to size a jockey pump that will make up the allowable leakage rate in 10 minutes or 3.79 lpm, whichever is larger. The head of the jockey pump is considered to be greater than the pressure of the main fire pump by a minimum of 69 kPa to avoid false starting of the main fire pump.

Types of fire sprinkler systems

There are four main types of fire sprinkler systems, as mentioned below:

1 Wet pipe
2 Dry pipe
3 Deluge
4 Pre-action

Special sprinkler systems are as follows:

1 In-Rack Sprinklers
2 Early Suppression Fast Response (ESFR) sprinklers

Wet-pipe fire sprinkler

This is the most common fire sprinkler system in use as a fire suppression system. In this system, the water supply piping is kept constantly filled with water under the desired pressure, as shown in Figure 5.7. When any sprinkler head is activated due to temperature rise supposed to be for any unwanted fire, since the system is kept filled with water under pressure, water becomes free to flow out of that sprinkler head, thereby pressure drops in the supply pipe. When pressure drops to a certain level, the main fire pump starts running, to supply water through the sprinkler heads. In wet-pipe type fire sprinklers, a special valve, as shown in Figure 5.8, is used to allow flow of water during need and from where water is drained out when water supply is stopped after dousing fire. They are also designed to automatically actuate electrically or hydraulically operated alarms when there is a steady

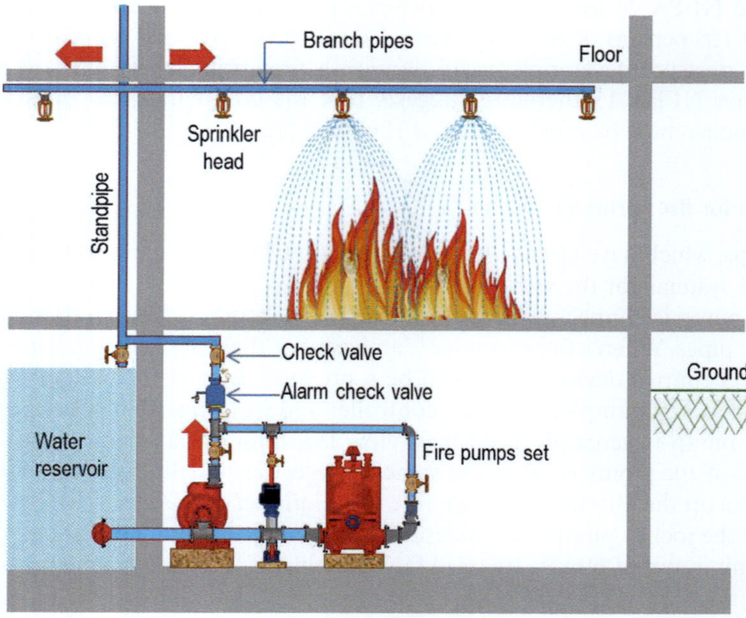

Figure 5.7 Typical layout of wet pipe sprinkler systems.

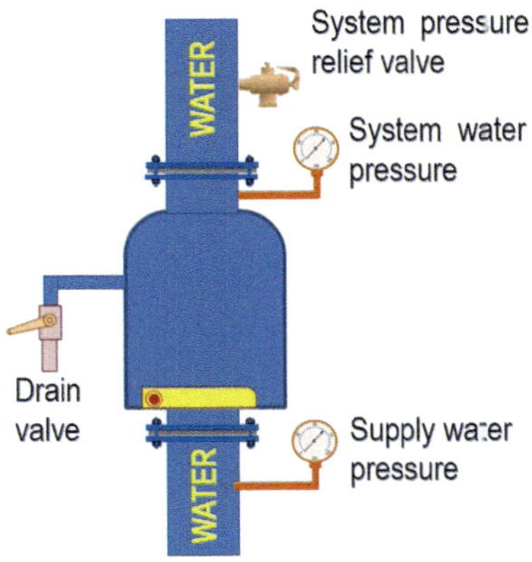

Figure 5.8 Alarm check valve for wet-riser.

flow of water into the system. These valves perform like a check valve, so termed an alarm check valve.

Since the wet-pipe type automatic sprinklers hold water under pressure for long, it should be regularly checked for sediment accumulation and loading in the piping of the system. Wet-pipe type sprinkler systems are the most reliable and cost-effective. But where there is a possibility of water becoming ice due to a fall in temperature less than 4°C, this system is avoided.

Dry-pipe fire sprinkler

A dry pipe system is considered to be the best option for the areas where the air is atmospheric, and so room temperature becomes 4°C or less and cannot be guaranteed to be maintained higher than this temperature to prevent water in a pipe from freezing.

In this system, the water is held under pressure behind a dry pipe valve, usually some distance behind the first branch pipe of sprinkler heads, and the rest of the piping above is kept filled with air under pressure, as shown in Figure 5.9. So, when a sprinkler head of any location is activated due to a temperature rise supposed to be for any unwanted fire, air in the pipe becomes free to flow out of that sprinkler head, resulting in a pressure drop in the system. A drop in pressure in the pipe will cause the dry pipe valve, as shown in Figure 5.10, to open and water to flow through the pipe and be sprinkled out through the open sprinkler head. In this system, there will be a delay in water flow through the furthest sprinkler head's operation when activated. So, the length of the dry pipe systems is limited to minimize the delay.

For effective operation during fire, the following measures are taken in the dry pipe fire sprinkler system:

1 The dry pipe valve must open when the pressure in the piping drops to 40 kPa.
2 A maximum delay of up to 60 seconds might be allowed to flow through the farthest sprinkler head.
3 An alarm shall be connected to the valve assembly to activate when the valve opens and flow occurs.
4 The dry pipe valve and associated pipe, equipment, appurtenances behind shall be installed in a space where temperature can be ensured above the freezing point of water.

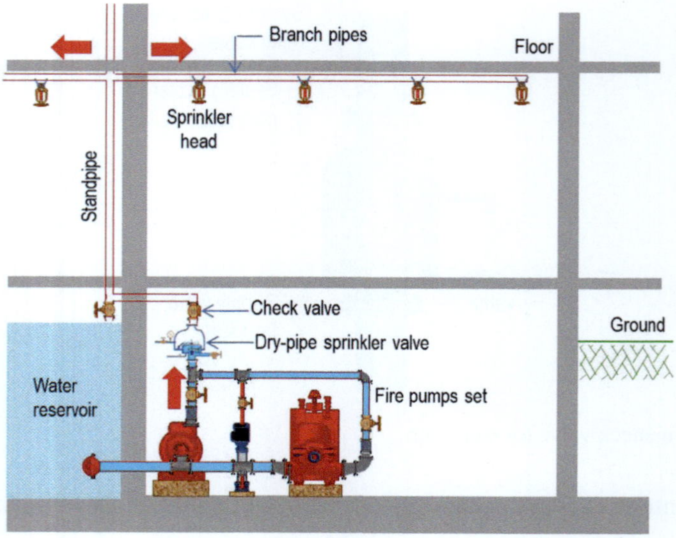

Figure 5.9 Dry-riser and its valve.

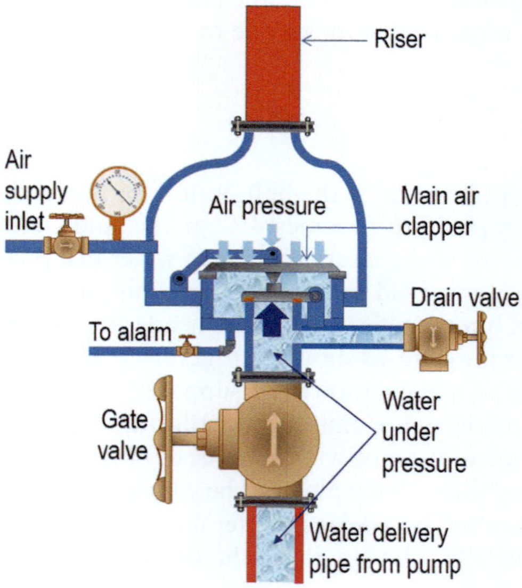

Figure 5.10 Dry-pipe sprinkler valve assembly.

In non-automatic dry pipe systems, all piping is normally kept dry, and water is supplied by firefighters through pumping water into the system via a fire department connection (FDC).

Deluge fire system

The deluge fire system is generally installed to protect high-fire-hazard areas that fall under fast-spreading fire with a very high rate of heat release, by activating a large number

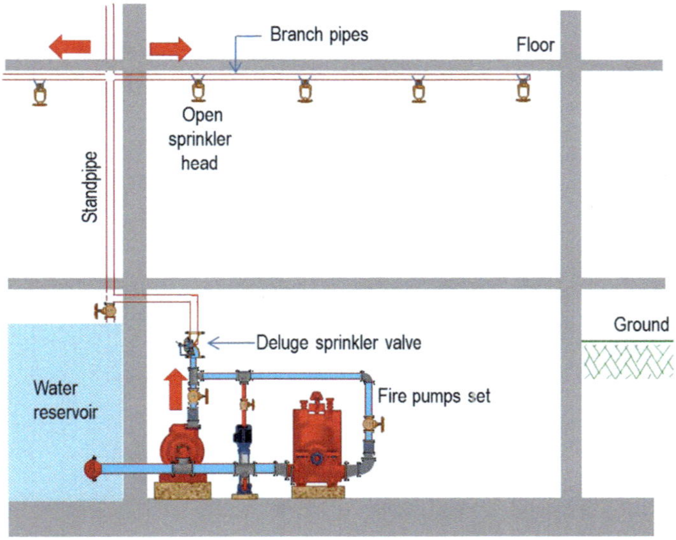

Figure 5.11 The deluge fire sprinkler system.

of open sprinkler heads simultaneously in case of a fire. These conditions require quick application of large volumes of water. So, in this system, a deluge valve is used for fast application of huge water flow through the riser and then the branch pipes and finally through the open sprinkler heads, thereby flooding the fire-affected area, as shown in Figure 5.11. The pipes, upstream the valve, are kept dry and unpressurized; the downstream pipe is connected to a pumping system that draws water directly from a water reservoir. The energized pump keeps water under pressure, but the water is held back by the deluge valve.

A deluge valve has an inlet, an outlet, and a priming chamber. The inlet and the outlet are separated from the priming chamber by the valve chamber and a diaphragm. In normal conditions, pressure is applied to the priming chamber through a restricted prime line. The pressure is trapped in the priming chamber and holds the clapper on the seat due to the differential of pressure. When the pressure is released from the priming chamber faster than it is supplied through the restricted priming line, the clapper moves; thus, allowing the inlet water supply to flow through the outlet into the system.

A deluge valve is basically a water flow control valve that remains normally closed but is activated by a fire detection system or manual release to open and allow water to flow into the system's piping. So, it is designed to operate manually or by actuation of an automatic fire detection system that is installed in the same area as the part of the water discharge system. Considering mode of operation, the deluge valve can be activated in the following ways:

1 Manually
2 Electrically, using a fire detector
3 Pneumatically, using a glass-bulb detector
4 Hydraulically, using a glass-bulb detector

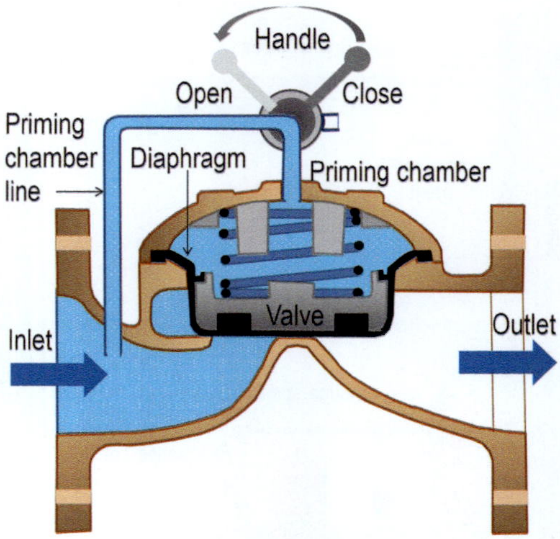

Figure 5.12 A manually operated deluge sprinkler valve.

Manually: Under fire conditions, water pressure in the deluge valve can be released from the control chamber by operating the manual handle to release the valve or, when fitted, by opening the additional manual release option incorporated in other types of valve release systems. By actuating these releasers manually, the deluge valve is opened to allow water to flow into the system's piping and to the concerned open sprinkler heads. Figure 5.12 represents a manual deluge valve in closed condition.

Electrically: When fire is detected by a smoke detector, it will energize a solenoid valve, releasing the prime water off of the top chamber of the valve and allowing the deluge valve to trip and deliver water to the sprinklers through riser and branch pipes.

Pneumatically: A pilot line of closed sprinkler detectors containing air under pressure is located in the area to be protected. It requires regulated dry air supply with a main supply point through a restricted orifice. The air pressure is to be maintained as specified for the dry pilot actuator. The pilot line is connected to the air inlet side of the actuator. The top chamber of the deluge valve is connected to the water inlet side of the actuator. When there is an air pressure drop, generally due to release of water through the sprinkler heads on detection of fire, the diaphragm of the actuator is lifted and allows the water to flow. In this condition, water pressure is released in the top chamber of the deluge valve, allowing the valve to get up from the seat and water to flow into the hydrant and alarm devices. For the dry pilot trim system, the recommended air supply pressure is 3.5 kg/sqcm. There must be a non-return valve at the air supply connection to deluge valve trim.

Hydraulically: Hydraulically, wet pilot operation uses a pilot line of closed sprinklers containing pressurized water, supplied through the upstream side of the Deluge valve through a restricted orifice. All the pressure release lines are connected to a common pressure release line connected to the upper chamber of the valve. Due to the release of any one of the sprinkler heads, the water pressure in the top chamber of the Deluge valve drops and the Deluge valve opens.

A deluged sprinkler system is though advantageous for some conditions, has some disadvantages as well. The advantages and disadvantages of this system are as follows:

Advantages:

1 The entire space is flooded with water, extinguishing the fire quickly and preventing further destruction.
2 The system gives quick response in water throwing.
3 Loss is minimized by preventing fire propagation, as the surrounding area near the fire cools.
4 It can be installed at comparatively low cost.

Disadvantages:

1 Water-sensitive items and electronic equipment might be damaged.
2 It needs longer cleaning and draining time for accumulated water than dust and gas systems.
3 It requires a comparatively large water reservoir.

Pre-action fire system

A pre-action fire sprinkler system is a two-step process before discharging water through the sprinkler head. The two-step process allows shutting off the system in case of a false alarm, minimizing the risk of water loss and related troubles, thus offering an additional layer of protection. This system is preferred for areas where accidental discharge of water through sprinkler heads needs to be avoided, generally to protect moisture-sensitive areas like data centers, art storage rooms, computer rooms, libraries, etc., and equipment. A pre-action system is relatively costly and more complex, in terms of installation and maintenance, than a traditional sprinkler system.

Pre-action fire sprinkler is basically a combination of dry and deluge sprinkler systems in which water is not contained in the pipes but is held back by a pre-action valve. Pressurized nitrogen or air is kept filled within the sprinkler pipes to keep the water in place, maintaining a minimum pressure of 0.5 bar. Pre-action valves are electrically operated and controlled valves that are activated by a heat, smoke, or flame detector's activation. When a fire is detected by these types of detectors, the pre-action valve is activated, allowing water to flow into the sprinkler pipes. Each fire sprinkler head is activated individually. The sprinkler head is activated to release water onto the fire-affected area only.

There are two common types of pre-action fire sprinkler systems, as mentioned below:

1 Single interlock
2 Double interlock

Single interlock: In a single interlock system, the supplemental fire detector has a lower temperature rating than the automatic sprinklers. The pre-action valve is released to flow water into the sprinkler pipes upon operation of fire detectors only. Single-interlock pre-action systems are usually employed in locations where there is less sensitivity to water damage.

Double interlock: The double interlock type has both the pre-action valve and a fail-safe. This failsafe, normally closed, prevents water from accidentally going through the valve. Water flows to sprinkler piping upon activation of both the fire detectors and automatic sprinklers. In this system, the deluge valve opens only when both pressures are reduced in the sprinkler piping and the fire detector is activated. There is provision for closing the deluge valve manually.

Double interlock pre-action systems, though more expensive, are preferred where the maximum degree of protection against an inadvertent sprinkler operation that could result in unnecessary flooding of water.

Sprinkler head

A fire sprinkler head, one of many types as shown in Figure 5.13, is basically one kind of automatic valve that prevents water from flowing from the pipe on which it is installed. A temperature-sensitive glass-bulb or link holds in place a "pip cap," which acts as a plug to prevent water from flowing. But when the temperature rises due to a fire, the glass-bulb expands until it breaks down to allow the free flow of water. Flowing water is deflected around in a sprinkling manner, being hit on a deflector of the sprinkler head. Sprinkler heads are mounted on pipes, generally at certain intervals.

Most fire sprinklers are available in two primary sizes: 12.7 mm and 19.3 mm. A 12.7 mm size is the most common size for office buildings, hotels, and schools, while 19.3 mm sprinklers are used in higher-challenge fire environments where larger quantities of water are necessary. One fire sprinkler head requires on average 30 lpm flow of water and typically requires a pressure of only 0.5 bar to operate.

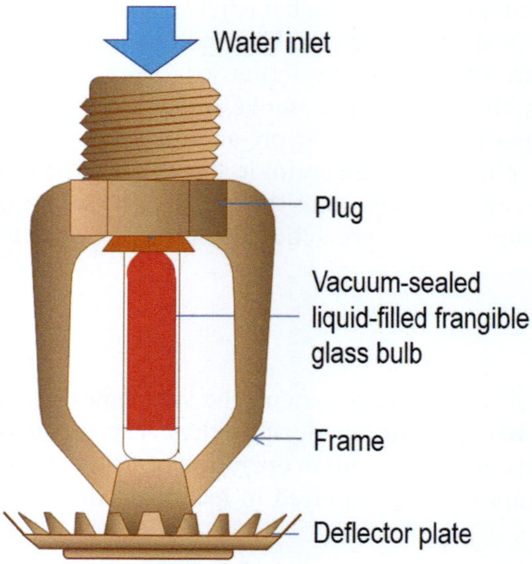

Figure 5.13 Sprinkler head, pendent type.

A typical sprinkler, those are used in manufacturing industries, discharges about 75–150 lpm. However, a typical Early Suppression Fast Response (ESFR) sprinkler at a pressure of 340 kPa will discharge approximately 380 lpm.

The sprinkler heads are composed of the following four main parts:

1 Cage frame
2 Quartzoid bulb
3 Valve assembly
4 Deflector

Sprinkler heads possess a steel cage with which a water deflector is fitted. The case houses a quartzoid bulb to actuate the valve assembly. These are commonly cast with a hex-shaped protrusion, called a wrench boss, to provide a contact surface for wrenches to securely grip to rotate while installing and uninstalling properly and easily.

The quartzoid bulb contains a capsule filled with highly expansible glycerin-based colored liquid that cause breaking of the bulb due to expansion when temperature rises to a certain level; thus, breakage of bulb, opens the valve and allow water to flow through the head. The upper end of the capsule presses against the valve assembly, which incorporates a soft metal seal. A small gas space is left inside so that, if the bulb is subjected to heat, the liquid expands and the gas space diminishes, generating pressure inside the capsule and causing the capsule to shatter once a predetermined temperature is reached. When the capsule bursts, it causes the opening of the valve, allowing water to flow through, and is deflected around being hit on the flower-shaped deflector in front, to make a wide spray of water, in a sprinkling pattern, on the fire below.

According to NFPA 13, the major characteristics that define the ability of a sprinkler head are the following:

1 Thermal sensitivity
2 Temperature rating

Thermal sensitivity

This sensitivity of a sprinkler head provides an indication of how quickly the thermal element of the sprinkler head is activated, which is measured with the Response Time Index (RTI). Sprinkler heads are categorized into fast or standard response, generally based on their RTI. But the major difference between standard and quick response sprinklers is the size of the bulb. In a standard sprinkler head, the bulb is slightly larger and takes a little bit longer time to work. A quick-response sprinkler head possesses a bit thinner and so has a faster response time. The Response Time Index (RTI) values for different categories of sprinkler heads have been shown in Table 5.2.

Table 5.2 Response Time Index (RTI) value for different categories of sprinkler heads [8]

Category	*RTI in (meters-second)$^{1/2}$*
Fast response	50 or less
Standard response	80 or more

Temperature rating

Generally, the operating temperature range at which the quartzoid bulb of the sprinkler head breaks is 68°C to 93°C, but bulbs operating at higher temperatures are also available. Quartzoid bulbs are manufactured in different colors to indicate the temperature rating of the bulb. In accordance with NFPA 13, the temperature rating of the sprinkler bulb and the corresponding color have been furnished in Table 5.3, and Figure 5.14 shows the different colors for the corresponding ratings.

On the basis of the way of actuating the sprinkler heads, these may be grouped into two broad categories, as mentioned below:

1 Open fire sprinkler head
2 Closed fire sprinkler head

Open fire sprinkler head: An open fire sprinkler head is a sprinkler head that is not equipped with any actuator or other heat-responsive elements to activate the sprinkler head for spraying water over a fire. Figure 5.15 represents an open fire sprinkler head.

Closed fire sprinkler head: Closed fire sprinklers are those that are equipped with an actuator or any heat-responsive elements and typically consist of a glass-bulb tube filled with very expanding liquid. In case of a fire, the air temperature exceeds a specified limit threshold; the expanding liquid bursts the bulb, thus allowing water to flow through and spray by the deflectors of the sprinkler head in a sprinkling manner over the fire below.

Table 5.3 Temperature rating of the sprinkler bulb and the corresponding color code

Temperature rating °C	*Temperature classification*	*Color code*
57–77	Ordinary	Uncolored or black
79–107	Intermediate	White
121–149	High	Blue
163–191	Extra high	Red
204–246	Very extra high	Green
260–302	Ultra high	Orange
343	Ultra high	Orange

Source: NFPA 13, 2007 edition [9].

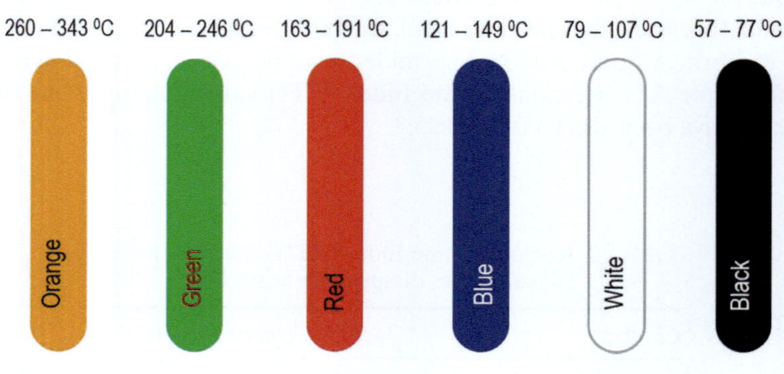

Figure 5.14 Temperature rating of the sprinkler bulb and the corresponding color.

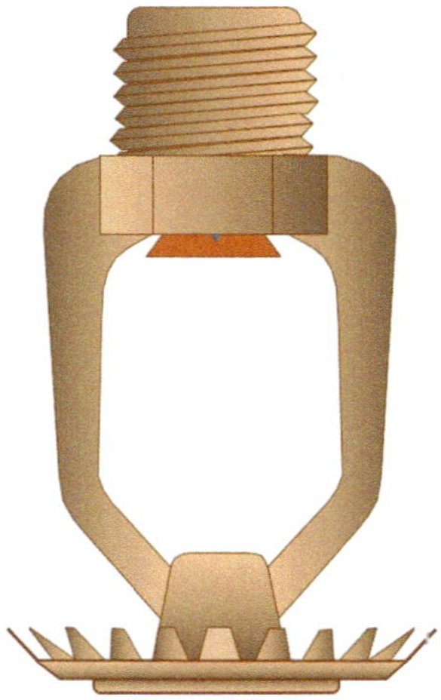

Figure 5.15 An open fire sprinkler head.

Types of sprinkler heads

There are four main types of sprinkler heads, as mentioned below:

1 Pendent sprinkler head
2 Upright sprinkler head
3 Sidewall sprinkler head
4 Concealed pendent head

Pendent sprinkler head

Pendent fire sprinklers are installed on branch pipes hanging down its head consisting of the deflector plate, as represented by Figure 5.13. Its heads feature a convex, almost umbrella-shaped deflector that guides water out and away from the center of the sprinkler head when properly installed. So, water is sprayed downward in a domed pattern to maximize coverage. These are ideal for offices, hotels, factories, etc. The most common pendent fire sprinkler has a 12.7 mm thread connection and a k-factor of 5.6.

There is a concealed type of pendant sprinkler head that is installed concealed, generally in the ceiling, and kept covered by a plate that falls off once it reaches to 9–19°C lower than the fire sprinkler temperature [9]. These are generally preferred where aesthetics is a major concern.

Upright sprinkler head

These sprinkler heads are installed on pipes, standing upright, as shown in Figure 5.16. In this head, a concave deflector is made to spray water downward in a hemispherical pattern. These up-right sprinkler heads are recommended for buildings having exposed ceilings like warehouses or industrial spaces and in rooms having deep beams, ducts, etc.

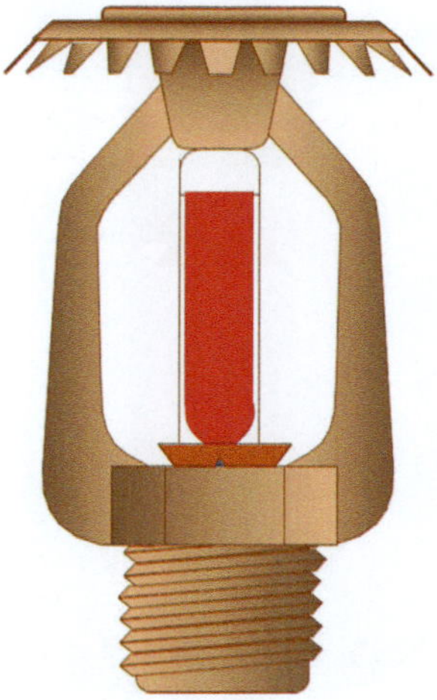

Figure 5.16 An upright sprinkler head.

For industrial, commercial, and storage spaces, up-right sprinklers of 19.05 mm thread connection with a k-factor of 8.0 are recommended.

Sidewall sprinkler head

These sprinkler heads are generally mounted on the side of a wall or beneath the beams. It has half of a deflector to spray water in a half circle in a crescent shape. These heads are mostly preferred for small rooms, hallways, and where sprinkler branch pipes run on the side of a wall. Most sidewall fire sprinklers utilize a 12.7 mm thread connection and feature a k-factor of 5.6. Figure 5.17 represents a sidewall fire sprinkler head.

Concealed pendent head

Concealed pendant sprinkler heads are basically sprinkler heads that have an additional decorative flat cover plate. During a fire event, when the temperature around rises to its operating temperature, the cover plate detaches first. Further rising of temperature causes the glass bulb to shatter, ultimately delivering a uniform water spray to fall onto the fire.

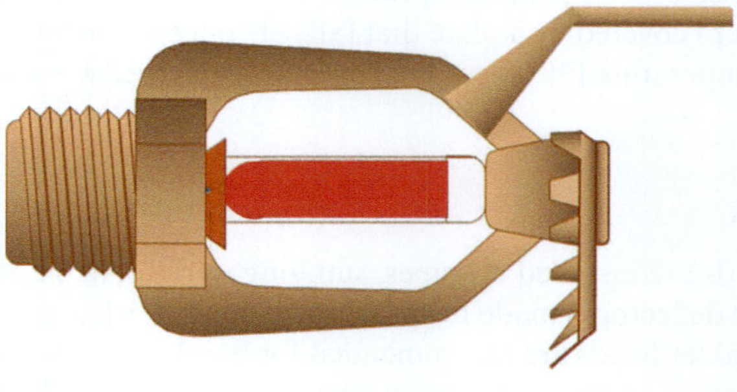

Figure 5.17 A sidewall sprinkler head.

The whole head unit is concealed for having a smooth ceiling. Sprinkler-head housing is threaded to fix the cover plate with the cup assembly to provide the required vertical adjustment. The sprinkler cover assembly can be removed and reinstalled. Pendant concealed sprinkler heads are also found in standard response or quick response types, maintaining standard coverage.

Automatic fire sprinkler system development

With a view to achieving an efficient automatic fire sprinkler system, the following five steps are to be outlined:

1 Determination of the building hazard classification
2 Determination of the number and spacing requirement of sprinkler heads
3 Selection of a piping configuration
4 Selection of a pipe sizing method
5 Determination of pump sizing
6 Water storage requirement.

Building hazard classification

From the perspective of fire hazard potential, buildings are generally classified into seven categories, as mentioned below, to provide relevant elements of the fire sprinkler system as required.

Class 1 – (Highest or severe hazard) occupancies

In these occupancies, quantity and combustibility of contents are supposed to be the highest, which under fire would tend to develop like an explosion and cause the highest rates of heat release. Examples of such occupancy buildings are explosives factories, oil refineries, petroleum product storage, paint or varnish factories, highly flammable chemical plants, Jute mills, cotton godowns, etc.

Class 2 – (High Hazard) occupancies

In these occupancies, quantity and combustibility of contents are supposed to be high, which under fire would tend to develop very rapidly and cause high rates of heat release. Examples are textile, paper, rubber, timber and plastics mills and its packaging, storage houses.

Class 3 – (Moderate hazard) occupancies

In these occupancies, quantity and combustibility of contents are supposed to be moderate, which under fire would tend to develop quickly and cause moderately high rates of heat release. Examples are engineering workshops, machine assembly plants with limited flammable materials, etc.

Class 4 – (Ordinary hazard) occupancies

These are the occupancies housing average fire risk materials, generally found in office buildings, academic institutional buildings, libraries, hotels etc.

Class 5 – (Low hazard) occupancies

In these occupancies, quantity and combustibility of contents are supposed to be low, which under fire would tend to develop slowly and cause low rates of heat release. Occupancies like hospitals, theatre or assembly halls with controlled furnishings, auditoriums, etc. fall in this group

Class 6 – (Very low or light hazard) occupancies

In these occupancies, quantity and combustibility of contents are supposed to be very low, which under fire would tend to develop very slowly and cause very low rates of heat release. Occupancies like museums with non-combustible construction and controlled contents, safe strong rooms etc. can be considered in this group.

Class 7 – (Lowest hazard or fire resistant) occupancies

Structural elements of buildings designed to be fireproof, containing little or no combustible materials are included in this class. Examples are: Rooms or buildings totally constructed as special fireproof structure and reinforced concrete structural members.

Fire sprinkler systems are generally developed considering a range of hazard categories or classes of buildings in the following ways:

1 Light hazard
2 Ordinary hazard (Group 1)
3 Ordinary hazard (Group 2)
4 Extra hazard (Group 1)
5 Extra hazard (Group 2)
6 Special occupancy hazard

Various occupations that fall under the above-mentioned classes of fire hazards have been furnished in Table 5.4.

Table 5.4 Types of occupancies under various classes of fire hazards [5]

Light Hazard	Churches, Clubs, Educational, Hospitals including veterinary facilities, Institutional, Kennels, Museums, Nursing or convalescent homes, Offices including data processing, Residential, Restaurant seating areas, Theaters and auditoriums, Unused attics.
Ordinary Hazard (Group 1)	Automobile parking and showrooms, Bakeries, Beverage manufacturing, Canneries, Dairy products manufacturing and processing, Electronic plants, Glass and glass products manufacturing, Laundries, Restaurant service areas, Porte cochère, Mechanical rooms.
Ordinary Hazard (Group 2)	Agricultural facilities (including barns and stables), Cereal mills, Chemical plant – ordinary, Dry cleaners, Libraries (large stack room areas), Machine shops, Metalworking, Mercantile, Paper and pulp mills, Piers and wharves, Stages, Textile manufacturing, Tire manufacturing, Carpentry areas.
Extra Hazard (Group 1)	Aircraft hangers, Combustible hydraulic fluid use areas, Die casting, Metal extruding, Plywood and particleboard manufacturing, Printing with inks with low flashpoints, Rubber-works, Sawmills, Textile work, Plastic foam upholstering.
Extra Hazard (Group 2)	Asphalt saturating, Flammable liquids spraying, Flow coating, Open oil quenching, Plastics manufacturing, Solvent cleaning, Varnish and paint dipping.

Sprinkler head requirement

The NFPA has guided specific requirements for sprinkler heads. Requirements of sprinkler heads are based primarily on hazard level and pump flow calculations as well. The number and spacing may need to be adjusted subject to any obstructions that may be present in the vicinity of the proposed sprinkler head.

Number of sprinkler heads

In the process of designing the automatic fire sprinkler system, the first job is to plan for locating the sprinkler heads determined for a particular area, based on the fire hazard classification of the area to be served. The sprinkler heads are generally positioned in particular alignment to get connected by a straight pipe to supply water to those. Majority fire department regulations suggest that a maximum of eight fire sprinkler heads should be installed on a branch pipe.

To find the total number of heads for a design area of particular hazard occupancy, the following formula can be used:

$$\text{Total number of sprinkler heads} = \frac{\text{Area to be served}}{\text{Area covered by one sprinkler head}} \tag{5.3}$$

Sprinkler heads spacing

Spacing of the sprinkler head depends upon multiple factors like the type of the head, the installation type, and the building hazard classification.

Based on coverage, there are two types of sprinkler heads, as described below:

1 The standard sprinkler head that is designed for a maximum coverage area of 15 sq m
2 The extended spray sprinkler head that has a maximum coverage area of 18 sq m

The installation types also determine the spacing of the sprinkler head. The wall-type sprinklers are spaced to attain an average coverage area of 4.2 m straight throw from their centerline. Recommend spacing between two wall-mounted sprinkler heads is 3.7 m.

Extended coverage sprinkler heads: If a building needs more protection but standard sprinkler heads cannot meet normal spacing requirements, then extended coverage sprinkler heads can be chosen. Extended coverage sprinkler heads sprinkler heads have increased diameter coverage.

To develop an efficient fire sprinkler system, NFPA recommends the maximum allowable spacing for a sprinkler head based on the hazard classification. The summary is furnished in Table 5.5.

NFPA also recommends an average of the recommended spacing as the maximum between sprinkler and wall or obstruction deeper than 350 mm, while special considerations should be given to custom ceilings.

Other sprinkler head spacing requirements for special cases include the following [11]:

1 Distance from walls: half the distance between sprinkler heads in standard condition; i.e., for 5 m spacing of sprinklers 2.5 m from walls; for 4 m spacing of sprinklers 2 m from walls
2 Minimum distance between sprinkler heads: usually 2 m

Table 5.5 Protection area and maximum spacing (extended coverage for pendent and upright spray sprinklers) [10]

Construction type	*Light hazard*		*Ordinary hazard*		*Extra hazard*		*High-piled storage*	
	Protection area (sqm)	*Spacing (m)*	*Protection area (sqm)*	*Spacing (m)*	*Protection area (sqm)*	*Spacing (m)*	*Protection area (sqm)*	*Spacing (m)*
Un-obstructed	37	6.1	37	6.1	–	–	–	–
	30	5.5	30	5.5	–	–	–	–
	24	4.9	24	4.9	–	–	–	–
	–	–	18	4.3	18	4.3	18	4.3
	–	–	13	3.7	13	4.6	13	4.6
Obstructed noncombustible (when specially listed for such use)	37	6.1	37	6.1	–	–	–	–
	30	5.5	30	5.5	–	–	–	–
	24	4.9	24	4.9	–	–	–	–
	–	–	18	4.3	18	4.3	18	4.3
	–	–	13	3.7	13	4.6	13	4.6
Obstructed combustible	N/A	N/A	N/A	N/A	N/A	N/A	N/A	N/A

3 Distance from ceilings: minimum 25 mm, maximum 300 mm. Most sprinkler heads shall be installed 25 mm from the ceiling. Concealed, recessed, or flush-mounted sprinkler heads may be listed for shorter distances away.

In addition, the sprinkler heads are to be provided in every building space other than the following, as specified by NFPA:

1 Bathrooms less than 5.1 sqm
2 Closets, pantries, and other non-dwelling spaces less than 2.2 sqm
3 Wall or ceiling surfaced with non-combustible material
4 Staircases, shafts, balconies, and other open building spaces

Pressure of sprinkler head

Without having a minimum water pressure, fire sprinklers cannot be used as an effective water sprinkler. They won't be able to sprinkle out enough water over a desired area to suppress fires. If the pressure at a sprinkler head is below a minimum of 0.5 bar (50 kPa), the water won't be able to push out the flow-holding cap even though the heat-sensitive bulb breaks due to temperature rise due to any fire.

Sprinkler piping

A fire sprinkler system being a water-based active fire protection method requires a water supply system maintaining adequate pressure and flow rate through a distribution piping system, to which fire sprinkler heads are installed to sprinkle water on fire, being activated by its heat. So, a planned piping network is to be developed following some methodologies.

Piping configuration

To develop an automatic fire sprinkler system, a network of piping is planned, designed, and installed to supply water from a source of water, mostly a water reservoir discussed earlier, with the help of a pump. The piping in the network for the sprinkler system may be grouped in two broad categories, as below:

1 Branch or lateral pipes
2 Stand or riser pipe

Branch or lateral pipe

Branch or lateral sprinkler pipes can be configured in three common patterns:

1 Tree piping layout
2 Loop piping layout
3 Grid piping layout

Tree piping layout

Fire sprinkler systems often use tree pipework configuration, although it is less efficient than other pipe configurations. For complex buildings such as schools, residential buildings, and systems that require a dry fire sprinkler installation to be installed, a tree piping layout, as shown in Figure 5.18, is preferable.

Although this configuration is not as hydraulically efficient as a loop or grid system, it is therefore required to select the most efficient piping configuration based on the room configuration.

Loop piping layout

A fire-sprinkler piping layout in which multiple cross mains are joined together so as to provide more than one path for water to flow to an operating sprinkler but branch lines are not connected together, as shown in Figure 5.19. Loop piping layout is highly preferred where seismicity is of great concern.

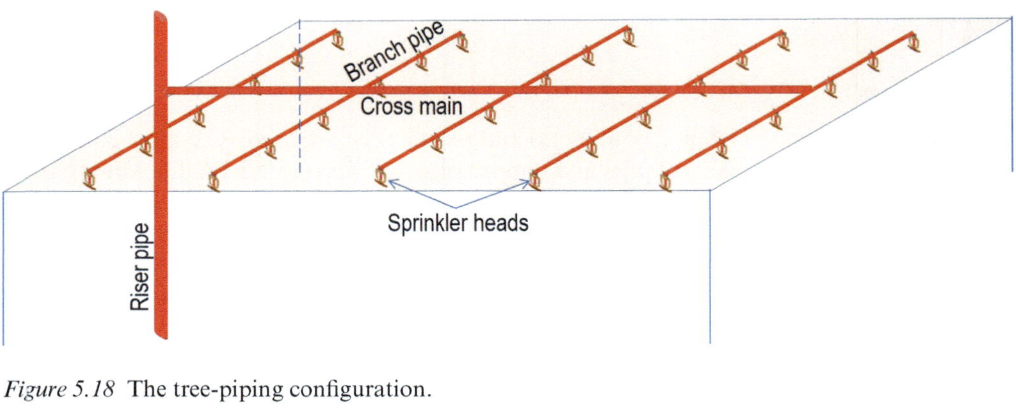

Figure 5.18 The tree-piping configuration.

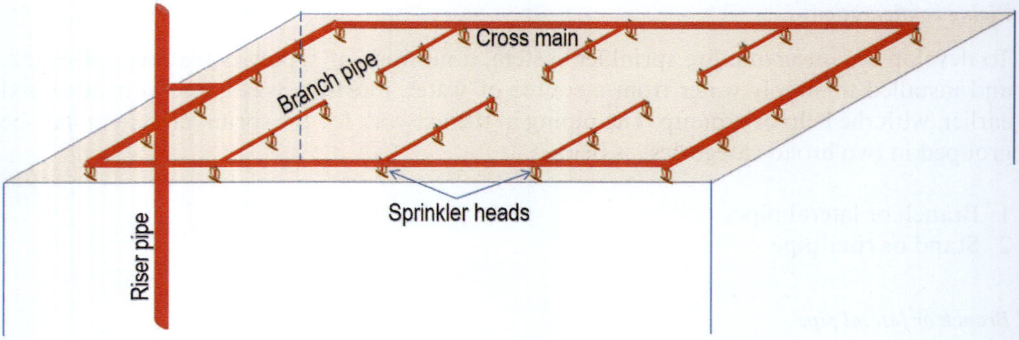

Figure 5.19 The loop-piping configuration.

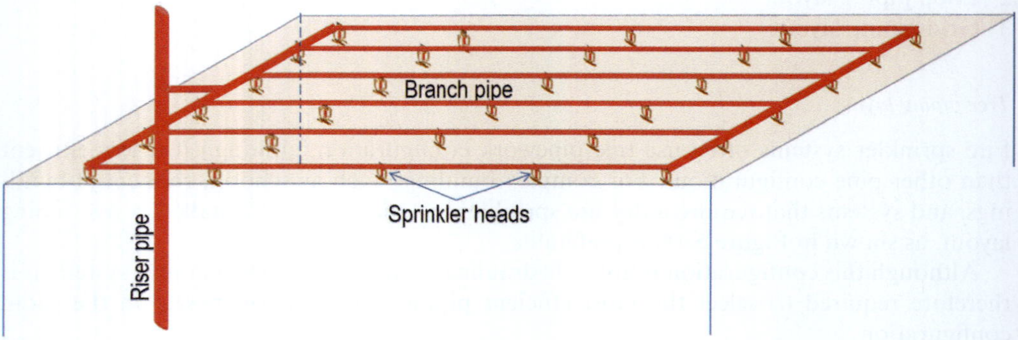

Figure 5.20 The grid-piping configuration.

Grid piping layout

A sprinkler piping layout in which parallel cross mains are connected by multiple branch lines as shown in Figure 5.20. An operating sprinkler thus receives water from both ends of its branch lines, while other branch lines help transfer water between cross mains. The grid piping layout is sometimes prohibited in some jurisdictions because of its difficult flexibility.

Piping material

In fire sprinkler piping systems, both metal and plastic pipe are used. The most commonly used metal pipes are black steel pipe and copper pipe, specified in NFPA 13. Among these two copper pipes are costly but good corrosion-resistant and smooth. In the group of plastic CPVC and PEX tubing is permitted for certain types of sprinkler systems, particularly in light-hazard occupancy areas. Some important features of the most popular piping materials recommended for fire sprinkler systems have been tabulated in Table 5.6 for comparison among the pipes.

Table 5.6 Important differences between the most popular piping materials recommended for fire sprinkler systems

Property	*Pipe or tube type*			
	Steel Sch. 40	*Typical light wall steel*	*Copper type M*	*CPVC SDR 13.5*
Color	Black	Silver	Copper	Bright orange
Weight of the DN25 size (kg/m)	2.5	1.8	0.7	0.4
Melting point (MP) or heat distortion temperature (HDT) (°C)	(MP) 1427–1538	(MP) 1427–1538	(MP) 1082	(HDT) 103
Damage susceptibility	Low	Low	Low	High with UV exposure and impacts
Corrosion susceptibility design C factor	High/120	High/120	Moderate/150	Low/150
Occupancy classification NFPA standards	Not limited	Not limited	Not limited	NFPA 13 light hazard. 13D. 13R. concealed and restricted exposure NFPA 90A
Maximum ambient temperature	Not limited	Not limited	Not limited	66°C
Flexibility/hanger spacing for the DN25 size (m)	Not flexible/3.7	Not flexible/3.7	Slightly flexible/2.4	Flexible/1.8
Expansion concerns/solutions	Negligible	Negligible	Negligible	Yes/offsets direction changes. loops
Fitting type	Threaded grooved flanged plain-type	Threaded grooved flanged plain-type	Soldering brazing grooved	Primer solvent cement
Compatible antifreeze	Not limited[a]	Not limited[a]	Not limited[a]	Glycerin[a]

Source: National Institute of Standards and Technology [12].

[a] NFPA recommends that only antifreeze solutions of glycerin or propylene glycol be used when the sprinkler system is connected to potable water.

Combined firefighting system

Combining firefighting systems involves integrating multiple fire protection strategies, generally incorporating sprinkler and wet riser systems, to ensure a comprehensive approach to fire safety in a building. Some guidelines, as mentioned below, are to be followed for this system:

1. Combining firefighting systems can be permitted for buildings of ordinary hazard category classification.
2. The water storage volume shall be based on the larger water storage demand among the two systems.
3. Separate pump sets for two systems shall be provided to allow for the simultaneous operation of the sprinkler and wet riser systems.
4. In the combined system, sprinkler riser and wet riser stacks cannot be interconnected by sprinkler system piping or a single riser that serves both systems.

Stand pipe size for combined system

If a building considers an automatic sprinkler system coupled with a fire hose system for special areas for its fire protection, then the system is termed a combined system. In the combined system, the minimum size of the standpipe can be 100 mm if the piping size is hydraulically calculated. The branch lines of the standpipe system are to be sized hydraulically but cannot be smaller than 65 mm.

The minimum size pipe for Class I and III standpipes is 100 mm. If the standpipe is part of a combined sprinkler system in a partially sprinkled building, the size of that stand pipe is increased to 150 mm [13].

Designing fire sprinkler system components

Designing the components of a fire sprinkler system involves sizing the following major components:

1 Determining the number of sprinkler heads
2 Sizing of pipes
3 Sizing of pumping system
4 Sizing of water storage tank

Determining number of sprinkler heads

In determining the sizes of major components of a fire sprinkler system, it is imperative to identify the remote area that is hydraulically most demanding, which is often the furthest area from the source, or the largest area to be served at a time. This is the area of floor space over which the required water density must be discharged. After calculating the area and selecting appropriate sprinkler types that meet the remoteness, the required number of sprinkler heads for the area is determined.

Based on the fire-hazard condition, a type of sprinkler head is selected, and its sprinkling coverage area is to be known. Dividing the area to be protected by the coverage area by the minimum required number of sprinkler heads for that area of a building, as discussed earlier, the actual number will be a little more to meet special requirements.

After finalizing the number of sprinkler heads for the design area, the sprinkler heads are to be arranged in a regular pattern, and the suitable piping layout is to be designed connecting the sprinklers, as discussed earlier.

Example: Number of sprinkler heads.

In a section of the piping network, the maximum number of sprinkler heads might be limited. In some cases, the maximum number might be limited to 200 numbers.

For a remote area of 140 sqm, when to be served by one sprinkler head covering 12 sqm,then total number of sprinkler heads required = 140/12 = 11.67. Say, 12 nos.

Sprinkler pipe sizing

Sprinkler pipe sizing is based on three important factors of empirical consideration, such as the pipe types, pressure requirements, and hazard classifications. Generally, two primary methods are followed in sizing sprinkler pipes: those are as follows:

1 The pipe schedule method
2 The hydraulic calculation method

Table 5.7 Sprinkler pipe sizing based on pipe schedule method [14]

Pipe size (mm)	*Number of sprinklers*			
	Light hazard	*Ordinary hazard*		*Extra hazard*
		Separation ≤ 3.7m	*Separation > 3.7m*	
25	2	2	2	1
32	3	3	3	2
40	5	5	5	5
50	10	10	10	8
65	30	20	15	15
75	60	40	30	27
82	100	65	60	40
100	> 100	100	100	55
125		160	160	90
150		275	275	150
200		> 275		–

Note. Protection area for light and ordinary hazard is 4830 sqm; and for extra hazard 3720 sqm.

Source: NFPA 13, 2019 edition.

Pipe schedule method

According to NFPA 13, Pipe Schedule method the pipe sizing is selected from a schedule that is determined by the occupancy classification. In this method a given number of sprinkler heads are allowed to be installed from specific sizes of pipe.

The pipe schedule method is mostly used for residential apartments or projects having an area 465 sqm or less. The minimum sprinkler pipe size is 25 mm in diameter.

The pipe schedule method thus involves computing each sprinkler pipe to the range pipe and eventually to the sprinkler main, using the sprinkler schedule Table 5.7. This process computation is based on the NFPA 13 standard, while different schedules were specified for different sprinkler pipes. The table summarizes the pipe schedule for different hazard degree.

Hydraulic calculation method

The hydraulic calculation method is an elaborated method based on the density-area curve furnished by NFPA, as shown earlier in Figure 5.4. These curves stipulate the required minimum densities and remote areas that establish minimum water requirements for sprinkler systems. The required minimum densities of water and remote areas to be served by a sprinkler head are very important to know to establish minimum water requirements for sprinkler systems.

Density: This is the amount of water that must be delivered for every square meter of floor space in every minute. The unit for density is liter per minute per square meter (lpm/sqm). In extra-hazard occupancy when designed for a density of 10 lpm/sqm, then the system must discharge 10 liters in every minute for every square meter of floor space over a specified area.

A preferable design density for Ordinary-Hazard Group-1 occupancy is 6.1 1 pm/sqm for a 140 sqm area. This means the sprinkler system operates effectively on the most hydraulically demanding 140 sqm. So, a portion of the sprinkler system is expected to operate during a typical fire, not all. If all sprinklers on a sprinkler system are to operate, then the water supply, most likely, may not be adequate to suppress the whole fire.

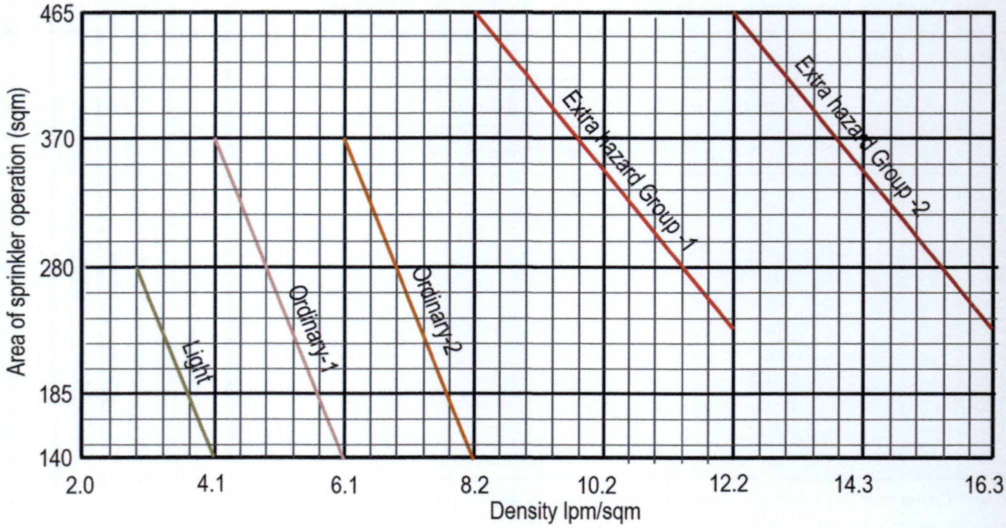

Figure 5.21 Density/area curves for various hazard classifications.

Example: Flow rate required

From Figure 5.21, it is found that in the extra hazard group-2 category, for an area of 230 sqm, the required density is 16.3 lpm/sqm. Therefore, the minimum required flow would be 16.3 lpm/sqm times 230 sqm, i.e., 3749 lpm.

Velocity in pipe

In hydraulic calculations for determining pipe size for fire sprinklers, the allowable velocity is to be known. Fire sprinkler design authorities like EN 12845 limit the velocity through pipes and valves in fire sprinkler systems; however, NFPA and FM do not have any restrictions in this respect. EN 12845 limits the velocity of flow to 6 mps through valves and flow switches and 10 mps in the sprinkler piping system.

Velocity of a particular flow in a pipe can be calculated using the Formula 5.4, from which the diameter of pipe needed can be determined when flow-rate is known and the velocity is assumed.

$$v = 21.22 \frac{Q}{d^2} \tag{5.4}$$

where,

v = velocity of flow in the pipe (mps)
Q = flow-rate of water in the pipe (lpm)
d = pipe's internal diameter (mm)

For a quick solution, the following Table 5.8 can be used, which provides the maximum flow rate in lpm through a particular diameter of steel pipework to EN 10255 specifications for velocities of 6 mps and 10 mps.

Table 5.8 Maximum flows through pipes for EN 12845 fire sprinkler systems [15]

Sl. no.	*Nominal diameter (mm)*	*EN12845*			
		Series M		*Series H*	
		For velocity (mps)			
		6	*10*	*6*	*10*
1	20	133	222	119	198
2	25	211	351	249	416
3	32	366	611	335	558
4	40	496	827	459	765
5	50	797	1329	744	1240
6	65	1342	2237	1273	2122
7	80	1851	3084	1760	2934
8	100	3135	5225	3029	5048
9	150	6802	11336	6732	11220

Example: Diameter of pipe required

From Table 5.7, it can be found that for a flow of 3749 lpm >3135 and <5225, the diameter of the pipe required is 100 mm for a velocity greater than 6 mps but less than 10 mps.

Fire pump sizing

Example: Let consider a fire sprinkler pump of flow 1600 lpm rated at 6 bar, which reveals that the pump will cause a flow of 1600 lpm at 6 bar when operating at 100 percent capacity. The pump can also produce a maximum flow of 2400 lpm (150 percent of 400 gpm flow) at 3.9 bar (65 percent of 56 psi) when operating at 65 percent capacity. So, the pump can be selected for the purpose, as the total head pressure is not less than 65 percent of the rated total head when the pump is operating at 150 percent of rated flow capacity.

Let consider a floor that require an estimated flow of 2000 lpm at 8 bar. From the pump curve if it is found that the pump will deliver approximately 5 bar when flowing 2000 lpm. When this discharge pressure is combined with the city water supply pressure of 5 bar, total 5 + 5 = 10 bar which is greater than 8 bar, it is evident that the selected pump can easily satisfy the hydraulic demand for the concern floor.

The cost of a fire pump is largely based on the horsepower rating of the pump and the type of controller. The popularly used fire pumps are horizontal split case, inline and vertical turbine. Vertical inline pumps are usually more cost-effective when compared to horizontal split-case pumps, when flow rate is comparatively small about 3800 lpm. Vertical turbine pumps are to be used in situations where the water level is below the suction of the pump, as NFPA 20 requires a positive suction pressure to a fire pump.

Jockey pump sizing

The most pump manufacturer recommend, as a rule of thumb, to size the jockey for about 1 percent of the rated flow of the main fire pump. Also, they recommend that the jockey pump be sized with 0.7 bar higher pressure than the system design pressure of the fire pump.

Table 5.9 Water flow rate and duration requirements for hydraulically calculated systems [16]

Occupancy	*Inside hose*	*Total combined inside and outside hose*	*Duration (minutes)*
	(lpm)	*(lpm)*	
Light hazard	0, 190, or 380	380	30
Ordinary hazard	0, 190, or 380	950	60 or 90
Extra hazard	0, 190, or 380	1900	90 or 120

Reservoir sizing

The basic approach to determining the minimum volume of water to be stored for fire suppression is to multiply the calculated fire flow requirement for the most hydraulically demanding area by the considered duration of flow, which gives the volume of water to be stored. A suitable internal dimension of the reservoir is then determined, in which a float depth which is about 15 percent of water depth, is added with the depth of water considered to find the internal height of the reservoir. Fire flow rate and duration of flow are primarily dependent on the hazard condition. For a combined system, the minimum water storage volume that shall be ensured for the minimum duration of flow and required flow rate based on hazard level have been furnished in Table 5.9; from which storage volume for the sprinkler system can be determined.

Example: For an extra-hazard area, if hose flow is considered 380 lpm, then only for the sprinkler system, the flow rate should be 1520 lpm. So, for 120-m duration, storage volume is required.

$$1520 \times 120 = 182{,}400 \text{ liters, i.e., } 182.4 \text{ cum.}$$

For an assumed area of width 5 m and length 10 m, the required water depth is 182.4/50 = 3.648 m; by adding 15 percent depth for float, the internal height of the reservoir is = 3.648 + 0.15 x 3.648 = 4.1952 m; Say, 4.2 m.

Drainage system

On the floors where there are sprinkler heads for fire suppression, adequate drainage shall have to be provided for draining sprinkled water from the heads subject to the occurrence of any fire event. Where there are flammable or hazardous liquids or the possibility of generating highly contaminated wastewater, if there is a sprinkler system to suppress any fire, there must be sufficient and separate drainage systems to contain the spreading of such hazardous wastewater to other floors of the building. The floors shall be well pitched, where necessary curbs or dykes shall be provided not to allow spreading such wasted water around, and trenches or sumps shall be created to facilitate runoff drainage from the floors.

Flushing

The sprinkler piping system needs to be flushed to clear it from debris and other residual materials that cause interruptions in the flow of water. During installation, sectional flushing can be done for very large projects, and after completion, final flushing is to be done. The lowest points shall be provided with appropriate-size drain points as provided

Table 5.10 Drain size for flushing

Sl. no.	*Riser or main size: mm*	*Drain size, mm*
1	Up to 50	20 or larger
2	65, 76, 82	32 or larger
3	100 and larger	50 only

SI conversion from source [2].

in Table 5.10. Draining shall be done through hose discharging into nearest floor drains available or made for the purpose.

Installation, testing, and commissioning

Automatic or non-automatic fire sprinkler systems must be planned, designed, and supervised by certified engineers from any recognized Institute for Certification in Engineering Technologies. The plumbers who will be installing and doing the maintenance work must have a competency license from an approved and recognized authority. Concerned regulatory authorities promulgate regulations that prescribe the credentials and experience requirements for the professionals involved in the design, installation, testing, and upkeep of fire sprinkler systems. Such requirements usually include licensing for engineers, plumbing and mechanical contractors, fire sprinkler installers, and other building trades providers.

Fire pumps must have provisions for installing a metering device or fixed nozzles to perform pump testing. According to NFPA the testing device must have a capacity of water flow not less than 175 percent of rated pump flow capacity [17].

References

[1] U.S. Department of Homeland Security (2006) "Structure fire response times" Topical Fire Research Series, Volume 5, Issue 7, U.S. Fire Administration/National Fire Data Center. Retrieved from https://apps.usfa.fema.gov/downloads/pdf/statistics/v5i7.pdf on 13 April 2024
[2] Kamal Ram (UD) "Design of automatic fire sprinkler system" Retrieved from https://www.kamal-mech-engr.com/design-of-sprinkler-system on 13 April 2024
[3] Balmoral Tanks Ltd (UD) "Fire fighting water storage solutions" Retrieved from https://www.balmoraltanks.com/sectors/fire-fighting on 14 April 2024
[4] Fire Sprinkler Services (2024) "Fire sprinkler systems" Retrieved from https://firesprinklerservices.com/service-areas/new-installations on 2 September 2024
[5] Quick Response Fire Supply, LLC (2020) "#358 – How much water pressure is required for fire sprinkler systems?" Retrieved from https://blog.qrfs.com/358-how-much-water-pressure-is-required-for-fire-sprinkler-systems/#:~:text=Two%20factors%20determine%20the%20required,there%20are%20minimum%20sizes%20 on 3 September 2024.
[6] www.tpmcsteel.com (UD) "Sprinkler nominal K-factor" Retrieved from https://tpmcsteel.com/sprinkler-nominal-k-factor on 3 September 2024
[7] Quick Response Fire Supply, LLC (2021) "#392 – How to size a fire pump and the basics of performance curves" Retrieved from https://blog.qrfs.com/392-how-to-size-a-fire-pump-and-the-basics-of-performance-curves on 3 September 2024.
[8] Valerie Ziavras (2021) "The basics of sprinkler thermal characteristics" Retrieved from https://www.nfpa.org/news-blogs-and-articles/blogs/2021/06/22/the-basics-of-sprinkler-thermal-characteristics on 19 April 2024
[9] Quick Response Fire Supply, LLC (2019) "#61 – fire sprinkler head types: pendents, uprights, sidewalls, and concealed" Retrieved from https://blog.qrfs.com/61-fire-sprinkler-head-types-pendents-uprights-sidewalls-and-concealed on 3 September 2024.

[10] Quick Response Fire Supply, LLC (2020) #380 — Sprinkler head spacing: extended coverage fire sprinklers, part 1" Retrieved from https://blog.qrfs.com/380-sprinkler-head-spacing-extended-coverage-fire-sprinklers-part-1 on 6 October 2025.
[11] Kauffman Co. (2024) "Fire sprinkler head location and coverage" Retrieved from https://www.kauffmanco.net/blog/fire-sprinkler-locations on 3 September 2024.
[12] National Institute of Standards and Technology (1994) "Comparison of fire sprinkler piping materials: steel, copper, chlorinated polyvinyl chloride and polybutylene, in residential and light hazard installations" 28, Retrieved from https://nvlpubs.nist.gov/nistpubs/Legacy/IR/nistir5339.pdf on 19 April 2024
[13] Shawn Mahoney (2021) "Standpipe system design and calculations" Retrieved from https://www.nfpa.org/News-and-Research/Publications-and-media/Blogs-Landing-Page/NFPA-Today/Blog-Posts/2021/11/19/Standpipe-System-Design-and-Calculations on 5 September 2024
[14] Fire Protection Specialists (UD) "Pipe schedule method for firefighting sprinkler system" Retrieved from https://www.fps-eg.com/2019/11/pipe-schedule-method-for-firefighting.html on 20 May 2025
[15] Canute LLP (2024) "Water velocity and limits for different design authorities" Retrieved from https://canutesoft.com/hydraulic-calculation-for-fire-protection-engineers/velocity-in-pipe#:~:text=EN%2012845%20limits%20velocity%20to,other%20point%20in%20the%20system on 5 September 2024
[16] Retrieved from https://up.codes/s/occupancy-hazard-fire-control-approach-for-spray-sprinklers on 5 September 2024
[17] WTWH Media, LLC (2024) "NFPA 20: fire pump design" Retrieved from https://www.csemag.com/articles/nfpa-20-fire-pump-design on 5 September 2024.

6 Fountain and waterfall

Introduction

Water is not only for the survival of life; it is also for beautifying and amusing life, as being one of the most beautiful and exciting materials on the planet. Fountain means the source or origin of anything; the source from which something originates. Water fountains are welling up of water from a source like a spring of water from the earth, and conversely, waterfalls are rolling down of water from a source at a higher elevation. Both water fountains and falls are special types of water streams originating from a source that exhibits special water features to add beauty to the environment and amuse humans' minds through enjoyment.

There are natural water fountains and waterfalls. Natural water fountains gush forth from beneath the earth's surface, whereas the waterfalls roll down to earth surface, almost vertically from the top elevations of the hills or mountains.

There are manmade fountains and waterfalls constructed mainly for beautification, entertainment, and as means of relaxation. But historically, these are constructed for many other purposes as well, such as climatic control in summertime cooling, community water supply, urban architecture, uplifting the social environment, tourism and aesthetics, etc.

In this chapter, the technology involved in the development of artificial manmade fountains and waterfalls is discussed elaborately.

Technological considerations

The basic principle of a water fountain and fall technology is to cause movement of water in a particular pattern with a view to creating eye-catching water features and joyful displays. So, both fountains and falls require a water source from where it conveys water, a pump to create pressure on water to cause water movement, and to direct water to flow from a particular point away from the source. To convey water, there is a need for a pipe and a special outlet element at the pipe end to make the feature of water spraying from the fountain or fall. For these purposes, the end of the pump-delivery pipe is fitted with a nozzle specially designed for creating particular features and displays of water, particularly in fountains. After the determination of nozzles for fountains and befitting outlets for falls, based on the display effect and pattern of water movement, the selection of the driving force or pump is done.

When the sprayed water is designed to fall on a solid surface, a drainage system is developed to direct water to the source to recycle and reuse water for continuous operation of the fountain. To avoid additional collection systems of sprayed water, most fountains are designed in such a way that the sprayed water falls on water contained in a basin, which also acts as the reservoir of water for the fountain.

DOI: 10.1201/9781003724056-6

After running for considerable time the fountains, mostly the outdoor ones, water gets contaminated mostly by absorbing pollutants in the surrounding air, where ultimately there occurs growth of algae in water. All these contaminants discolor the water; and they also affect the internal parts of pumps and nozzles and the interior of pipes. As a result, the fountain becomes inoperative, and so the repairing and maintaining costs become high. To mitigate these problems, fountain technology involves a filtration system to maintain a trouble-free longer operation of the fountains. Maintaining good quality of water through regular maintenance of the filtering system, pumps, nozzles, and other accessories will ensure longer efficient operation of the fountains and water falls.

Structural components of a fountain and waterfall

A fountain and waterfall are built comprising a number of major elements depending upon the grandness, type, and feature of the water spray designed. Water fountains are made with a multiple number of major and minor components. Among those, the major components are as follows:

1 Nozzle
2 Pool or basin
3 Pump
4 Piping
5 Valve
6 Overflow armature
7 Water filter unit
8 Underwater lighting or illumination (optional)
9 Multi-directional drives 1D or 2D (optional)

Water fountain

Water has multiple characteristics. Some of the characteristics of water, such as its plasticity, movement, sound, and reflectivity, are considered important aspects of water fountains or falls. Water is plastic because its shape and appearance are determined by the size, color, texture, and location of the container that holds water. Water's movement can be classified as either static, as in a placid pond, pool, or still container, or dynamic whenever it is acted upon by a force. Sound is an unnoticed or noticed quality of water depending on the velocity of water and turbulence created during the movement of water involved, which generates a diverse nature of sounds of different intensity, rhythm, and tone. The audible aspects of sound generated due to flowing or by the thrust of water pressure are very important to humans mental peace; too little dripping of a faucet at deep, quiet night can be irritating too much and might also be harrowing, while too much dripping in smaller, close spaces can be overpowering. Reflectivity of water is enjoyed during its calm and quite setting of large water bodies, during its static state under almost no flowing wind. In this condition, the reflection of its environment overwhelms the onlookers.

Humanity has conducted and manipulated water primarily to control its potential for survival and livelihood, but also to enjoy its beauty more or less. Fountains are built with the objective of enjoying the natural beauty of water in general.

In the historical manmade fountains, the water flow of fountains is mostly made by gravity. Around the 20th century, the availability of electric pumps and later on the advancement of flow-control technology helped more ingenuity to the design of fountains with special effects in water flowing features in the fountain. Now the designers make water fountains very elegantly, combining state-of-the-art technology and artistic features very successfully, to attract the public more than before.

Modern water fountains involve light and sound in the movement or flow of water in the fountain. Combining light and sound with water dynamics in an intelligent way creates a more lively fountain and gorgeous landscapes. Mechanical equipment in fountains cause water to perform in dynamic features mostly in the open atmosphere, with special aesthetics being stimulated by sound and glowed by lightings of diverse color; specially designed electro-mechanical systems for the fountains help create amazing scenarios. All the water fountains are not the same; depending on their particular features, location, and technology, different types of fountains can be created, employing interesting design principles of architectural, structural, electromechanical, and plumbing knowledge and knowhow. Due to the inherent multidisciplinary nature of fountains and their appeal to the general public, there exists a great potential for enhancing special technical knowledge, both theoretical and practical, in this regard.

Water fountains not only beautify the urban landscape and bring enjoyment to the city dwellers; they also cool down the surrounding environment, reduce dust, and thereby reduce the surrounding air pollution.

A range of various interchangeable nozzles are available for different fountain effects as well as different fountain height sprays.

Types of water fountains

Based on the location of a fountain with regard to the building aspects, fountains can be grouped into two types:

1 Indoor fountains
2 Outdoor fountains

Based on the water display shape and performance, fountains can be categorized as:

1 Stable
2 Sequencing

Stable fountain

Stable fountains are those in which the displaying shape of water of the fountain is kept stable throughout the entire time of operation. Adjustable nozzles and their installation allow water streams to fall inward, outward, upward, shooting, or even crossed streams. The majority of jet fountains are stable fountains.

The stable fountains are generally supported by underwater color lights to glorify their nighttime display, along with the must-use electromechanical device— a centrifugal pump.

Sequencing fountain

A sequencing fountain, as opposed to stable fountains, changes the displaying shape that cycles through a sequence of dynamically changing water display patterns and features of water discharge by possible logical combinations in configurable speed, being controlled by a programmable sequence control unit. Multiple nozzles are set in various ways—linear, circular, zigzag, etc. Changing water shape and features is more technological based on the incorporation of various electro-mechanical devices in addition to pumping.

They are mostly large ornamental fountains, typically installed outdoors in public places and in amusement parks, as well as smaller-scale ones made indoors. Individual nozzles create patterns of water that vary in height, direction, and discharging. Wind velocity is the major concern in maintaining the desired pattern and shape of discharge of water in a sequencing fountain.

The sequencing fountains are supported by underwater color lights, rhythmic music, or sounds complying with the changing shape and feature of water display to glorify, beautify, and make the fountains more attractive during their display time, both day and night.

Sequencing fountains are generally developed in three categories:

1 Jumping jet
2 Musical
3 Dancing

Jumping jet

Generally in fountains, water is continually sprayed from the nozzles, but in jumping jet fountains, water jets magically shoot through the air in broad arcs and disappear, just as unexpectedly as they appear. Jumping jets are produced and shot by an electro-mechanical device fitted on the pump delivery pipe end to pop, burst, and jump again and again while the fountain is in operation.

Musical fountain

The musical sound turns a fountain into a lively, fascinating spectacle of water shapes and features. The fountain nozzles discharge to the music in real time with absolute synchronization and no delay. The function of a basic musical fountain includes the sequencing of the water performance of every nozzle in relation to the pitch and wavelength of musical sound integrated with the fountain display. This function makes water sequentially rise and fall at various speeds and in all possible logical combinations based on the amount of water discharge.

Over the past number of years, fountain designers have been experimenting with interactive water features, enticing the public to not only watch but also participate in the dynamism of the fountains. Fountains, those are sequential and are choreographed, additionally supported with music and color-changing lighting, are quite sophisticated in terms of planning, design, installation, and maintenance.

The nozzles, in particular, are part of a more sophisticated system and require complex control panels and programming. Skilled manpower is to be involved in developing customized-programmed musical fountains with synchronized light shows to meet public ever-changing needs.

Dancing fountain

In the dancing fountain, water is made to dance by changing direction and height of spray following a rhythm. In these fountains, swivel nozzles, as shown in Figure 6.1, are used instead of fixed nozzles, through which the water spray is varied by electro-mechanical actuation of the nozzles and controlled hydraulically; both are synchronized, programming logically.

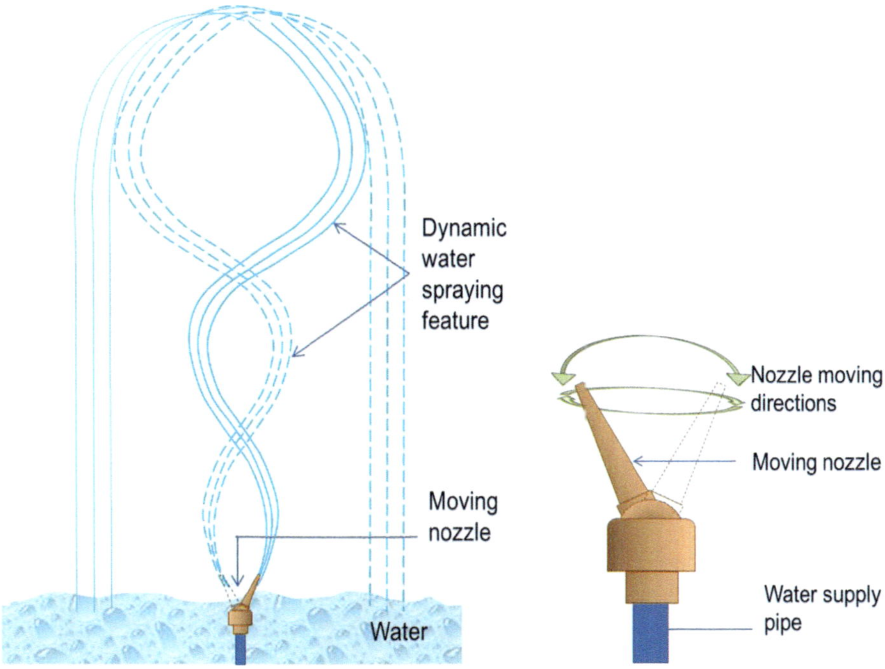

a. A dancing fountain

b. A dancing fountain nozzle

Figure 6.1 Dancing fountain and its nozzle.

With a view to creating a mesmerizing display, with the display of fountains, music is played to make the dancing of fountains rhythmic and synchronized to the rhythm of music. In addition to the music, multicolor lights are installed with the fountain display. By changing the intensity and color of the flashing of light, the fountain display is made more lively, attractive, amusing, and enjoyable.

A series of effects and choreography are programmed in which the movements of the water and the changes in the lights are coordinated with the rhythm of the music, thus creating the dancing-characteristic of these fountains. Movement of dancing fountains is powered by electricity of low voltage to minimize the possibility of any accidental occurrences.

Technologically, dancing fountains are the most complex dynamic water features, which are made moving, maneuvering the nozzle actuating electro-mechanically, and the spray is controlled hydrologically. All these performances are controlled from a control panel that performs by a precise hydro-electro-mechanical system actuated by software programming.

Some software allows the control of all the fountain's systems from a remote position, which allows a controller to see the status of the fountain in real time and control the light and water features remotely.

Indoor fountain

Indoor water fountains are specially designed with a view to bringing tranquility, relaxation, and beauty to indoor special spaces. The indoor fountains are basically interfacing a sculpture supported by a hydrological feature in an indoor environment, which makes not only an aesthetic spectacular element but also functions as a therapeutic device for the dwellers.

Figure 6.2 An indoor water fall [1].

The indoor fountains are relatively smaller, in which the aesthetic spectacular value of the sculpture gets the prominence with respect to the feature of the water movement. The generation of sound in these indoor fountains is one of the vital aspects of enjoyment, particularly at quiet moments of late night. Indoor fountains mostly have smaller water management elements that let virtually the enjoyment of the presence of water indoors, in a very controlled way within the sculpture designed with a highly decorative touch.

Indoor fountains, as shown in Figure 6.2, are also known as residential fountains, as the majority of water fountains built outdoors are usually found as part of urban beautification structures.

Outdoor fountain

Outdoor fountains are built outside the building, inside the building premise, mostly in front of the building. Fountains are also built by the concerned authorities as a part or element of the urban built environment. Though there are a good number of varieties of outdoor fountains, from technical point of views, outdoor water fountains can be grouped in the following three types:

1 Dry fountains
2 Floating fountains
3 Water jet fountains

Dry fountain

Dry fountains, also called dry deck fountains, are composed of one or multiple numbers of water jets that spring up at ground level. Dry fountains are primarily designed as playful water features rather than as an aesthetic element of outdoor spaces.

In a dry fountain, a centralized water supply system uses one or several large pumps that supplies water from a centrally located water source and then transports water to multiple nozzles through a distribution piping system. Dry fountains are preferred for large public spaces or landscapes, such as plazas, parks, or large outdoor recreation areas.

Dry fountains can offer multifunctional opportunities for using them. They allow diverse uses of the same space when the fountain is turned off, the space can be for various purposes as planned. If these fountains include devices that let people control the water jets, they are known as interactive fountains.

Floating fountain

Floating fountains are fountains generating water features that do not have a grounded anchor. The floating fountains are built on surface of water bodies that are equipped with fastening cables that ground the devices to the bottom, so they remain perfectly controlled while keeping their water jets sprinkling. Floating fountains stay on the water surface without tipping. Only the nozzle and the surface of the float element termed a buoy are found visible; the pump and other accessories remain submerged underwater. The fountain may consist of multiple nozzles mounted on a circular header or distributor with the nozzle tips just protruding above the water level, as shown in Figure 6.3.

Position of floating fountains is unaffected due to changing water depths. The varieties of water spraying are varied by the use of diverse types of nozzle sets, and the floating fountains are also named accordingly.

In addition to offering an impressive effect on water, it is used for artificial aeration of stagnant water bodies where installed, as stagnation of water causes its contamination in diverse ways. Moreover, the dissolved oxygen level in water increases significantly.

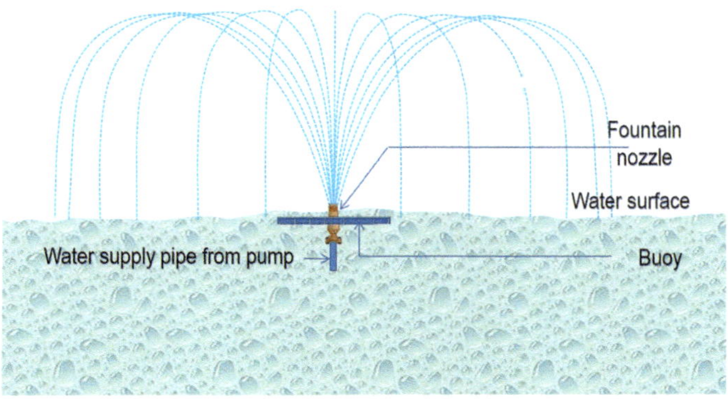

Figure 6.3 A floating fountain.

Water jet fountain

Water jets of a water fountain are the mechanisms, created by pushing water ultimately trough nozzles that control how water is ejected or sprayed from the fountain system. They shape the water flow into various patterns, directions, and effects, creating the visual display. Various types of nozzles are used to discharge or throw water in various shapes and patterns, created due to the design of nozzle tips differently.

Based on the number of water-jet of a fountain, the fountains can be categorized in the following way:

1 Single jet
2 Multi jets

Based on the clarity and effect of water display fountains can be categorized in the following way:

1 Crystalline jet
2 Frothy or foam jet
3 Mist or spray jet

Single-jet water display

Single-jet fountains are small tapered clear stream fountains that display with a minimum of distortions. Single-jet fountains may be displayed as vertical columns or trajectory patterns as desired. In a single-jet of high and slender column of water stream, an attractive plume shape is sometimes created at the top.

A simple, vertical column of clear water is ideal for small- and medium-sized displays. Single-jet fountains are mostly arranged in an array of smaller, vertical jets to create a delicate upward spray, but angular spray is also made by installing the nozzle at varied angles as designed.

Dramatic displays of water can be made by single jet fountains by grouping multiple single jet fountains in a straight line or in a traditional arch or circular pattern.

Fountains of various patterns and shapes can be created in a single-jet discharge, as discussed below, using single special nozzles designed for the purpose:

1 Column jet
2 Spout
3 Fan
4 Mushroom or bell
5 Tulip
6 Foam
7 Film
8 Curtain

Column jet fountain

It is the simplest single-jet fountain, which displays a vertical column of clear water of varying or same height as designed. Multiple small column jet fountains arranged in

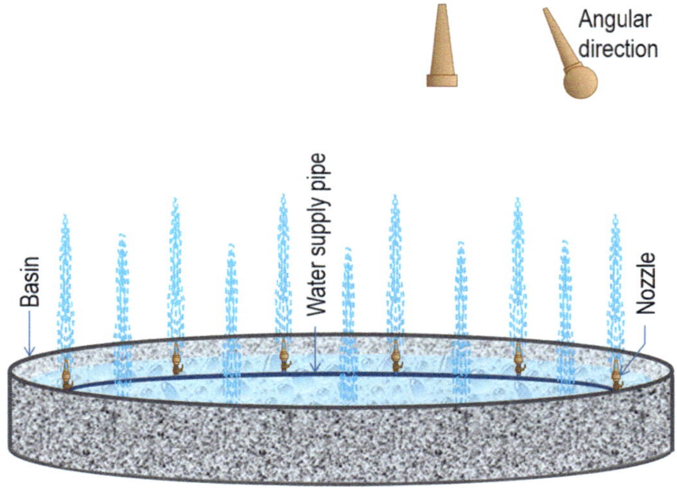

Figure 6.4 Fountain of upright single lance jet nozzles arranged in a circular way.

various patterns, circular or linear, small- and medium-sized water displays are created. Sometimes clustering of high and small column jet fountains, or of the same heights as shown in Figure 6.4, is designed for varying the shape and enlarging the size of the fountain.

There is also burst jet single column fountain that is made for high single shot or "burst" effects of up to as high as 15 m. The design of the jet needs to meet high pump pressure and flow rate of the water. The jet creates a speedy stream that rises rapidly but disperses at its peak to fall gently in a controlled manner. This type of fountain is also termed a high vertical jet fountain. This jet requires a nozzle of very high abrasion resistance to withstand high velocity.

Spout fountain

A spout fountain is one that uses one or multiple spouts, mostly circular but may be square or rectangular, a piece of pipe protruding from a wall or any other structure, through which water is discharged continuously, striking on the water surface remaining below. Generally it is found that this type of fountain is paired up with a statue or a sculpture, as if water is being sprayed from a particular spot, mostly the mouth if there is one, on the statue or sculpture, where there is a nozzle embedded that sprays the water in a specific pattern designed for which necessary piping is also embedded into the structure, and a pumping arrangement is made as usual to supply water. Figure 6.5 features a classic spout fountain paired with a sculpture.

The pouring of water from a structure on a water body below, through a spout at a high altitude, creates huge sound by the falling water while striking water below and also aerates water greatly.

As the piping and the nozzle are generally kept embedded within a concrete or any masonry structure, its material shall be long-lasting in terms of corrosion resistance primarily. In this perspective, spouts of bronze or rustic iron fitted on stainless steel pipe would be a good choice.

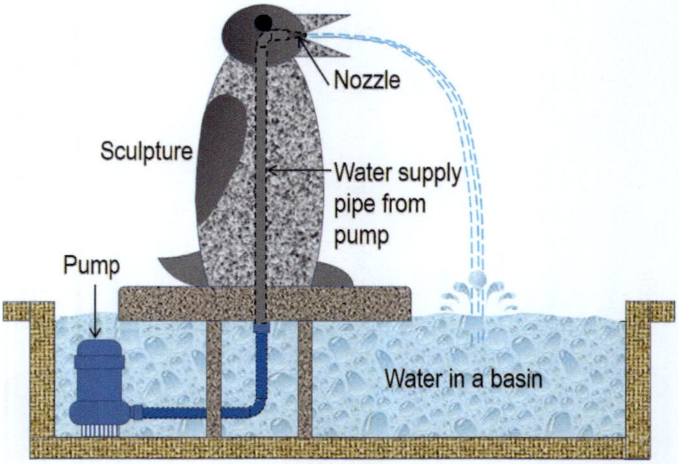

Figure 6.5 Spout fountain paired with a sculpture.

Multi-jet water displays

There are fountains that are designed to display multiple jets of water spray from a single point. For this purpose, a multi-jet-producing single nozzle is used. Multi-jet water display that produces the unique feature of multiple-jets of clear water streams can also be formed using multiple single nozzles. Multi-jet fountains are popular for use in small- and medium-sized fountains and play-decks.

Various types of multi-jet fountains are designed as mentioned below:

1 Finger jet
2 Palm tree jet
3 Crown jet

Crystalline jet water display

A crystalline jet water fountain is called so because the water it projects appears to be clear, smooth, and glass-like transparent. The visual clarity, purity, and cohesion of the stream of water are key features of crystalline appearance of the water jet. This type of fountain water display is the most commonly used in the world of fountains.

Creating a crystalline jet involves maintaining laminar flow through nozzles where no turbulence is developed. For this purpose, constant pressure is to be maintained and water should be kept clear.

The crystalline water jets are also made hollow to reduce water flow rate. Water jet being hollow, it requires comparatively less water and low power in a pump, which ultimately saves significant energy in operation due to the low power consumption of the pump.

Crystalline jet-producing nozzles may be again of different types, producing different features of water jets; some very common are as mentioned below:

1 Lance jet
2 Fan jet
3 Bell or mushroom jet

Frothy jet water display

This fountain is a type of unique water feature producing an aerated jet that appears as a white foamy column or burst of water. This frothy appearance is created by entraining air into the water, resulting in a highly aerated, vigorous water jet. These fountains are also called foam jet, aerated jet, or geyser jet fountains.

The water display of such a fountain has low wind sensitivity due to producing a cohesive plume. The display appears aesthetic even when wind disperses fine droplets, unlike a crystalline jet display.

Mist jet water display

These types of fountains generate a large volume of water spray in the form of fine, cloud-like water droplets, resulting in a mist-like effect, which is light, so it drifts along the wind flow.

In this fountain water is forced through a very small orifice of a special type of nozzle under high pressure. Flow of water through a very small orifice under high pressure atomizes the water into ultra-fine droplets; spraying of what in open air creates a mist or cloud of water particles.

Nozzle

Nozzles are the main component of a fountain, which actually causes the patterns of the water spray to create different types of water fountain discussed so far. In a fountain, there might be the use of a single nozzle producing a particular type of water jet or multiple nozzles projecting the jet in the same or multiple directions. Nozzle tips are designed or shaped to produce water spraying features of different natures.

Some nozzles also help create a water jet up to certain heights depending on the pressure employed and size of the nozzle. Incorporating a smooth ball joint, the nozzles can be tilted at a maximum angle of 20° from vertical, thereby enabling the throwing of parabolic jets that can be designed with no additional parts. These nozzles also might have a valve inbuilt for regulating and adjusting the flow rate. Small and medium sizes having 4–10 mm outlet are suitable for indoor architectural fountains, while larger than 10 mm diameter outlets are most suitable for outdoor fountains.

Lance jet nozzle

The nozzle has a tapered end that provides a single sharp water jet. The water jet is basically a single-dimensional drive that may be vertical or at an angle toward a particular direction, making a parabolic line of the water jet. All sizes of single jet nozzles may have a ball joint with which the jet can be pivoted up to 15° from the vertical plane. Figure 6.4 also represents a fountain incorporating multiple upright single lance jet nozzles in a circular way.

Palm tree jet nozzle

These fountain nozzles comprised of a set of small jet nozzles, arranged in such a way that on spraying the jets look like the trunk of a palm tree and on falling like its leaves, as shown in Figure 6.6.

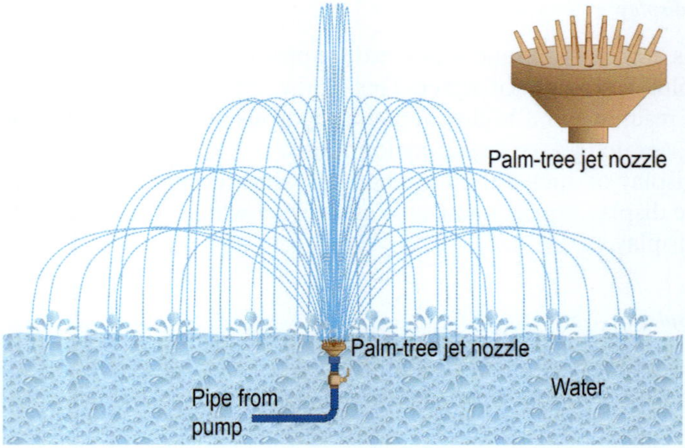

Figure 6.6 A palm-tree jet fountain and the nozzle.

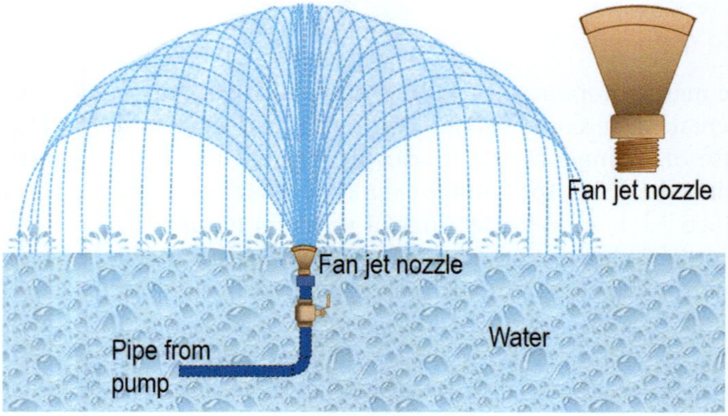

Figure 6.7 A fan-jet fountain and the nozzle.

Fan jet nozzle

Fan jet nozzles are made conical-shaped. A fan-shaped water veil is generated by the gushing water from these nozzles, as shown in Figure 6.7. Depending on the installation of the nozzle for a particular direction of spray, the fan nozzles generate either a vertically ascending water veil or a diagonally spouting water veil of 6 or 8 mm in thickness. A clear, fan-shaped display with narrow-based spread can be made to reach heights up to 3 m.

Finger jet nozzle

Finger jet nozzles generate multiple clear single stream jets from individual fingerlike individual nozzles arranged in the same plane. In this nozzle multiple fingerlike nozzles protruding from a header cause the finger jet water spray, as shown in Figure 6.8.

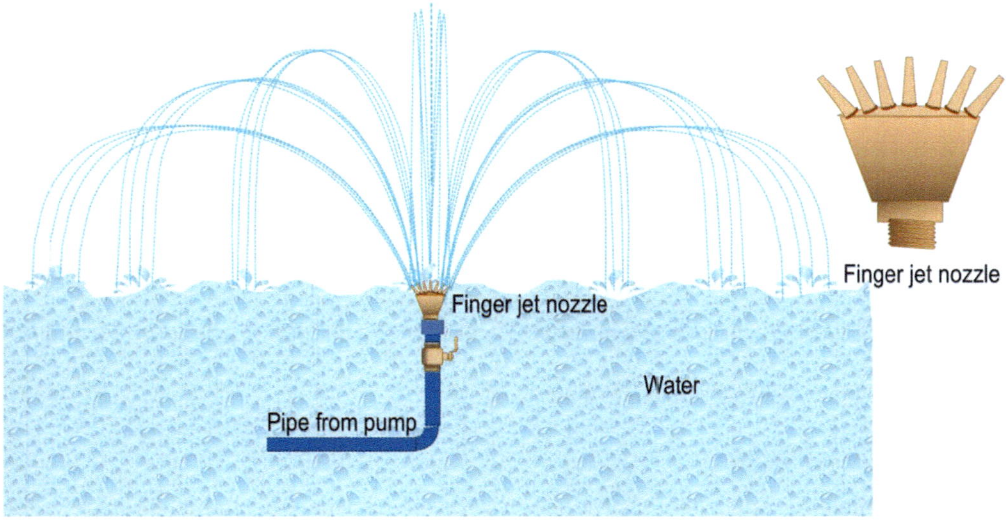

Figure 6.8 A finger-jet fountain and the nozzle.

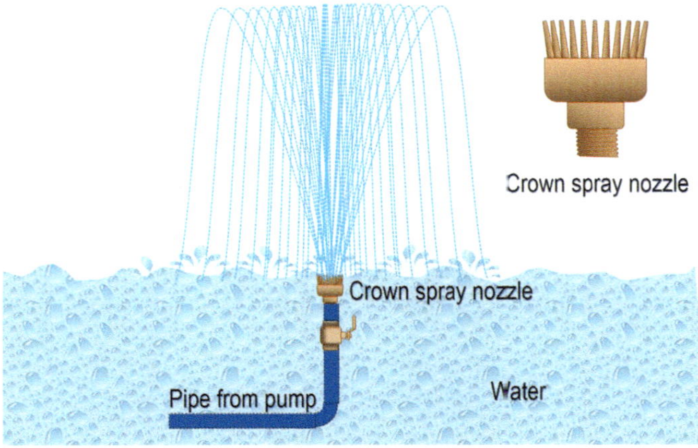

Figure 6.9 A crown spray fountain and the nozzle.

Crown jet nozzle

A number of single jets are arranged in a circle on a nozzle header. The middle one is set up-right, and the others in a particular circle are installed tilted at a particular angle, as shown in Figure 6.9. There may be multiple circles of nozzles on the header. Thus the jets create a calyx. The water pattern looks like a calyx in which a strong water jet in the middle can be produced if designed. These nozzles produce a variety of water jets at different heights where jets of a particular circle are of similar heights.

The nozzles may be in a single header or ring-pipe header of particular diameter designed. When nozzles on a ring header are found generating different natures of jet streams in terms of height and flow rate of water, the nozzle can be equipped with a ball, globe, or a pressure-reducing valve, which would be necessary.

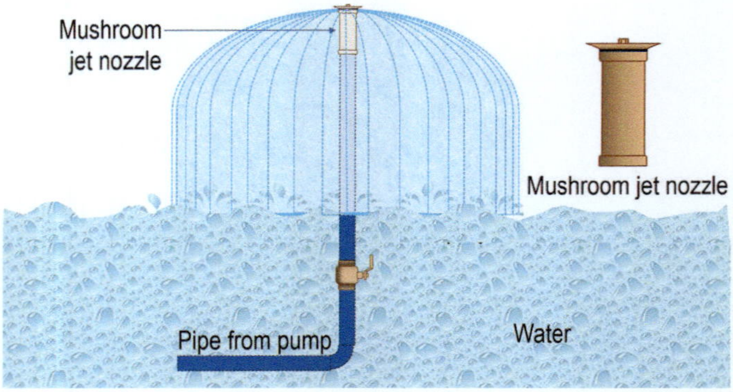

Figure 6.10 A mushroom jet fountain and the nozzle.

Bell or mushroom jet nozzle

The bell or mushroom jet nozzle produces a sheet of transparent laminar flow of water, creating a bell or mushroom shape from the nozzle-outlet being at the center, as shown in Figure 6.10. The name comes from the bell- or mushroom-shaped sheet of water it produces. The breaking of the water sheet primarily depends on the thickness of the water film; air or wind velocity is another factor. Its operating speed and noise level are comparatively low.

In addition, there are also special types of single-jet nozzles for spraying water, producing different features with respect to volume of water, density of spray, and height of spray, as mentioned below:

1 Film spray
2 Geyser jet
3 Foam column
4 Mist or Fog

Diffuser sheet nozzle

Not only are the sprays made in jet shape, they can also be made in sheet shape using a very different type of nozzle known as a diffuser sheet nozzle, as shown in Figure 6.11. These nozzles produce sprays of clear water sheets in various sizes as designed.

Minimum sound or splash is created by this sheet of clear water. For wind-free locations in small- and medium-sized pools, these fountains are created, to avoid breaking of water-sheet.

Geyser jet nozzle

Geyser jet nozzles produce a fountain jet that combines a mixture of air with water from the pump immersed in water in the basin or pond. The water jet pattern is a column of white water, created by a mixer of air with water together. These nozzles generate a slender water column and fall in slightly conical form, as shown in Figure 6.12. Geyser jet fountains are usually installed as floating fountains.

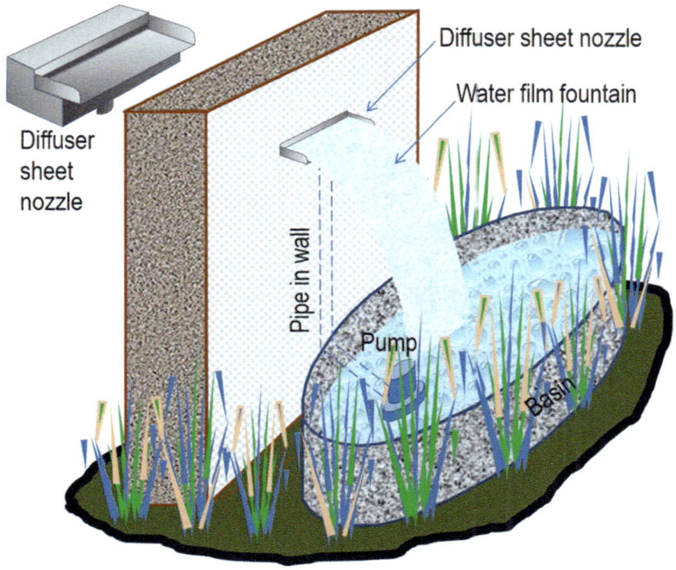

Figure 6.11 A water film fountain and a diffuser sheet nozzle.

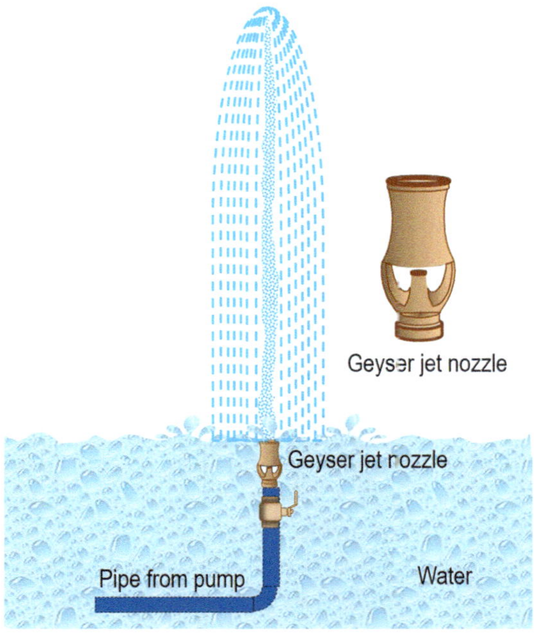

Figure 6.12 A geyser jet fountain and the nozzle.

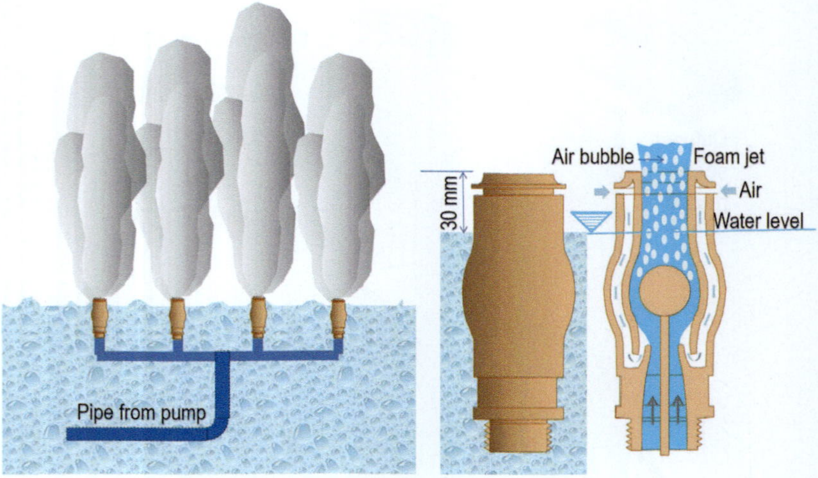

Figure 6.13 A foam jet fountain and the nozzle.

Foam jet nozzle

Foam fountains are water fountains, but the display appears voluminous foamy water instead of the common narrow, thin water appearance, as shown in Figure 6.13. By mixing air with water, the gushing water pattern appears to be extremely high-volume of foamy nature. The specially designed aerator nozzles actually have a low water requirement.

The discharge of foam from the nozzle creates a mass of frothy, aerated water that generally reaches lower heights and becomes a splashy, frothy column at levels around 2.5 m. Foamy shape is better achieved at lower heights. These foam fountains are ideal for small-, medium-sized, and large displays. The beautiful soft and foamy white column-jet offers a strong contrast to its surrounding environment.

Mist or spray jet nozzle

As the name implies, the fountain is made to appear as mist or fog from a nozzle, as shown in Figure 6.14. In this type of fountain, a special nozzle is designed in which innumerable water jets are generated from many tiny single holes, jets originating from a header of the nozzle, thus creating very fine droplets of mist or fog, as if being sprayed in one full jet of fog or mist from a source. Although these types of fountains add special beauty to landscaping, they offer many other unique benefits, such as air purification by dust removal, cooling down, and humidifying the environment.

The specially designed fog-nozzles are found made of brass and stainless steel to work under an operating pressure of 70 kg/cm to produce a flow rate of 29–355 ml/min through innumerable orifices of diameter 0.1 mm–1.0 mm [2].

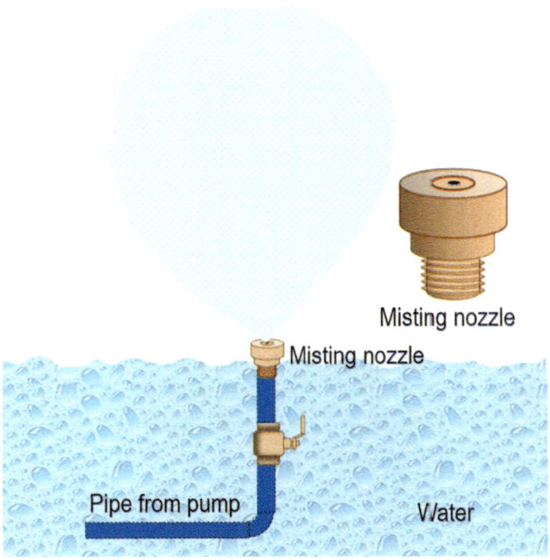

Figure 6.14 A mist fountain and the misting-nozzle.

Fountain development planning

Planning for placing and locating the major components of a water fountain plays an important role for its efficient development, operation, and maintenance. The ways of planning the development of a fountain irrespective of its spraying water features have been discussed below:

1 The simplest type of fountain can be developed by placing the basin directly on a floor, as shown in Figure 6.15. This type of basin placement eliminates the need for excavation of soil. The pump is submersible type; so, it is installed on the basin floor, and the cable is led over the basin edge. Water depth should be sufficient to submerge the pump and nozzle in some cases. Though this is a low-cost option, all the engineering and mechanical components remain almost visible, which may not be aesthetically pleasing.

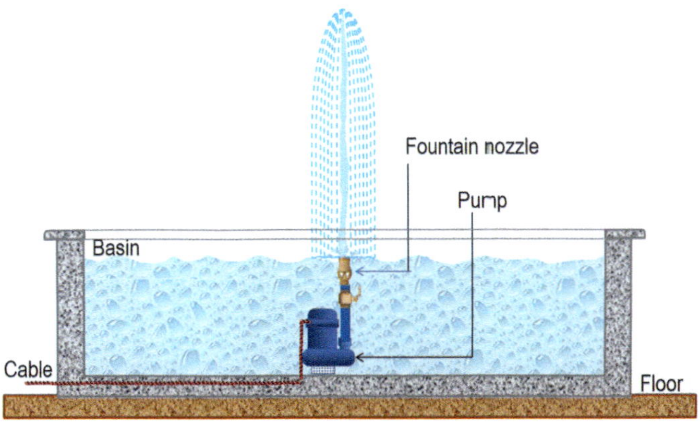

Figure 6.15 A simplest fountain.

Maintenance cost is relatively high as there is little opportunity of providing filtration or water treatment systems.

2 In this planning, the basin is placed in-ground, partly or fully. A concealed space for the submersible pump and piping is created in the floor as shown in Figure 6.16. This produces a more clear appearance of the basin floor, but construction costs go higher. Water depth in the basin can be kept low by increasing the depth of the sump for the pump to keep submerged.

3 In this type of planning, the pump is located outside the basin of the pool, incorporating a dry-mount pump, as shown in Figure 6.17. The water basin can be constructed on or in the ground, as discussed earlier. The suction and delivery pipes lead to a remote but nearby location where the pump, filtration, and electrical controls are located. A spring check valve and strainer shall be fitted at the suction end. This planning is best used when water volume exceeds 3786 liter [3]. A dedicated space or room is required near the fountain basin for accommodating pumps and the filter units.

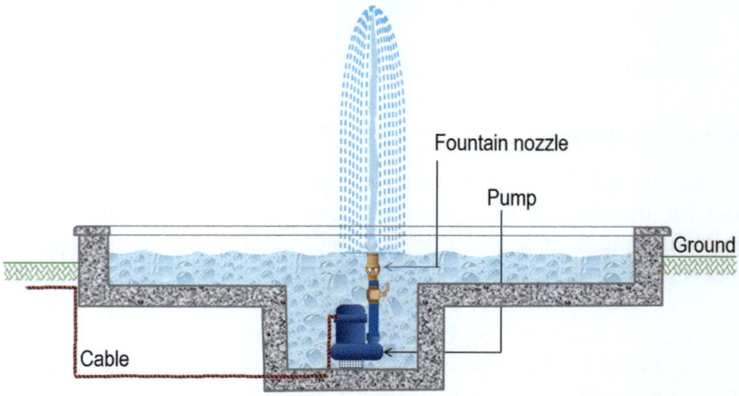

Figure 6.16 Fountain having the pump in a concealed space in the basin.

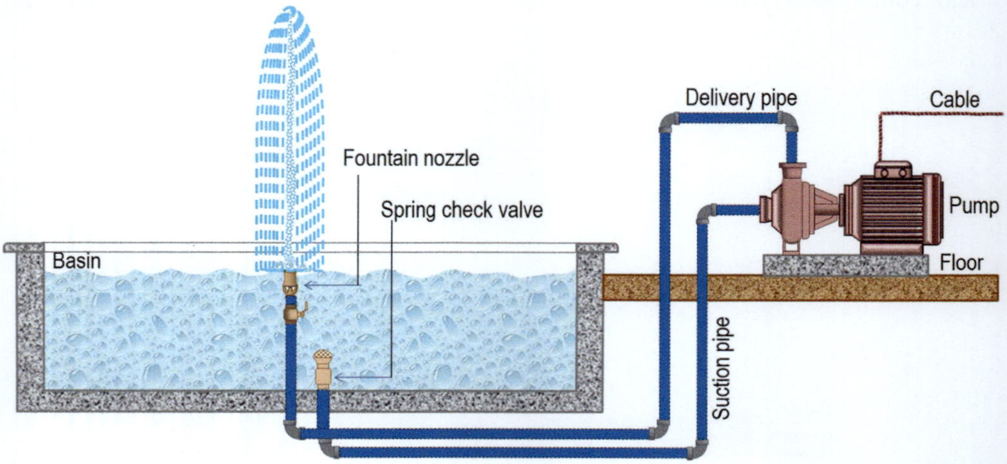

Figure 6.17 Fountain having the pump outside the basin.

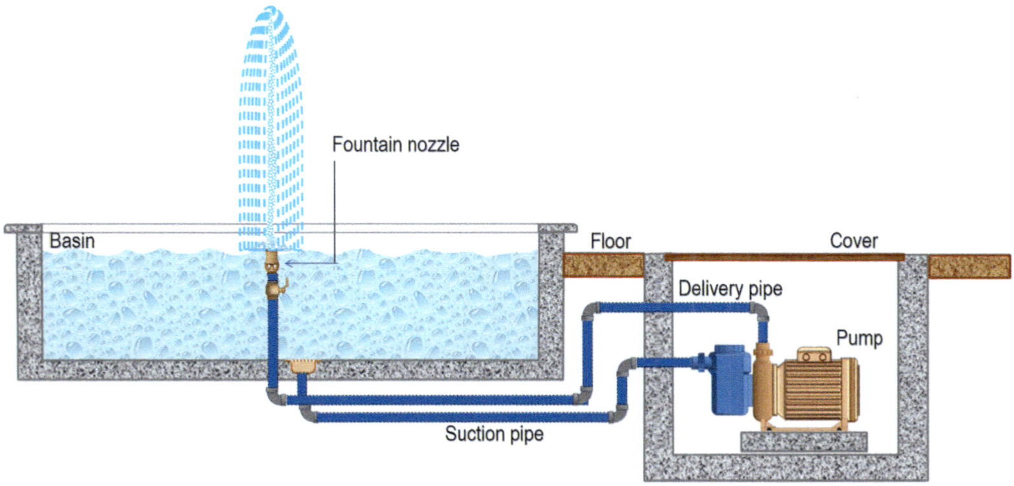

Figure 6.18 Fountain pump in an underground chamber outside the basin.

4 This planning is more or less similar to the previous planning, but the difference is that the pump is placed in an underground chamber, thereby concealing all the pumps as shown in Figure 6.18. Where possible, the filtration unit can also be located by the pump side. It is the most expensive installation planned mainly for commercial large fountains.

Waterfall

Waterfalls are one kind of fountain in which water either flows down or falls from an upper elevation freely or is supported by some structural elements. Theoretically, waterfalls are fountains in which water flows over a weir and then falls below. A weir is a surface, or edge, over which water flows by a force, mostly under gravity. Weirs may be designed in various shapes and sizes to control the shape and feature of the waterfall. In addition to shape and size, the level and surface condition of the weir also cause changes in the feature, and most importantly, the texture of the waterfall.

Based on the texture of the water fall, it may be divided into two types:

1 Clear-sheet waterfall
2 Aerated waterfall

Clear-sheet waterfall

When a volume of water, after flowing over a long weir, falls, it creates a sheet of waterfall. The clarity of the sheet of waterfall depends upon the thickness of flow at the tip of the weir and disturbance in terms of undulation on the weir surface; it encounters turbulence in the flow while just falling and wind conditions around the fall. So, to ensure a very clear sheet of waterfall, the surface of the weir shall be horizontally truly leveled, very smooth, and free from zero undulation, creating no wave in flow, and having very calm flow with a minimum possible velocity of flow under very calm airflow.

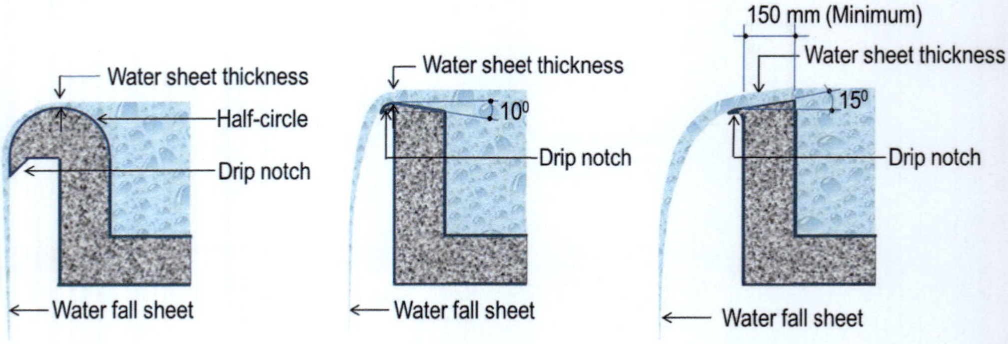

Figure 6.19 Various natures of clear water sheet falls.

The falling of water from the weir can be differently shaped to create different natures of sheet-fall, as mentioned below and shown in Figure 6.19:

1 Drop sheet
2 Normal or clear sheet
3 Accelerated sheet

Drop sheet

A drop sheet of clear waterfall is that nature of waterfall that falls truly vertical from the weir tip. To ensure this nature of water fall, the weir is configured perfectly quarter-rounded, as shown in Figure 6.19a. Perfect roundness and the tip angle being zero with the vertical shall be the key factors in this regard. There must be a perfect drip notch at the tip of the weir, which prevents the dropping of some unwanted water separately beyond the sheet of the waterfall.

Normal or clear sheet

When the weir is horizontal, theoretically allowed up to a maximum angle of 10 degrees from the horizontal, the water sheets fall at an angle more than zero from the vertical, depending on the initial velocity of flow on the weir, which is again dependent on the angle of the weir with the horizontal and the pressure at the start of water flow. This configuration of weir can produce the maximum clear sheet of water fall.

A clear drop-sheet waterfall can be made free falling and falling by flowing over a surface. When it falls freely, it is termed a water curtain, and when it flows over a surface—generally on wall—is termed a water wall, as shown in Figure 6.20.

A water curtain can be designed in different shapes, as mentioned below, by shaping the weir accordingly.

1 Flat or curtain type
2 Curved type
3 Circular or cylindrical type

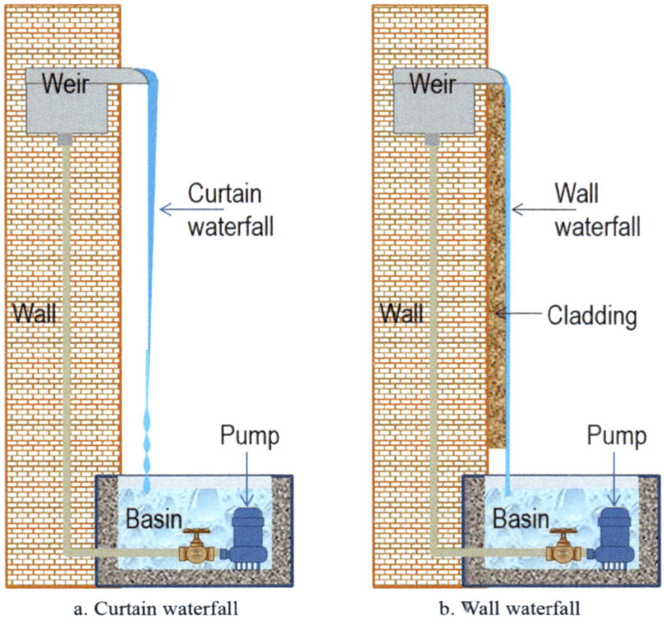

Figure 6.20 Clear sheet waterfall: a. Curtain type and b. wall type.

Water curtain

A curtain of water is basically transparent film-type water falling tumbling down generally into a basin, forming an elevated position. The curtain water falls are made in varying widths and heights. The thickness of the water film at the start of the fall depends on these two factors: width and height of fall.

The curtain waterfall is an un-breaking thin sheet of water, sensitive to wind, and the thickness of the sheet narrows the farther it falls. When the thickness of the sheet of water becomes too thin, the surface tension may not be sufficient to hold it together, and then it will break apart. Additionally, when it falls from high, it gains velocity that also causes breakage of the water sheet. At fall heights above 1 m, the sheet effect begins to break apart and to produce an annoying buffeting noise. So, a maximum mounting height of 1 m for the sheet effect is recommended for a quiet environment.

To make the water fall down very clearly and remain away from the wall face, the front lip of the water blade must have a protrusion of at least 30 mm out of the wall face.

Water is pumped to circulate water from the basin below to the weir at a higher level. Water is pumped through one riser and header pipes into a trough, which being overflown, water flows over the weir and rolls over the tip to fall down as film or sheet into the basin from where water is pumped, and thus water is circulated. In the wide water curtains, the lip should be made perfectly level to prevent the water film from breaking up at the upper levels, which is not desirable. For a perfect transparent appearance, wind flow shall be very calm and the pump is correctly sized.

Water wall

Water wall is mostly indoor water fall made by allowing water film to run down on a vertical surface, generally a tiled smooth wall or wall of glass, acrylic, stainless steel, etc. No splash of water is caused by the steady down-flow of water. A water wall is also very quiet. The waves on the sheet surface can be created by making the wall surface very light-wavy.

Accelerated sheet

When the angle of weir is greater than 10 degrees, the texture of water sheets tends to change, losing its clarity [4]. The angle of falling at the start will be more subject to the angle of weir from the horizontal, which is recommended to be 15 degrees.

Aerated waterfall

Aerated waterfalls are another type of waterfall in which water texture is changed by causing aeration while flowing by rolling down on a surface. Where aeration is done, water texture turns foamy and white due to the production of huge water droplets therein.

Aerated waterfalls can be created in three different techniques, as mentioned below:

1 Stair-step
2 Boil-over
3 Rough wall

Figure 6.21 illustrates three different aerated waterfall weir configurations.

Stair-step

The most common method for creating an aerated waterfall is to cause a water flow over a series of stair-steps; a minimum of three steps are required to induce full aeration. The stair-step-induced aeration is caused by allowing the water to tumble down the steps. To create the tumbling action, the steps must be correctly sloped and sized, and a minimum flow depth is to be maintained. The stair slope should be 1:1, having steps of 75 mm. both tread and rises, and 12–19 mm depth of flow is considered optimal in creating maximum aeration. In addition, it is important that the first or top step be half the height and width of the second and third steps. This slows the velocity of the water enough as it starts down

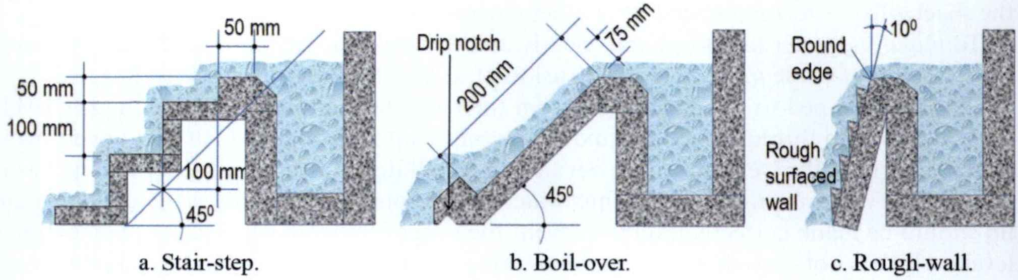

a. Stair-step. b. Boil-over. c. Rough-wall.

Figure 6.21 Typical aerated waterfall weirs.

the steps to allow it to tumble. If this first step is too wide, the water will accelerate as it crosses the step, causing it to miss the second and third steps. If for structural reasons the top step must be wider than 38–50 mm, then the step should be sloped down in the direction of the source trough at a minimum 15 degrees to increase the depth of water on the step and to minimize velocity [4]. Figure 6.21a illustrates stair-step aerated waterfall weir configurations.

Boil-over

The "boil-over" technique of aeration is done first by accelerating water down a short ramp. At the base of the ramp, a weir is created, as shown in Figure 6.21b. The accelerated water crashes into a small trough of water, being hindered by the weir, thus creating turbulence and a rolling boiling action. The flowing water then surges over the weir edge, creating a highly pulsating, aerated waterfall. This type of water fall is preferred when a broken, pulsating, aerated waterfall appearance is desired after water is going to fall down a significant distance from the start of the water fall.

Rough wall

In the "rough wall" technique, an inclined wall having its surface finished with a uniformly created very rough surface, as shown in Figure 6.21c, is used. As water slides down the rough-surfaced wall, it begins to tumble and roll very closely. Because the wall is inclined, the water flow continues to interact with the rough wall, creating significant aeration in the water.

Cascading

A cascade waterfall is basically a waterfall of aerated water created by flowing water over a series of steps made on an inclined slab of desired size. Water flows gently over, at multiple levels from one step to the next, as shown in Figure 6.22. Cascading fountains are usually self-contained, which incorporates all the necessary plumbing housed inside the unit.

Water is lifted from a basin built under the cascade up to the highest level of the cascade and flown over the weir made along the length of the cascade.

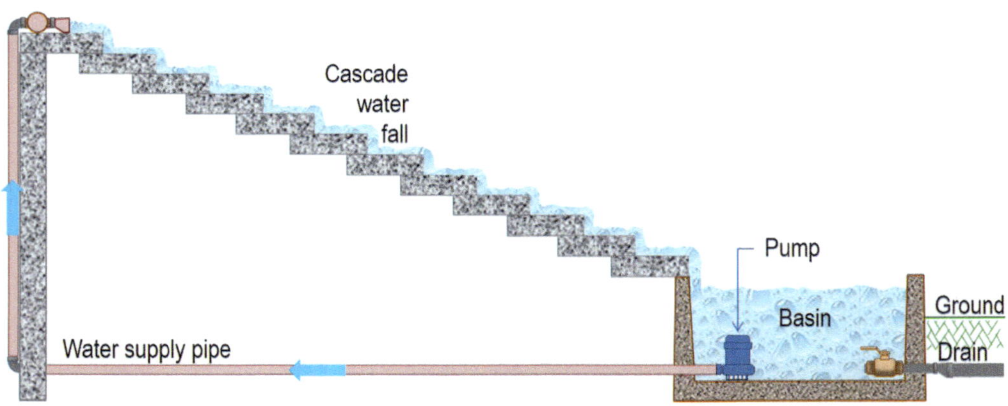

Figure 6.22 A cascade waterfall.

Pool or basin

A pool or basin is the container of water that also receives sprayed water from the fountain or fall, from where water is pumped up to supply water through the nozzles to display the feature of a water fountain or fall.

The water pool not only just holds water but also serves several other purposes, depending on the design and context. The following are the major uses and benefits of a water pool:

1 It increases the aesthetic appeal of the structure.
2 It enhances the visual and sensory experience of the waterfall.
3 It reflects the waterfall and surrounding landscape, adding beauty and depth.
4 It helps absorb the energy and noise of falling water, reducing splash and excessive noise.
5 It helps maintain water clarity and supports selected aquatic life to grow for eco-friendly landscape design.
6 It may be used for wading or swimming if designed safely and hygienically.
7 The combination of water movement and evaporation helps cool the surrounding environment, especially in hot climates.

The size and shape of the pool are determined considering the water spraying effects of the fountain. The common practicing rule of thumb is that the water feature effect height is the equivalent of half of the radius of the pool. Water depth in the pool should be no more than 450 mm, to avoid the safety issues related to water depth.

Typical splash zone extends from the nozzle as far out as the water feature is high or tall. A water feature 1 m tall may splash out 1 m in every direction; so, a 2 m diameter splash zone is considered [5]. In the form of a formula it can be said that if the height of a tallest fountain is H_t meter then the size of the basin will be at least 2 H_t meter. So, if the nearest fountain is H_n meter high, the basin boundary wall shall be at least H_n meter away from that fountain, as illustrated in Figure 6.23.

In order to prevent wave amplification in pool water, wave suppression measures would be required, mostly when fountains are installed in circular pools.

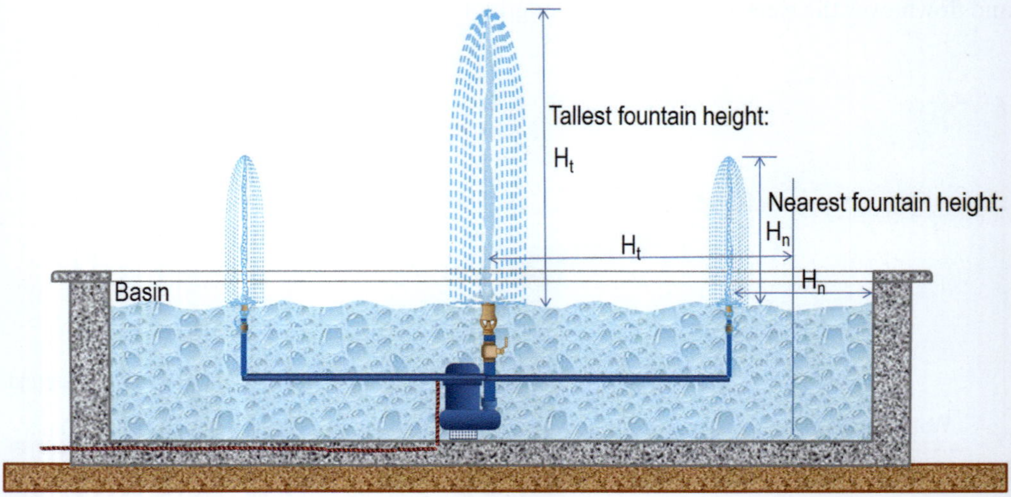

Figure 6.23 Fountain splash coverage.

Inside of side walls should be very smooth for prohibiting any microbial growth and easy cleaning. It is also suggested to select particular colors like blues, grays, and black, as they create an illusion of depth and help conceal pumps, pipes, etc. placed on the basin floor. Yellow and brown colors should be avoided for inside pool surface finishes, as they tend to make the water appear dirty. Greens appear as an artificial intervention that clashes with the surrounding greenery where it exists. Uniform coloring of basins magnifies imperfections; so, patterned or dappled finishes would be better appealing.

A basin must have a drainage system to drain out its contaminated water when the water loses its treatability. Drain pipe shall be discharging to an approved point of disposal, in wastewater or combined building drains or sewers available nearby.

Water pool below a water fall

As provided in the water fountain, a befitting pool is also constructed under the waterfall for various reasons, mainly for collecting water to be pumped back up to the top of the waterfall, making a closed-loop water circulation system. It is also essential for maintaining water levels and continuous operation. A water pool also helps absorb the energy and noise of falling water from a high level, reducing splash and excessive noise.

The dimension, width, and length of the water pool below a water fall mainly depend on the width of the blade of water fall and also how high the lifting of the water is. As the sheet of water cascades downward, it will finally hit the water in the pool, which could potentially create a splash. If the pool is too small to catch the water splash, it would cause very quick loss of water and may damage the submersible pump put in the pool.

As a general guide, it is suggested that the reservoir is around 150–300 mm wider than the blade on both the left and right hand sides. If a cross wind tends to drive the water sheet, this amount of enlargement of the pool dimension should be enough to cater for deviated falls due to wind. From the wall facing coming forward, it is suggested that the pool should be around 600–800 mm. This surface configuration would be able to capture the majority of splashes. With regard to water depth, it is ideally recommended a depth of 500–600 mm where possible [6]. This will allow the pump to remain submerged on the floor of the pool with sufficient water above.

Pool basin accessories

Suction drains and debris screen

A debris screen should be installed on suction drains to prevent solid matter from entering the pump, thereby preventing damage to pumps and clogging of spray jets.

Overflow drains

Overflow drains remove excess water received in the basin; this generally occurs by rainfall or accidental overfilling of the fountain basin. An overflow drain performs two functions: it provides a means for removing excess water in case water makeup should malfunction and a means for draining rainwater. The preferable location for an overflow drain is in the basin wall of the pool and at the water level to be maintained. All overflow drains are fitted with screens to prevent large debris and are connected to the drain pipe for disposal.

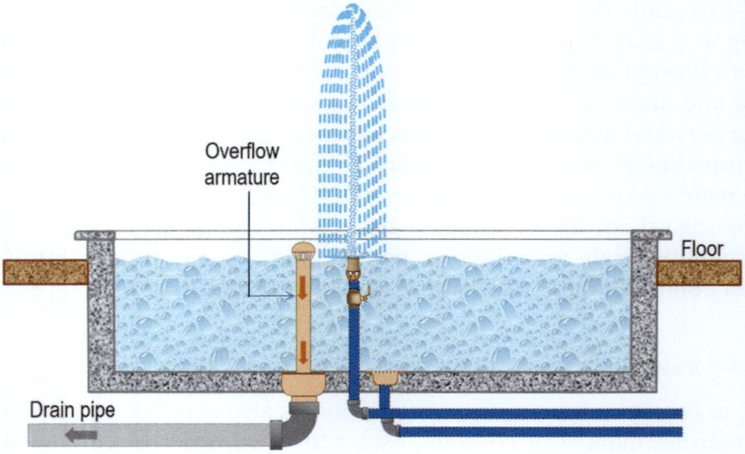

Figure 6.24 Fountain drain including overflow armature.

To maintain water height an overflow or drain armatures is also installed, as shown in Figure 6.24. The overflow or drain armatures is basically an overflow pipe with an overflow crown positioned at the water level to be maintained, and the bottom part is fixed on the drain pipe leading to the disposal point.

Sleeve fitting

Pipes that penetrate pool basin walls and floors are the major source of leaking. Sleeve fittings, or stub-ups, are to be used to penetrate pipes through the wall or floor to achieve a mechanical means of sealing between the pipe and the structure.

Water level control

Electronic and non-electric water level sensors are used to monitor water level in the basin and help replenish water lost through evaporation or splashing.

Water level controls provide both a cut-off for safe operation of submerged fountain lights as fountain lights operate 50 mm below the water surface and a regulating effect for proper submersion of jets. Pump suction pipe must have sufficient water depth to prevent vortex occurrence. Some jet nozzles draw additional water or air from their surrounding water to increase their visual impact. These jet nozzles' performance is water-level dependent and requires very accurate water-depth controls. So, evaporation loss is to be managed well in the pool.

Pump

Pumps for supplying water are selected based on flow rate of water and pressure requirements for the combined fountain effects to get the desired size or volume of water features.

Mostly submersible surface-mounted pumps are placed in the basin of the fountain, but surface-mounted pumps at remote locations outside of the basin can also be placed.

Submersible pumps in the basin must be placed in a vertical position. Where needed, a sump might be needed to construct to house the pump in vertical position.

There are also dry pumps used for fountains. Dry pumps are located outside the basin. Dry pumps should be equipped with suction strainers.

Piping

In water fountains or waterfalls, selection of the appropriate type of pipe is essential to ensure durability, water quality, and ease of maintenance. Among plastic pipes, HDPE pipe will be more long-lasting and impact- and chemical-resistant than uPVC pipes. Among the metal pipes, steel pipe is preferable to galvanized iron pipe, as it may corrode much earlier due to having constant water contact. Steel pipe, though costly, is used for more aesthetics and longer life in exposed conditions.

Filtration system

A suitable filtration system is required for a fountain or waterfall to maintain the quality and clarity of its water. Filtration through a sand media traps fine suspended particles present or accumulated in fountain water during operation, to prevent the water from looking murky.

Filter type and sizes are based on the pool capacity and flow rate desired that must be calculated in order for proper operation and to ensure clear water. For this purpose, pressure filters are mostly opted for considering limitations for space and high filtration rate requirements. Filtration for small water bodies less than 1900 liters may be too expensive to install and not worth the investment. Draining the full water and cleaning regularly may be the best solution for such small fountains. Bacteria and algae will soon develop and multiply in a fountain if not countered with any disinfection system like chlorination or bromination. Fountains should be employed with water treatment, maintaining 4-hour turnovers and proper recirculation.

Lighting and illumination

Well-designed fountain-lighting adds additional exciting dimension in the dark, mostly at night, by illuminating the water feature effects. Fountain lighting is desired to illuminate sprayed and moving water. They are two different illumination solutions practices that should not be used interchangeably. One or maximum two lights of different colors per single nozzle is recommended for proper illumination. The type and wattage of lamps are determined by the size and height of the water sprayed and whether colored lenses are used.

To illuminate a fountain of spray-circle less than 1.2 m, one fixture with a wide flood lamp is to be used. When using colored lenses, keep in mind they will produce different light intensities. Amber and turquoise require 50 percent more light than clear. Red requires 100 percent more light than clear. Blue and green require 250 percent more light than clear. High ambient light will have an impact on lighting selection.

All lighting fixtures should be kept submerged before lighting and must be installed by a qualified electrician. For safety, all electrical installations must meet the concerned local Electric Code.

Junction box: Junction boxes are used in underwater watertight cable connections for lights, pumps, sensors, etc. and disconnections when needed. It should have corrosion resistant housing. Junction boxes have one inlet and multiple outlets for cables or wires, which should be determined by the number of lights and electrical circuits to be served.

Control panel: Electrical control panels are installed to control the pumps, lights, and water level equipment. Ground fault interruption hardware safeguards people from the contact of water and electricity. Control panels can also be equipped with operating timers and, where needed, timers and relays for sequencing valves.

Designing fountain elements

Sizing a pump

As usual, pumps for water fountains and falls are sized by their flow rate and pressure requirements.

Pump flow rate: This is the flow rate of water described in liters per minute (lpm) that the pump will be discharging through the nozzles of fountains or over the weir of waterfalls. To determine the flow rate, determining the velocity at the nozzle is very important.

To find the flow rate of a fountain using one nozzle, which is the flow rate of the pump delivery pipe, the following formula is used:

$$Q = AV \tag{6.1}$$

where,

Q = Flow rate (cum/s)
A = Cross sectional area of the pipe to which the nozzle is fitted

$$\text{Again,} \quad A = \pi d^2 / 4 \tag{6.2}$$

So, the diameter of the pipe connecting the nozzle is calculated.

The velocity at the nozzle is determined by the following formula:

$$\text{It is known that velocity } V = \sqrt{2gh} \quad \text{i.e. } h = \frac{V^2}{2g} \tag{6.3}$$

where,

g = Acceleration due to gravity = 9.8 m/sec.sq
h = Height of jet projectile (m)

For a fountain of multiple nozzles connected to a pipe, the total flow rate of the branch or main pipe supplying the pipes connected to the nozzle is the summation of the corresponding connected nozzles flow.

Pump head: It is the total dynamic head of the pump obtained by adding the static head from the water level of the basin to the nozzle top, frictional head loss in the piping and its accessories, and the nozzle pressure or head loss. It is needed to consider the head loss of one nozzle requiring the highest pressure to operate.

The total head of the pump H is thus expressed by the following formula:

$$H = H_s + H_f + H_v + H_n \tag{6.4}$$

where,

H_s = Static head (m)
H_f = the pipe friction head (m)
H_v = Velocity head (m)
H_n = nozzle head (m).

Pump power requirements can be found from the following formula:

$$\text{Pump power requirement}(\text{kW}) = \frac{Q \times H}{360 \times E} \tag{6.5}$$

where,

Q = flow rate in cum/h
H = Total dynamic head in m
E = Efficiency ranges between 60 and 85 percent.

Pipe sizing

In an outdoor fountain of multiple nozzles, four types of pipes are to be sized for different functions and should be sized considering the appropriate governing factors. In sizing the pipes, the most important factors to be considered are the flow rate, which was discussed earlier, and the next factor to be considered is the velocity of flow to be maintained in the pipe. The considerations for these factors for different pipes are discussed below:

1 Nozzle pipe: The pipe at the end of which a fountain nozzle is installed. This pipe is to be sized based on the velocity to be maintained in the flow through the nozzle.
2 Branch pipe: The pipe supplying the nozzle pipes is to be sized based on the flow required on all the nozzle-pipes connected to a branch pipe. The maximum velocity for a branch pipe can be considered 3.0 mps.
3 Pump pipes: The pump delivery pipe shall be sized considering the total flow requirement for all the nozzles of a fountain. The suction pipe should be one size bigger than the delivery pipe.
4 Drain pipe: The drain pipe shall be sized on its maximum allowable velocity of 0.9 mps.

Maintenance of fountain

Water fountains, outdoor fountains in particular, must go under regular maintenance. With the following routine maintenance measures, fountains can be run smoothly without having major trouble with system failure:

1 Cleaning basin: Water fountains need regular cleaning. Generally, the inside of the basin wall shall be cleaned to remove attracting dirt and dust. Scrub away any lime-scale or algae growth with a soft or hard bristled brush, using some detergent. Such cleaning is to be done to empty the basin every six-month interval at least.
2 Cleaning pump: The fountain pump should be regularly drained, cleaned, and refilled. There must not be any blockages or debris build up in the strainer basket and the pipes. Pumps should be cleaned out at least once a year.
3 Prevent algae: To stop unsightly algae building up in the basin, algae prevention solution can be added to the water. To avoid algae growth, drain out basin water and refill with fresh water, adding in an algae repellent every month.
4 Cleaning water: To keep water clean, circulate pool water through an ideal filtering unit.

References

[1] Shutterstock, Inc. Royalty-free image. Stock Illustration ID: 1700631700 Retrieved from https://www.shutterstock.com/image-illustration/3d-rendering-interior-design-fountain-isolated-1700631700 on 15 September 2024
[2] www.fibigreen.com "Misting nozzles" Retrieved from https://www.fibigreen.com/en/product-585332/Complete-Stainless-Steel-High-Pressure-Misting-Nozzles-3001-SS.html on 13 September, 2024
[3] Crystal Fountains Inc. (2002) "How to build a fountain" A brief technical overview of simple fountain design, p. 11, Ontario. Retrieved from http://www.vodne-fontane.si/f/docs/Manjse_fontane/Fountains_101.pdf on 02 June, 2025
[4] The Fountain People (1999) "Fountain design guide" Texas Retrieved from https://cms.esi.info/Media/documents/SSPFo_designguide_ML.pdf on 9 May 2024.
[5] Best Planet (2024) "Pump sizing for overflowing fountains" Retrieved from https://best-planet.com/pages/pump-sizing-for-overflowing-fountains on 14 September 2024.
[6] Water Garden Ltd (2024) "Water blade installation guide" Retrieved from https://www.water-garden.co.uk/tech/product-selection-guides/water-blade-guide on 14 September 2024.

7 Sprinkler irrigation system

Introduction

A land or soil is needed to be irrigated for cultivating or even maintaining it covered by green grass. For small pieces of land, irrigation by water can be done manually, but for very large areas of land, manual irrigation by water may not be effective because it is very labor intensive and time-consuming, where one or more of the various water conveying and distribution systems are adopted. Irrigation by sprinkling water using pumps, pipes, and accessories in various ways is one of the most efficient irrigation systems for particular land cultivating or maintaining particular greeneries. Sprinkler irrigation systems are classified as mobile, semi-mobile, or fixed based on the degree of mobility and permanence of their components, especially how easily and frequently the pipes, pumps, sprinklers, and other major components can be moved or relocated. A fixed sprinkler irrigation system involves items and works that fall under the purview of plumbing and may be considered of a special nature. So, in this chapter, among all the irrigation systems, only the fixed-type sprinkler irrigation system will be discussed elaborately.

Irrigation

Irrigation is the artificial method of applying water uniformly to the soil, generally for the purpose of cultivation or to increase the average yield and quality of certain crops, plants, or even grass, adopting various systems. Irrigation is necessary in areas where irregular rainfall or dry times or drought is expected.

To irrigate, the application of water throughout the land can be done naturally, manually, and mechanically, depending upon the source of water, land size, and its topography. Sprinkler irrigation system is a mechanical way of applying water, generally adopted for very large and having farmable slope, whether uniform or undulating.

Sprinkler irrigation

Sprinkler irrigation is a method of applying irrigation water on the ground in a controlled manner and in the form of fine droplets, more or less similar to natural rainfall. Generally, in sprinkler irrigation, water is sprayed by high-pressure sprinklers or guns from a central location in the field or parts of a large field.

DOI: 10.1201/9781003724056-7

Applicability

Sprinkler irrigation systems are generally recommended for sandy soil with high infiltration rates, although they can be planned for other various soil types. Sprinklers are not suitable for soils that might form a crust or in soils where there is a risk of salinization. Moreover, it should not be adopted where there is a chance of clogging due to the presence of sediments or debris. In these situations, if sprinkler irrigation is unavoidable, then to minimize the negative implications, very light and fine spray of water should be maintained in the sprinkler irrigation system.

Types of sprinkler irrigation systems

Sprinkler irrigation systems may be of the following three types, as discussed below:

1 Mobile system
2 Semi-mobile system
3 Fixed system

Mobile system

In this system, the motor and pump unit, the main pipe, and the laterals can be shifted. This system is widely used for crop cultivation and nurseries of irregular-pattern land areas with diverse natures of plants.

Semi-mobile system

In this system, the motor and pump unit and the main pipe are kept fixed in particular positions and locations while the laterals are made movable. This system is widely used for irrigating comparatively larger plants.

Fixed system

In this system, the motor and pump unit, main pipe, and laterals are installed fixed in particular positions and locations. This system is widely used for frost protection and as a permanent system for crop cultivation in large fields. Figure 7.1 represents a fixed-type sprinkler irrigation system in which the major elements of the system have been shown.

Sprinkler irrigation system components

A typical sprinkler irrigation system is developed incorporating the following main components:

1 Source of water
2 Pump
3 Piping
4 Sprinklers

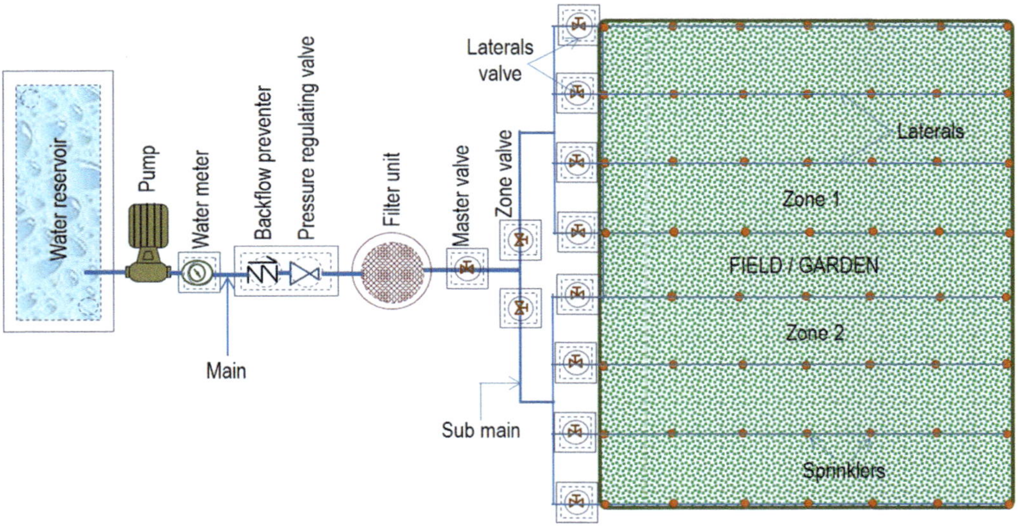

Figure 7.1 Schematic diagram of a simplified fixed-type sprinkler irrigation system.

Accessories and appurtenances

In addition to the main components, the following accessories, appurtenances, and controllers are incorporated for better and more efficient performance of the sprinkler irrigation system:

1 Water meter
2 Pressure gauge
3 Gate valve
4 Check valve
5 Pressure-regulating valve
6 Coupler

Source of water

The most common source of water for any irrigation of large land areas are the very common surface water bodies like ponds, lakes, canals, or rivers that are found available near the land and have potential in terms of quality and quantity.

If the surface water sources are not found at a cost-effective distance to collect, then ground water is considered the second option as a source of water for irrigation when its availability and abstraction are found to be reasonably cost-effective in comparison to other optional sources and permissible in the local context.

When both the surface water and groundwater sources are found to be either limited or costly, the third option is envisaged, which is mostly rainwater if it is found to be sufficient and regular for the concerned area.

Based on the condition of the source of water, there might be a need for a water reservoir to store water for this purpose. In those cases, water in a constructed reservoir is considered the source of water for irrigation.

Pump

A pumping unit is required to draw water from the source of water and convey the water under pressure through a pipe network to spray water to various locations of the field to irrigate. The pressure forces the water through sprinkler heads installed on pipelines, which form a spray of water around them. A constant-speed centrifugal pump is used for this purpose in individual fields. The driving unit for the pumps may be either an electric motor or an internal combustion engine.

Based on features and performance, various types of centrifugal pumps are designed; among those, the following four types are found in sprinkler irrigation systems:

1 Centrifugal pump
2 Vertical turbine or diffuser pump
3 Propeller pump
4 Submersible pump

Centrifugal pump

A centrifugal pump, as shown in Figure 7.2, is preferred when it is to be mounted on a surface and when water needs to be drawn from reservoirs, surface water bodies, or shallow wells, and the required suction head of the pump from the deepest water surface is comparatively low. Though theoretically, the maximum suction depth is 10.33 m, practically it can be limited to 3–6 m. For pumping water from deep tube wells, submersible deep well pumps are used.

Advantages of centrifugal pumps

1 Highly efficient over a range of operating conditions
2 Simple, economical, and adaptable to many situations
3 Electricity, diesel engines, or tractor-mounted power can be used
4 Does not overload within the increased total operating head (TOH)
5 Vertical centrifugal pumps may be submerged and need no priming
6 Installation is easy

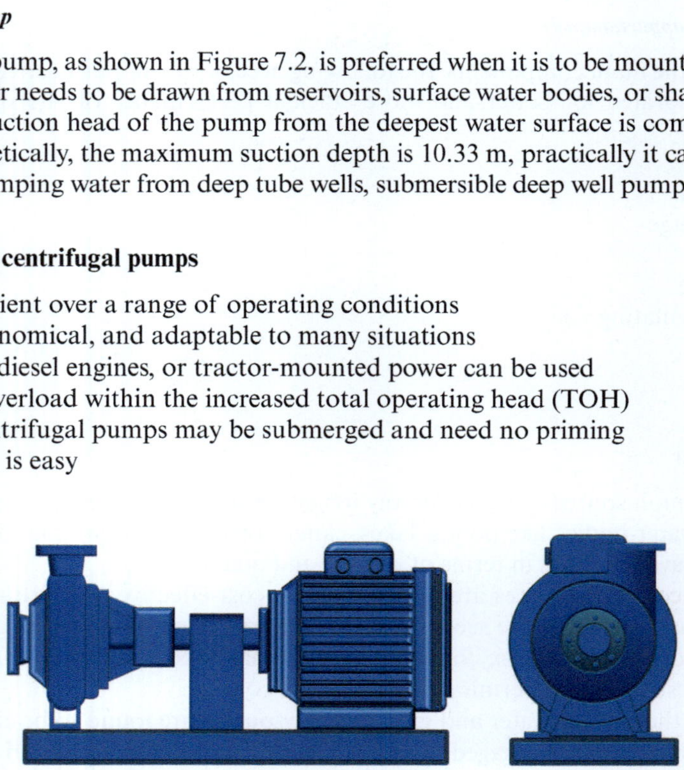

a. Side view   b. Front view.

Figure 7.2 A centrifugal pump.

Disadvantages of centrifugal pumps

1 Operating suction lift is restricted; it is recommended to be below 3 m.
2 Priming is required when installed over a water source.
3 Loss of prime can cause cavitation, thereby damaging the pump.
4 The motor is overloading when the head is much lower than the design value.

Deep-well turbine pump

These pumps are preferred for use in cased deep wells or where the water level is below the practical limits of a centrifugal pump. Turbine pumps are also used to draw water from surface water bodies.

The turbine pump has three main components: head assembly, shaft and column assembly, and pump bowl assembly, as shown in Figure 7.3. The pump motor is generally found installed on a concrete base on a surface and connected to the pump assembly, which remains submerged in water.

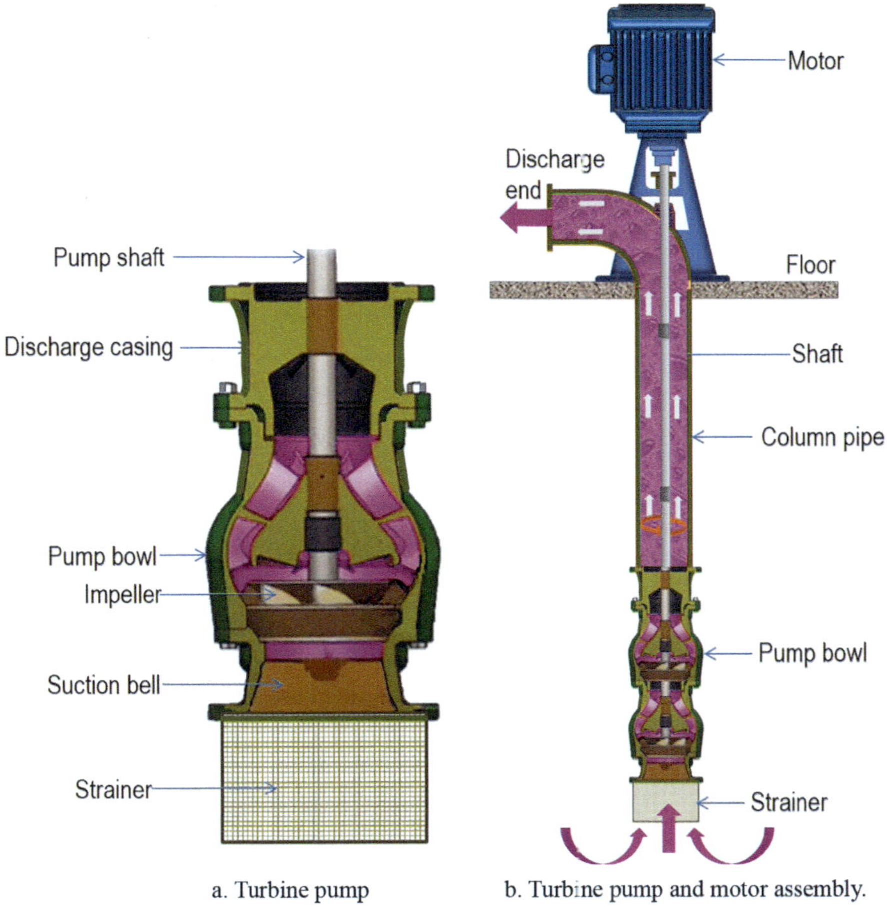

Figure 7.3 A deep-well turbine pump.

Advantages of turbine pumps

1 It is adapted for use in tube wells, so no floor space is required.
2 At high efficiency, it provides high head and flow rates.
3 Electric or diesel engine power can be used.
4 Priming is not needed.
5 It can be used where water level fluctuation is comparatively high.
6 Easily modifiable subject to changed hydraulic conditions.

Disadvantages of turbine pumps

1 Comparatively difficult to install, inspect, and repair
2 Higher initial cost of investment than required for a centrifugal pump
3 Impellers are required to be adjusted periodically, to maintain high efficiency
4 Repair and maintenance are more expensive than that for centrifugal pumps

Propeller pump

Propeller pumps are basically axial flow and mixed flow types of centrifugal pumps, used generally for large flow rates of water at low heads generally ranging between 3 and 5 meters. The impeller consists of a central boss with a number of blades mounted on it. The impeller is rotated within a cylindrical casing with fine clearance between the blade tips and the casing walls, as shown in Figure 7.4. Impeller is driven by means of a standard motor. Its impeller is placed completely submerged in water to be drawn.

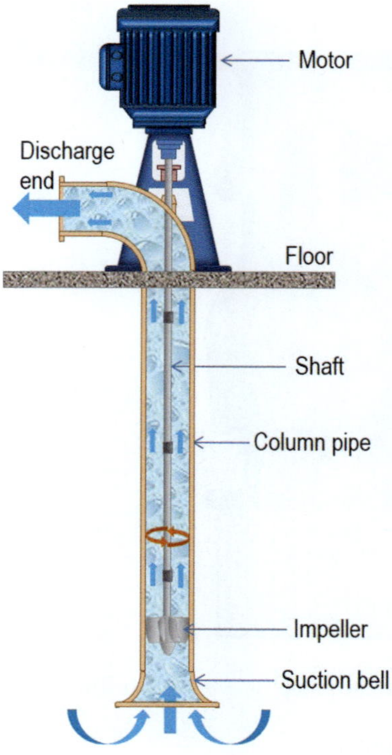

Figure 7.4 A deep well propeller pump.

Advantages of propeller pumps

1 It is simple in construction and so in maintenance.
2 It can pump coarse sand.
3 Priming is not required.
4 It is efficient for large flow rates and low-head operations.
5 Both electric and diesel engines can be used to drive the pump.
6 It is easy to install; and therefore, it is suitable for portable operation.

Disadvantages of propeller pumps

1 Not operable for suction lift
2 Cannot be valved to reduce flow rate
3 Intake submergence depth is very critical
4 Limited to low TOH which is less than 25 m

Submersible pump

A submersible pump is a centrifugal pump close-coupled to a submersible electric motor as shown in Figure 7.5, with both pump and motor remaining submerged in the water of a well, reservoir, or water body. Submersible motors may be designed as dry or wet motors.

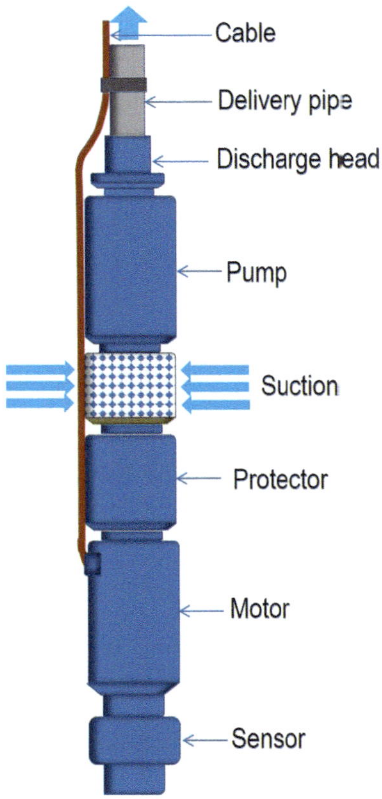

Figure 7.5 A deep tube well submersible pump.

Dry motors are hermetically sealed with high-dielectric oil to exclude water from entering the motor; conversely, wet motors are somehow exposed to the well water, allowing the rotor and bearings of the motor to operate in the water. Submersible pumps are basically vertical-type pumps, so they should be installed vertically, particularly in reservoirs. Water-lubricated submersible pumps are to be used for irrigation purposes.

Advantages of submersible pumps

1 Suitable for use in deep wells and tube wells
2 Priming is not required
3 Can be used in crooked wells
4 Very easy to install
5 Smaller diameters are less expensive than comparable-sized vertical turbine pumps

Disadvantages of submersible pumps

1 Larger sizes are more expensive than deep well vertical turbine pumps.
2 Only electric power is used.
3 These pumps are more susceptible to lightning.
4 A leak-proof motor is required.

Selection of pump

There is a wide range of pumps available for use in irrigation purposes. In some specific condition, there might be a need for special pump requirements, but in general, there are some common considerations that should be addressed in the selection of an appropriate pump. The common considerations are as follows:

1 Irrigation system
2 Source of water and water level
3 Discharge and pressure (or head) required
4 Suction conditions
5 Source of power available
6 Operation cost
7 Potential for corrosion and wear
8 Facilities for maintenance

Based on the required pump capacity in terms of flow requirements for a particular operating head, the preferred pump can be selected from the values furnished in Table 7.1.

Piping

The piping system consists of main line, sub-mains, laterals, and risers. The main line delivers water being driven by a pump drawing water from the source. Then it distributes water to the sub-mains. The sub-mains convey water to the laterals, which in turn supply water through the sprinkler heads on the riser pipes installed on the laterals.

Table 7.1 Preferred pump types for sprinkler irrigation, based on range of flow rates and the delivery heads

Sl. no.	*Flow (lpm)*	*Total dynamic head (m)*		
		15 or less	*16–152*	*153 or more*
1	0–1135	Propeller Centrifugal	Centrifugal Vertical turbine Vertical turbine Submersible	Centrifugal Vertical turbine Submersible
2	1136–18927	Propeller	Centrifugal Vertical turbine Submersible Vertical turbine submersible Vertical turbine submersible	Centrifugal Vertical turbine Vertical turbine Submersible
3	18928 or more	Propeller	Centrifugal Vertical turbine Propeller Submersible Vertical turbine submersible Vertical turbine submersible	Centrifugal Vertical turbine

Converted to SI unit from source [1].

Aluminum or PVC pipes are generally used for portable and mobile sprinkler systems, while steel pipes are usually used for center-pivot laterals. PVC and wrapped steel pipes are usually used for fixed-type buried irrigation piping systems.

Main: The main pipe is that which originates from the pumping system and distributes water throughout the sub-mains and lateral distribution piping of an irrigation piping network.

Sub-main: The sub-mains are the portion of a pipe network that is designed to convey irrigation water from the main line to the laterals on which the irrigation sprinklers are installed. Sub-mains are found smaller than the mains but bigger than the laterals.

Lateral: Laterals are the horizontal pipework that carries irrigation water to the risers supporting the sprinkler heads or any outlet faucet.

Riser: Risers are the segment of piping that is required for connecting the sprinkler heads or drip lines to the lateral pipes in order for the irrigation system to work. Risers are mostly installed vertical.

Irrigation sprinkler head

Irrigation sprinkler head causes flowing water through it into a spraying of water uniformly over the field without causing runoff or loss of water due to deep percolation. Though there are various types of sprinkler heads available, there are three basic categories of sprinkler heads, as mentioned below. Each of these categories again includes several different variations:

1 Fixed heads
2 Rotary heads
3 Bubblers

Some of the different variants are as follows:

1 Micro or misting
2 Pop-up

Fixed head sprinkler

The head of the sprinkler, which sprinkles water, remains fixed in its position when in operation. Fixed-head sprinklers are generally used in irrigating small land areas. These require comparatively less pressure to operate; so, coverage area per sprinkler head is also smaller.

Fixed-head sprinklers release comparatively more water per unit area than rotating sprinklers. Hence, fixed-head sprinklers are the good choice for soils with a high percolation rate.

Sprinkler heads are also designed for part-circle coverage, as shown in Figure 7.6, for placing at particular locations of the land area, as mentioned below:

1 Quarter circle heads: for placing at the corners of the land area. The horizontal angle of spraying is around 90 degrees.
2 Half circle heads: For placing between corners, i.e., on the sides of a land area, following recommended equidistant spacing. The horizontal angle of spraying is around 180 degrees.
3 Full circle heads: For placing each perimeter within the land area, the number will depend on the perimeter's overall dimensions requiring covering the land area. The horizontal angle of spraying is around 360 degrees.

Rotary head sprinkler

There are sprinkler heads that can rotate around its axis. The head of the sprinkler starts rotating when in operation. Rotary head sprinklers are commonly used to irrigate comparatively large land areas. These require comparatively high pressure to operate; so, coverage

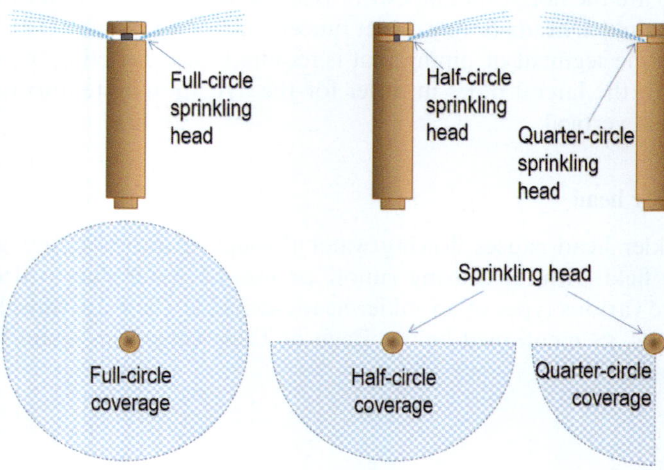

Figure 7.6 Fixed spray irrigation sprinkler heads and their coverage.

area per sprinkler head is also larger. The heads rotate in a circle delivering a single stream of water. It is the oldest type of sprinkler head still in use.

These sprinkler heads are progressively rotated by making the oscillating impact on an oscillating arm with a counterweight on one end and a drive spoon on the other that repeatedly engages the flow-stream from the nozzle. This flow-stream from the nozzle momentarily strikes the body of the sprinkler head, causing the whole body to rotate each time.

The rotary-type heads can be selected for a wide range of flow rates, applications, and spacing depending upon their size and pressure range. This type of sprinkler is available in sizes 12 mm up to 32mm threaded inlet sizes with nozzle flow rates between 11.35 lpm and 473 lpm at operating pressures between 207 and 552 kPa. These rotary sprinklers usually cover a 10–20 m radius with an application rate between 5 and 12.5 mm per hour [2]. There are smaller rotary heads that can cover distances of 5–12 m (Figure 7.7).

Bubbler

These sprinklers are designed to satisfy water demand for particular landscapes, as shown in Figure 7.8, generally at side or end strips of land area, planter boxes, etc. Bubbler-type heads are used for delivering a considerably huge amount of water at a fast rate, in small areas like tree basins or around shrubs, etc. These heads work by flooding water to the ground and around the vegetation's root zone instead of spraying the foliage. Spray nozzles deliver water to areas that are up to 1.5 m wide or less. Bubblers are commonly known as spot sprinklers.

Bubblers are mostly small water emitters that discharge water at flow rates of 100–250 lph at operating pressures of 1.0–3.0 bar. Each basin for a tree can have two bubblers if required. Adopting bubbler irrigation application, efficiency of up to 75 percent can be achieved [3].

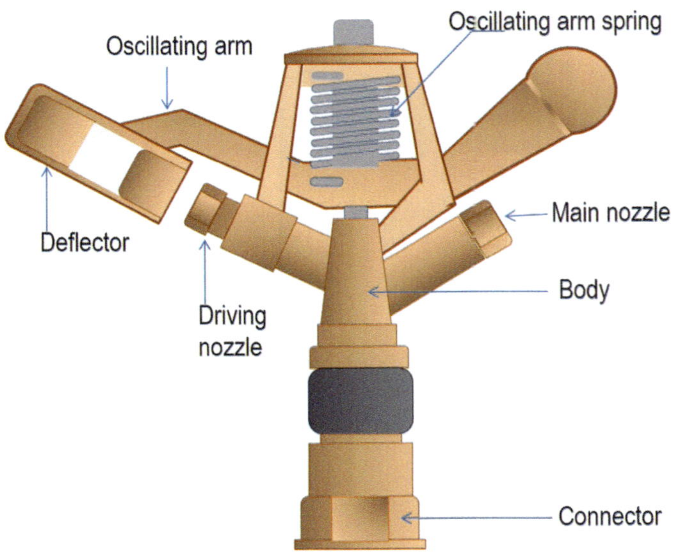

Figure 7.7 A rotating irrigation sprinkler head.

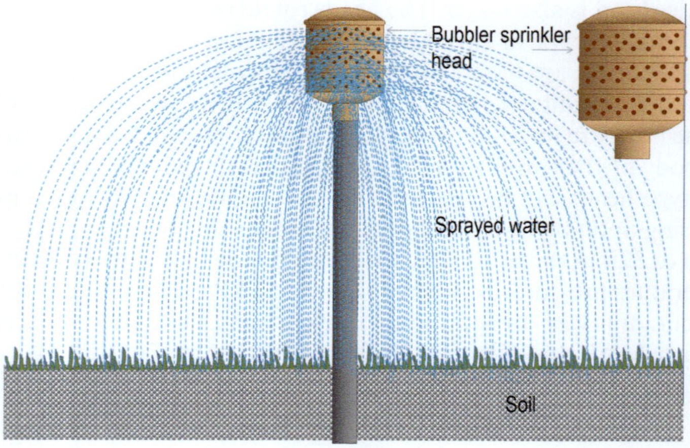

Figure 7.8 Watering by a bubbler.

Spot sprinkler

Spot sprinkler heads are preferable for isolation watering in plant beds. These are basically small sprayers, perfect for targeting individual trees or plants over others.

Pop-up sprinkler

Within the groups of fixed and rotary heads there are pop-up sprinklers, which are installed even with the grade, and riser-mounted shrub heads, which are installed above grade.

The pop-up sprinklers and the pipework are all installed buried underground as shown in Figure 7.9, keeping its top at soil level and hidden from view, thus eliminating obstruction and providing a great aesthetic in the landscape.

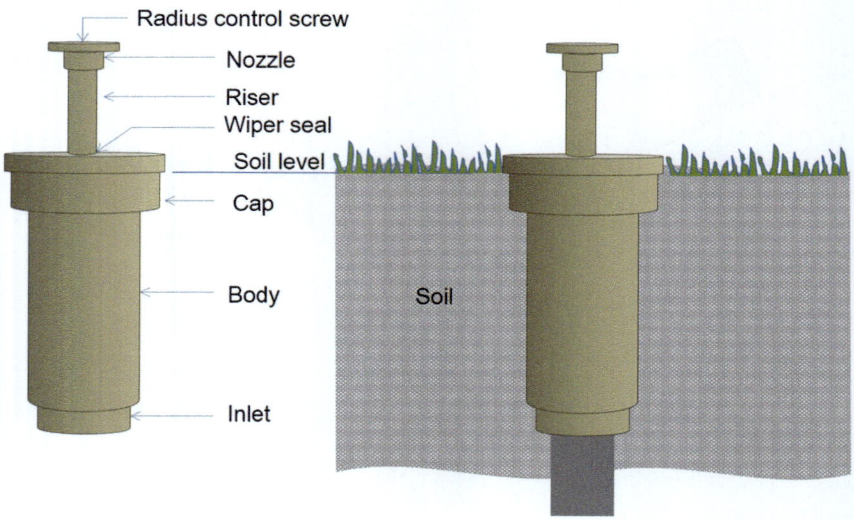

Figure 7.9 A pop-up irrigation sprinkler head.

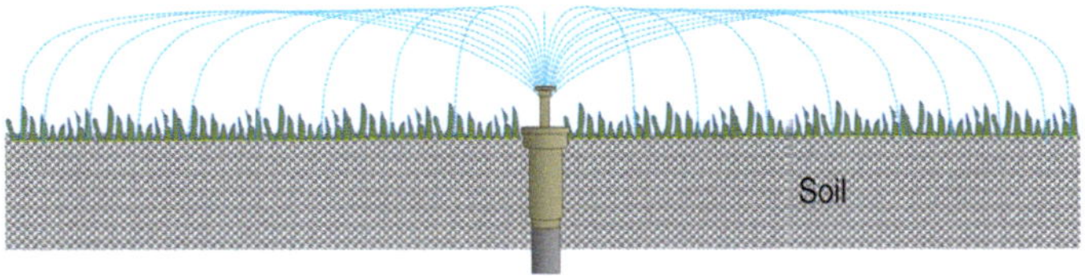

a. Fixed spray irrigation sprinkler.

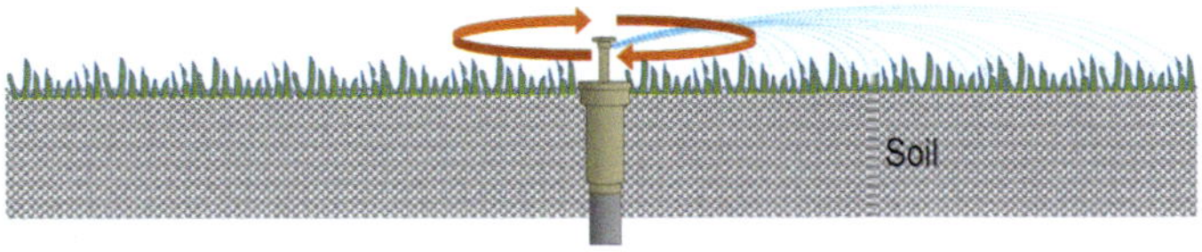

b. Rotating spray irrigation sprinkler.

Figure 7.10 Sprinkler heads installed above ground: a. fixed type and b. rotor type.

In a pop-up style sprinkler, there is a piston that contains the nozzle that lifts up from the sprinkler body when the sprinkler is activated and retracts back when the system is turned off. There is a check valve assembly inside the body that prevents water from draining out from the sprinkler, the lowest sprinkler of a zone or lateral in particular. This helps activate the sprinklers faster whenever they need to be turned on and also helps avoid issues with pooling and leaking water around the sprinkler heads, thereby saving the landscape from overwatering damage.

Pop-up sprinklers are mostly chosen for irrigating the grass of large sports fields. These are also preferred for shrub areas of gardens. An important benefit of choosing a pop-up sprinkler system over a traditional lawn sprinkler system is the considerable reduction in water use, as the pop-up sprinklers, when installed and function properly, deliver water precisely and efficiently, eliminating wastage of water. Pop-up sprinklers may be a fixed spraying type or a rotating spraying type, as shown in Figure 7.10.

Sprinkler sizes

When water is sprayed out from a sprinkler head, it breaks up into droplets of sizes between 0.5 and 4.0 mm [4]. The small water drops fall close around to the sprinkler, whereas the larger water drops fall near to the edge of the wet sprayed circle, as shown in Figure 7.11. Large drops may damage delicate plants and soils, and so in such conditions it is best to use the smaller sprinklers.

Drop size can be controlled by pressure and nozzle size. When the pressure is low, droplets tend to be much larger as the water jet does not break up easily. So, small-diameter nozzles operating at or above the normal recommended operating pressure are to be used to avoid plant and soil damage. The flow rate of particular diameter irrigation sprinkler nozzles at different pressure levels has been furnished in Table 7.2.

Sprinklers arrangement pattern

The effective positioning of the sprinklers to irrigate the required land is one of the most vital concerns for irrigation. There are two popular patterns: the square pattern and the triangular pattern are in practice, as shown in Figure 7.12.

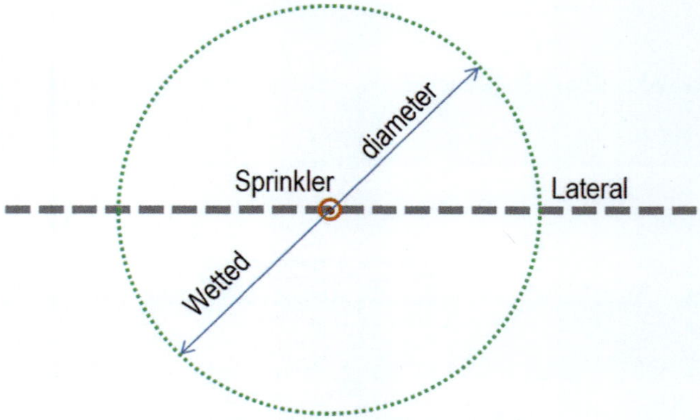

Figure 7.11 Wetting pattern for a single sprinkler (top view).

Table 7.2 Sprinkler flow rate with respect to pressure

Sl.no	*Sprinkler system design capacity*							
	Static pressure	*Bar*	*2*	*2.76*	*3.45*	*4*	*4.83*	*5.52*
	Water meter mm	*Service line*	*Max. lpm*	*Max. lpm*	*Max. lpm*	*Max. lpm*	*Max. lpm*	*Max. lpm*
1	12	12	7.57	15.14	18.93	22.71	26.50	26.50
		19	15.14	22.71	30.28	30.28	37.85	45.42
		25	15.14	26.50	30.28	37.85	49.21	56.78
2	19	19	15.14	22.71	30.28	34.07	37.85	45.42
		25	18.93	26.50	37.85	53.00	64.35	75.71
		32	18.93	45.42	64.35	75.71	83.28	83.28
3	25	19	15.14	26.50	30.28	34.07	45.42	45.42
		25	18.93	30.28	53.00	68.14	75.71	75.71
		32	18.93	53.00	90.85	98.42	113.56	128.70
Working pressure		Bar	1.724	2.0	2.413	3.103	3.45	3.792

SI conversion from source [5].

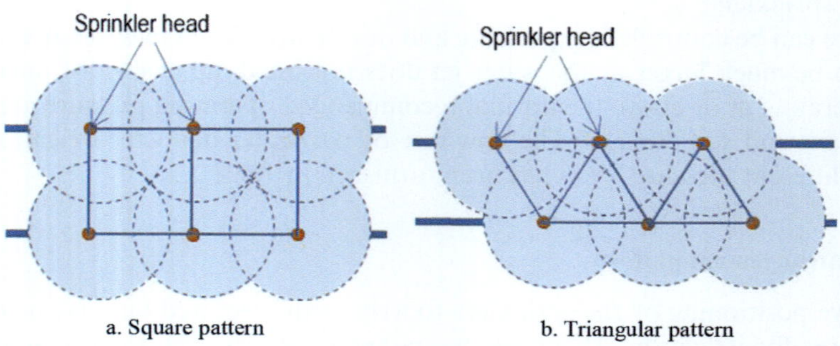

Figure 7.12 Sprinkler arrangement patterns.

Square pattern

In the square pattern, sprinklers are positioned in such a way that a group of four of them forms a square, which is placed parallel in each row.

In rectangular land, the square pattern is easier to plan, but it requires more sprinkler than required in another triangular pattern. In the square pattern, the spacing between the sprinklers is 50 percent of the sprinkler-throw diameter, whereas in the triangular pattern, the sprinkler spacing can be made up to 60 percent of the sprinkler-throw diameter.

Triangular pattern

In the triangular pattern, sprinklers are positioned in such a way that a group of three of them forms equilateral triangles, which are placed, offset in each row.

The triangular pattern is considered to be the most efficient pattern in terms of water consumption, but the placement of the sprinkler may be difficult. This pattern is popular for non-rectangular areas, especially for circular areas. In a triangular pattern, the sprinklers can be spaced slightly further apart than in the case with the square pattern.

Accessories and appurtenances

A good number of accessories and appurtenances are used in sprinkler irrigation systems. Among those following are some major items:

1 Valves
2 Pressure gauge
3 Quick coupler
4 Water meter
5 Timer controller
6 Drain

Valve

Valves are kinds of accessories in any water supply piping, including irrigation piping, that are provided to control the flow of water through the system as they open and close to perform as intended. Various types of valves are used in sprinkler irrigation systems. The following valves are most common:

1 Gate valve
2 Globe valve
3 Check valve
4 Backflow preventer
5 Pressure regulating valve

Gate valves

Gate valves are used to start and stop the supply of water to the sprinklers. They also help isolating a part or zone from another part or zone of the irrigation system.

Globe valves

Globe valves are used to reduce the flow of water by throttling the flow by the valve. It can also be used to start and stop the supply of water to the sprinklers.

Check valve

A check valve in sprinkler irrigation is installed to stop water from draining out when the sprinkler zone is cut off for any reason, which will cause wastage of huge amounts of water remaining in the piping network.

Backflow preventer

If any sprinkler irrigation draws water from a potable water supply system, then the potable water supply system must be protected against backflow of non-potable irrigation water by using an atmospheric-type vacuum breaker, a pressure-type vacuum breaker, or a reduced-pressure-principle-type backflow preventer. At downstream from an atmospheric vacuum breaker, no valve shall be installed. Where there is possibility of chemical contamination, the potable water supply shall be protected by a reduced pressure principle backflow preventer [6].

Pressure-regulating valve

A pressure-regulating valve controls pressure regardless of variation in flow in a pipe. A pressure-regulating valve is primarily used in sprinkler irrigation to maintain high distribution uniformity by reducing pressure fluctuation in the system. There is less water requirement for irrigation where higher distribution uniformity is maintained, thereby resulting in lower pumping costs, minimizing the risk of chemicals' leaching, and reduced runoff. Uncontrolled pressure fluctuations into sprinklers result in flow rate deviations and spraying pattern distortions. So, the total pressure variation in a sprinkler irrigation system should not exceed 20 percent; if pressure difference is more than 20 percent, pressure-regulating valve is to be used for its management [7].

Valve box

A valve box or pit with operable cover is constructed or installed to house the valves installed underground, which provides protection and easy access to the valves for operation or maintenance.

Pressure gauge

It is necessary to know whether the sprinkler irrigation system is running under the desired pressure to ensure water distribution uniformity. To know the working pressure of the system, a pressure gauge is installed. It is recommended to monitor pressure at some critical locations in a sprinkler irrigation system, such as the pump outlet, both sides of the filtration system, and at the inlet to each irrigation zone.

Quick coupler

Quick couplers are the devices used for connecting two pipes and uncoupling quickly and easily. A quick coupler essentially possesses the following special features:

1 Remarkable flexible connection
2 Leak-proof joint
3 Simple and easy coupling and uncoupling
4 Light, non-corrosive, durable jointer

Water meter

They are used to measure the volume of water delivered to the irrigation field. This is necessary to operate the system for a period to provide the required quantity of water. To monitor water flow in irrigation, water meters play an important role, which should cause low head loss even while addressing water with a high content of impurities.

Timer controller

With the help of a timer, an irrigation sprinkler system is designed and scheduled to determine when, how often, and how long each valve of a particular irrigation zone is to be kept open to supply water. It controls how much water goes where and when in the landscape based upon the design of the system. For these purposes, a timer controller is programmed. The more programming flexibility the controller will have, the more efficient water distribution programming can be done for optimum use of water in irrigating a landscape.

Drain

The unused water or the remaining water in irrigation pipes, the laterals in particular, should be drained out. To facilitate draining of such unused water, manual or automatic drain-cocks are installed at the end of sprinkler pipes and at any low points in the piping. Draining water from pipes helps it keep it free from being damaged by corrosion.

Material of items

The present conventional understanding is that any metals of the best quality are more durable than any best quality plastic, and so, metal products being more durable are costlier than plastic items. Up until the late 1970's, metals like brass, steel, zinc, etc. were the standard common materials from which almost all sprinkler items were made. However, times have changed humans mindset and attitude, mostly governed by money, which diverted them to go for plastic as the most common material for almost all the items and so for sprinklers. Manufacturers are now reluctant to make an all-metal sprinkler anymore as it is not cost-effective. Any machine-made metal parts are enormously expensive in comparison to injection-molded similar plastic items. Fortunately, most of today's plastic items are very well engineered and so perform better than the old metal items. Some manufacturers make some plastic sprinkler items that embed metals at critical parts or locations for longevity.

Planning sprinkler system

The most important job of developing a sprinkler irrigation system is to plan the system first. To make better planning, the following steps are to be followed:

1 At the onset of planning, it needs to have an actual sketch to scale of the area where the system is to be installed. In the drawing there must be mention of trees, flower beds, or other vegetation, as it requires different water requirements than turf-grass, in addition to marking hardscapes like walkways, driveways, hard courts, etc. where irrigation is not required.
2 The next job is to mark on the map the area where the piping and the control box for the system could be placed.
3 Designate zones according to the water requirements for the specific vegetation growing, like turf-grass, shrubs, trees, etc., as each type has different water requirements.
4 Taking into consideration any obstructions like a slope, mark on the map the proposed position of the sprinkler heads based on the capacity of heads to be chosen, following some basic guidelines in this respect, as mentioned below:

   i Sprinklers of the same capacity and specification should be chosen for particular hydro-zones of the whole irrigation system.
   ii Different sprinkler heads should be chosen to meet the watering needs of various hydro-zones of different natures of vegetation.
   iii For the area to be irrigated, the spraying pattern of each head must overlap the spraying area by its neighboring heads by 50 percent, which is recommended but must not be less than 65 percent of the wetted diameter.
   iv Place quarter-circle heads at the corners, half-circle heads between corners, and full-circle heads within each perimeter, as shown in Figure 7.13. Following recommended equal distant spacing, subject to the capacity of the heads chosen.

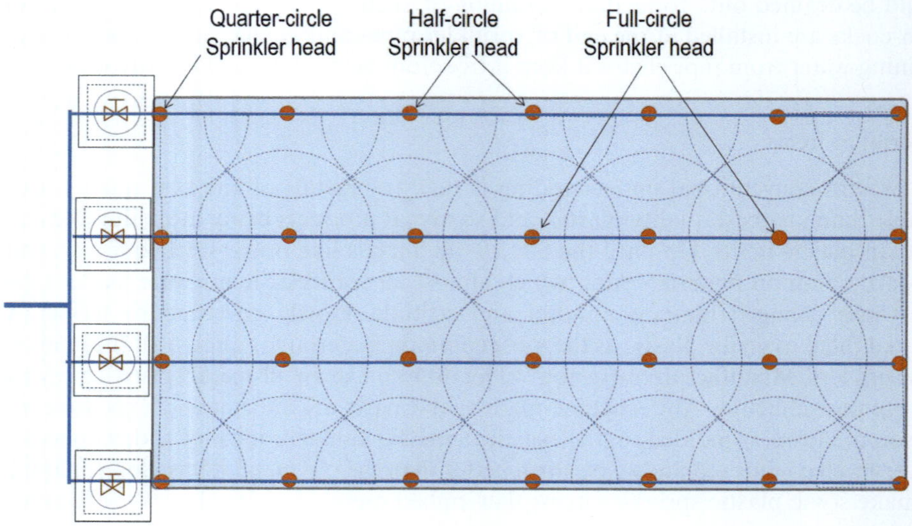

Figure 7.13 Layout of irrigation sprinkler heads and their spraying circle.

v On sloping land there may be considerable pressure differences between sprinkler heads on high and low ends of the line, depending on gradient of slope, and so there will be varying spray rate among those spray heads. In this situation, flow control spray nozzles should be used to ensure the uniformity of water application. By flow control nozzles, water can be sprayed at the same rate, regardless of water pressure.

5 For large lands, enough water may not be available to irrigate the entire land at a time; then, the area can be divided into zones, which may be called hydro-zones.
6 Group up the similar sprinkler heads into clusters of different circuits each controlled by one gate valve, i.e., every zone must have its own gate valve to on or off the flow of water to a zone. Otherwise, there might not be enough pressure to operate all the sprinklers at a time. For very large areas, a number of valves are grouped together in an assembly, termed a valve manifold that can then be attached to an irrigation controller placed at an easily accessible location for easy maintenance. The manifold should be placed close to the area of the valves, which is not sprayed when activating the system manually. Then the pipes, fittings, sprinkler heads, and valves are organized to work together.
7 Sketch the irrigation system on paper, showing the exact location of the sprinklers, zones of different sprinklers, and distance between the various parts of the sprinkler system, such as valves, controllers, piping, manifolds, etc. It is good practice to draw the spraying circles of the sprinkler heads while preparing the drawing during planning so that the whole lawn is ensured watered. In final drawings for execution of the work, the spraying circles should be deleted.

Designing sprinkler system components

The sprinkler irrigation system must be designed in such a way that it ensures application of water uniformly without wastage from runoff or erosion of soil. The application rate of water by the sprinkler system must be compatible with the infiltration rate of the most restrictive soil in the field. A higher application rate than the soil intake rate will cause runoff from the field, and lower application is never acceptable to attain the objective of irrigation. So, a well-designed sprinkler system helps provide water uniformly to the soil surface and apply sufficient water to meet the peak demands of the plants without producing runoff.

The sprinkler system capacity is the flow rate needed to adequately irrigate an area, which is expressed in liters per minute per acre (lpm/acre). The system capacity depends upon the following:

1 Peak water requirements during the plant growing season
2 Effective plant rooting depth
3 Texture and infiltration rate of the soil
4 Available water-holding capacity of the soil
5 Wind, air temperature, and humidity (day versus night)

The potentiality of the source of water must be sufficient enough to satisfy the pumping capacity to meet the sprinkler irrigation system capacity designed.

Precipitation rate

The precipitation rate (P_r) of the sprinklers is the flow rate at which water is delivered from a sprinkler head to an area covered by it, which is averaged as millimeters per hour (mm/h). The precipitation rate should be less than the soil percolation rate to avoid surface run-off. The precipitation rate is calculated using the following formula:

$$P_r = 1000 \times Q_n / A \tag{7.1}$$

where,

Q_n = Nozzle flow rate (cum/h)
A = Area covered (sqm) = $S \times L$ (sqm)
S = sprinkler spacing (m) and L = row spacing (m).

Application rate

The application rate (A_r) is the average rate at which water is stored in the soil, measured in mm/h. The application rate depends on sprinkler application efficiency based on the size of sprinkler nozzles, the operating pressure, and the distance between sprinklers, which varies from 60–80 percent; an average value of 70 percent is considered reasonable. The average application rate should be less than the basic infiltration rate of the soil. The application rate is as follows:

$$A_r = 0.7 P_r \tag{7.2}$$

Slope of land

A correction of the precipitation rate is recommended in order to avoid runoff in sloping land. Values furnished in Table 7.3 are commonly used to assess the maximum sprinkler application rates based on average soil, slope, and tilt.

Table 7.3 Suggested maximum application rate for average soil, slope, and tilt [8]

Soil texture and profile	*Slope*			
	5%–0	*8%–5*	*12%–8*	*16%–12*
	Maximum application rate			
	mm/hr	*mm/hr*	*mm/hr*	*mm/hr*
Coarse sandy soil to 1.8 m	50	38	25	13
Coarse sand soils over more compact soils	38	25	19	10
Light sandy loams to 1.8 m	25	20	15	10
Light sandy loams over more compact soils	19	13	10	8
Silt loams to 1.8 m	13	10	8	5
Silt loams over more compact soils	8	6	4	2.5
Heavy textured clays or clay loams	4	2.5	2	1.5

Nozzle size and pressure

Sprinkler application rate varies considerably with respect to size of the sprinkler nozzle and pressure available at its inlet. Sprinkler application rate with respect to size of the sprinkler nozzle and available pressure have been furnished in Table 7.2.

Irrigation duration

The duration of irrigation needed to store sufficient water in the plant root zone is as follows. For this purpose, the water-holding capacities of various types of soils and typical crop rooting depths have been furnished in Table 7.4.

$$I_d = I_r / A_r \tag{7.3}$$

where,

I_d = Irrigation duration (h)
I_r = Plant irrigation requirements (mm)
A_r = Application rate (mm/h)

Wind effect

In strong wind, droplets of sprinkled water travel long through the air; so, evaporates more. Evaporation increases if the droplet remains airborne for an appreciable period. Wind prolongs the time that the droplet remains airborne. As droplets evaporate, they become lighter and travel further in the air, allowing further evaporation and causing more drift before reaching the intended area. So, irrigation under calm conditions should be practiced. In strong wind coverage patterns change, as shown in Figure 7.14, for which closer spacing of sprinklers is required than the spacing required for calm conditions. Based on wind velocity, the suggested reduction in spacing in percentage has been furnished in Table 7.5.

Table 7.4 Available water-holding capacity of various types of soil and typical crop rooting depths

Available water-holding capacity of soils		*Typical crop rooting depths*	
Soil texture	mm of available water per foot of moist soil	Crop	Typical active root zone depth, m
Sands and fine sands	12.5–18.75	Alfalfa	1.5
Very fine sands, loamy sand	20–25	Corn	1.2–1.5
Sandy loam	30–37.5	Small grains	0.9–1.2
Loam	48.25–50	Dry beans	0.9
Silt loam, silt	50	Pasture	0.45–0.75
Silty clay loam	48.25–50	Potatoes	0.6–1
Sandy clay loam, clay loam	43.18–50	Turf	0.3–0.6
		Vegetables	0.45–1

Allowable depletion to avoid crop water stress is usually about 50 percent of available water holding capacity for most field crops.

Converted to SI unit from source [9].

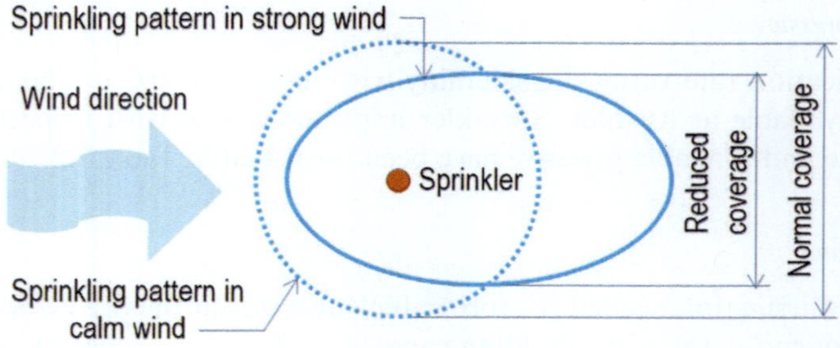

Figure 7.14 Sprinkler coverage pattern under strong wind.

Table 7.5 Maximum allowable spacing of irrigation sprinklers, based on wind speeds [10]

Sl. no.	*Average wind speed (km/hr)*	*Spacing as the percentage of the diameter of the water spread area of a sprinkler*
1	No wind	65
2	0–6.50	60
3	6.50–13.00	50
4	More than 13.00	30

Sprinkler head sizing

Sprinkler heads shall be sized based on the land coverage for irrigation per head. It should be minimum 65 percent and maximum 70 percent of maximum coverage by a sprinkler head. This is the maximum spacing between the sprinkler heads. Based on allowable diameter or radius, i.e., spacing of sprinklers, the size can be chosen from Table 7.2. The required pressure is to be maintained to get the expected flow of the sprinkler.

Verticality of the Sprinkler Head: If the sprinkler head is in an inclined position, it will create an acentric watering application. Thus, sprinkler heads must be kept vertical while operating.

Pipes designing

The pipes installed in a sprinkler irrigation system may be grouped into the following:

1 Riser pipe
2 Lateral pipe
3 Sub-main
4 Main

Riser pipe

The riser pipe size is provided as the same size of the sprinkler head chosen. If the sprinkler head size is 20 mm, then the riser pipe size should be 20 mm in diameter.

Lateral pipe

The lateral pipe is sized based on the flow requirement at a particular velocity of flow. The total number of heads connected to a pipe operating simultaneously is counted to determine the total flow requirement for that particular part of the pipe.

Begin the sizing of the pipe segments from one sprinkler head to the following one. Start calculating from the head farthest from the zone valve. The pipe connecting the last head to the second-to-last head should be the same size as the size of the sprinkler head chosen. The lateral pipe segments are to be sized based on the cumulative flow requirement of that segment of pipe for supplying the upstream sprinkler heads.

Sub-main pipe

Sub-main pipes of each separate zone are sized based on the number of sprinklers connected on the lateral pipes connected to that particular sub-main, operating simultaneously.

In this way, the sizing of pipe segments is continued until the segment of pipe connecting the main pipe is done. On the start pipe of a hydro-zone, an on-off valve of the same size is installed. This procedure is repeated for other zones.

Main pipe

The main pipe serves sub-mains to supply the manifolds. So, the water flow requirement for all manifolds is summed up to find the flow requirement of the sub-main pipe to serve and then to main. The corresponding required diameter of the main pipe can be determined from Table 7.6, corresponding to the flow requirement at velocity 1.5 mps. For economy, the main pipe size should be made of varied diameter from nodes to nodes of supplying pipes to the manifolds, as shown in Figure 7.15. The size of the last segment of the main pipe would be equal to the size of the sub-main pipe supplying the farthest manifold. The segment of main pipe following the farthest manifold supply pipe should be sized for the cumulative flow requirement for that farthest manifold. In this way, the first segment of the main pipe feeding water to all sub-mains shall be sized for the water requirement for a

Table 7.6 Pipe sizing table based on velocity limitation for Schedule 40 plastic pipes [12]

Nominal Size	*Flow at different velocity*			
	1.2 mps		*1.5 mps*	*1.8 mps*
Mm	*lps*	*lpm*	*lpm*	*lpm*
12.7	0.23	13.80	17.25	20.70
19	0.42	25.20	31.50	37.80
25.4	0.68	40.80	51.00	61.20
31.8	1.17	70.20	87.75	105.30
38.1	1.60	96.00	120.00	144.00
50.8	2.63	157.8	197.25	236.70
63.5	3.77	226.20	282.75	339.30
76.2	5.8	348.00	435.00	522.00
102	10.00	600.00	750.00	900.00

Friction loss p corresponding to the flow rate Q for piping having fairly smooth surface condition after extended service, from the formula $q = 4.57p^{0.546}d^{2.64}$.

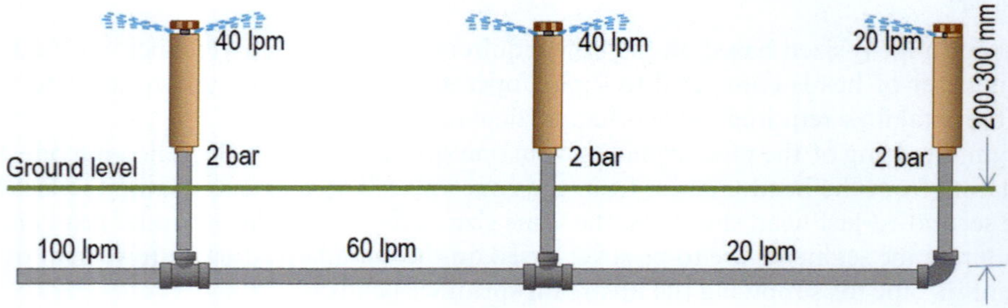

Figure 7.15 Cumulative flow in different sections of a lateral pipe.

manifold. It is advised to design the pipe for all the sprinklers operating at a time on a manifold serving the maximum number of sprinklers, if any. If it is planned that one or multiple manifolds among many manifolds will be operating at a time, then the main pipe should be sized for the water requirement of those manifolds to be operated at a time as planned, having a maximum water requirement only. In this way, the economic size of the main pipe can be determined.

Velocity of flow

Velocity of flow is the distance flown by water through a medium per unit time, which is expressed in meters per second (mps). Velocity indicates how fast the water is moving through a pipe. Higher velocity causes higher friction loss, resulting in pressure drop; again, lower velocity is not desirable. In sprinkler irrigation, the main or lateral pipe shall be sized based on a maximum water velocity of 1.5 mps [11], but maintaining a velocity of 1.2 mps would be acceptable.

Pipe sizing

To determine the diameter of a segment of pipe of a lateral pipe, its cumulative flow requirement is calculated first. The corresponding diameter of pipe based on velocity of 1.5 mps has been selected from the Table 7.6 developed for plastic pipes.

Pump sizing

Correct pump sizing is one of the important jobs for developing an efficient and energy-saving sprinkler irrigation system. It is already mentioned that a centrifugal pump is the best choice for sprinkler irrigation. An irrigation pump is not kept running 24 hours; the runtime of the pumps between 12 and 18 hours per day is typical [13].

In designing an irrigation pump, it is needed to calculate the following factors to determine the required pump capacity:

1 Flow required for the pump
2 Head of the pump

Table 7.7 Water requirement for various vegetables [14]

Vegetables category	*cm*
Shallow rooted	
Cabbage	30
Lettuce	45
Onion	40–60
Spinach	25
Sweet corn	45
Medium rooted	
Bean	30–45
Beet	45
Carrot	40
Cucumber	45
Summer squash	45
Pea	45
Pepper	45

Water flow requirement for the pump, which is denoted by "Q", is the flow for a main, to be applied during the peak period of irrigation. The flow needed in the main pipe of irrigation piping is the flow of water to be delivered by the pump, depending upon water requirement for the type greeneries, plants etc. to be irrigated.

For irrigating grass on lawns or fields, provide 120–200 mm of deep water per weak; for the majority flowering beds, it is general practice to provide 25 mm of water every seven days. For vegetable gardens, use Table 7.7 to find the water requirement for various vegetables.

The head of the pump, denoted by "H," required for sprinkler irrigation, is the total dynamic head to overcome the following four elements of pressure to spray the correct amount of water on a distant land area from the pump.

$$H = H_e + H_s + H_p + H_f + H_v \tag{7.4}$$

where,

H_e = Elevation head, also called the static discharge head, which is the elevation from the center of the impeller to the highest sprinkler on the system, as shown in Figure 7.16, (m).
H_s = Suction head for the pump, which is the elevation from the center of the impeller down to the water level while pumping, (m).
H_p = Highest pressure required for the water application devices (such as sprinklers, spray heads, drippers, etc.), and pressure required for the farthest sprinkler from the pump, whichever is higher (m).
H_f =Friction loss while flowing water through the piping system, including pipes, fittings, valves, screens, etc., (m).
H_v = Velocity head due to velocity of flow in a pipe.

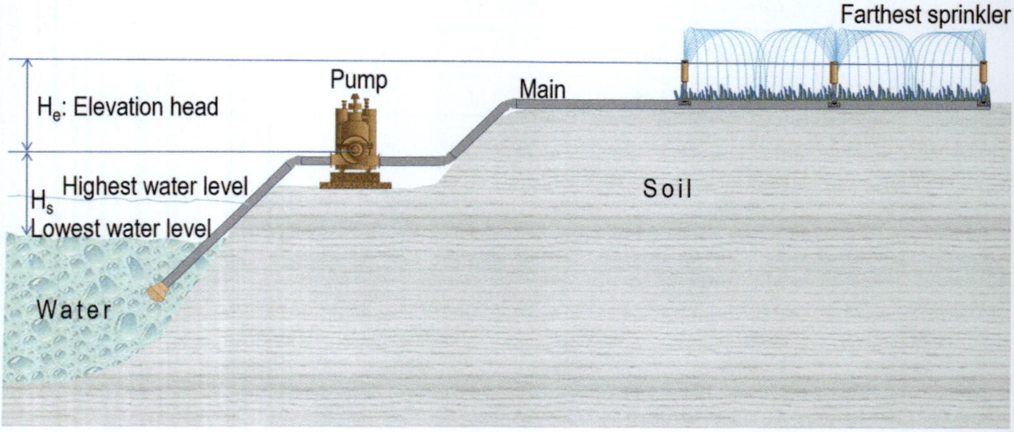

Figure 7.16 Pump elevation and suction head.

Pressure required: Pressure required for the water application devices like sprinkler spray heads or drippers can be found from the manufacturers' data sheet; if not available, then it can be considered from the following criteria:

1 Drip irrigation requires a head of 207 kPa (21 m).
2 Spray type sprinkler requires heads of 276 kPa (28 m).
3 Rotor-type sprinklers require heads of 310 kPa (31 m).

Friction loss

Total friction loss while flowing water through the piping system, including pipes, fittings, valves, screens, etc., can be determined by calculating the equivalent pipe length for all the fittings and accessories on the pipe and adding to the existing pipe length to determine the total equivalent pipe length, and then finding the total friction loss "*f*" with respect to the corresponding flow rate from the chart furnished in Figure 7.17. Table 7.8 furnishes the equivalent length in meters of straight pipe for fittings like bends, returns, tees, and valves having screw joints.

Velocity head: The velocity head for a pump in meters can be computed from the following formula:

$$H_v = \frac{V^2}{2g} \tag{7.5}$$

where,

g = Acceleration due to gravity 9.8 m/sec.sq and
v = velocity of flow (mps)

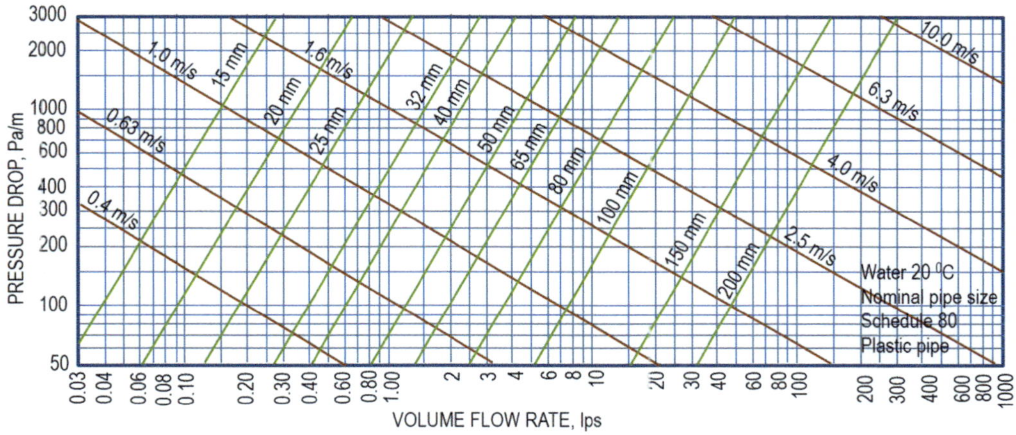

Figure 7.17 Friction losses in plastic pipe.

Table 7.8 Equivalent length (meter) of straight pipe for fittings like bends, returns, tees, and valves by making screw joints [15]

For valves and fittings, consider the equivalent length of straight pipe (m)										
Screwed fittings		Pipe size (mm)								
		12	19	25	32	38	50	63	75	100
Elbows	Regular 90 deg	1.1	1.3	1.6	2.0	2.3	2.6	2.8	3.4	4.0
	Long radius 90 deg.	0.7	0.7	0.8	1.0	1.0	1.1	1.1	1.2	1.4
	Regular 45 deg.	0.2	0.3	0.4	0.5	0.6	0.8	1.0	1.2	1.7
Tees	Line flow	0.5	0.7	1.0	1.4	17	2.3	2.8	3.7	5.2
	Branch flow	1.3	1.6	2.0	2.7	3.0	3.7	4.0	5.2	6.4
Return bends	Regular 180 deg.	1.1	1.3	1.6	2.0	2.3	2.6	2.8	3.4	4.0
Valves	Globe	6.7	7.3	8.8	11.3	12.8	16.5	18.9	24.1	33.6
	Gate	0.2	0.2	0.3	0.3	0.4	0.5	0.5	0.6	0.8
	Angle	4.6	4.6	5.2	5.5	5.5	5.5	5.5	5.5	5.5
	Swing check	2.4	2.7	3.4	4.0	4.6	5.8	6.7	8.2	11.6
Strainer		1.5	2.0	2.3	5.5	6.1	8.2	8.8	10.4	12.8

Pump power requirements can be found from the following formula:

$$\text{Pump power requirement}\left(\text{kW}\right) = \frac{Q \times H}{360 \times E} \tag{7.6}$$

where,

Q = flow rate in cum/h
H = Head in m
E = Efficiency ranges between 60 and 85 percent.

Operation and maintenance of sprinkler system

Only the proper design and installation of a sprinkler system cannot ensure the successful performance of the system. The key to success lies in maintaining a good operation and maintenance program for the system. For this purpose, some guidelines are to be followed, as discussed below:

Operation of sprinkler system

In the cases of tractor-driven and movable pumping units, it should be ensured that the prime mover and the pump are in true alignment, wherever it is placed. To achieve this, the driver shaft and the pump shaft should lie at the same elevation.

While assembling and laying the main and lateral pipes, it should always begin from the pump end. This necessarily helps the correct connection of all quick coupling pipes. While joining the couplings, it is to be ensured that both the couplings and the rubber seal rings are clean.

In starting the sprinkler system, the motor or engine is started with the gate valve on the main keeping closed. The pump must attain the pressure stated on the type-plate, or otherwise there is a fault in the suction line. After the pump reaches the regulation pressure, the delivery gate valve is opened slowly. Similarly, the delivery valve is closed after stopping the power unit.

To maintain good water distribution uniformity, the inlet pressure must be maintained at an optimum level. Leakage and breakage of the pipelines can drastically reduce the discharge and pressure, resulting in poor sprinkler performance. Regularly checking for leakages should be ensured and repaired.

It is recommended to operate the sprinklers during periods of low wind, as water distribution uniformity is greatly influenced by stronger wind.

Maintenance of sprinkler components

General principles regarding the maintenance of the pipes and fittings and sprinkler heads are given below:

1 Pipes and fittings: The pipes and fittings require virtually no maintenance, but attention must be given to the following procedures:

   i Occasionally clean any dirt or sand out of the groove in the coupler in which the rubber sealing ring fits. Any accumulation of dirt or sand will affect the performance of the rubber sealing ring.
   ii Flush piping after installation and every few months, or more often if water in source is fairly dirty. If algae growth within the lines is a problem, chlorine should be used during flushing.
   iii Do not lay pipes on new damp concrete or on piles of fertilizer. Do not lay fertilizer sacks on the pipe.

2 Sprinkler heads: The sprinkler heads should be given the following attention:

   i Sprinkler heads should be cleaned periodically to insure proper functioning.
   ii When moving the sprinkler lines, make sure that the sprinklers are not damaged or pushed into the soil.

iii Do not apply oil, grease, or any lubricant to the sprinklers. They are water lubricated, and using oil, grease, or any other lubricant may stop them from working.
iv Sprinklers usually have a sealed bearing, and at the bottom of the bearing there are washers. Usually it is the washers that wear. Check the washers for wear once a season or every six months, which is especially important where the water is sandy.
v After several seasons' operation, the swing arm spring may need tightening. This is done by pulling out the spring end at the top and re-bending it.
vi In general, check all equipment at the end of the season and make any repairs and adjustments so that the equipment is in perfect condition to start in the next operation.

References

[1] Thomas F. Scherer (2022) "Irrigation water pumps" North Dakota State University Retrieved from https://www.ndsu.edu/agriculture/extension/publications/irrigation-water-pumps on 16 May 2024.
[2] Ed Butts (2020) "Irrigation fundamentals" *Water Well Journal*. Retrieved from https://waterwelljournal.com/irrigation-fundamentals-11/ on 18 May 2024.
[3] H. A. Mansour, E. F. Abdallah, M. S. Gaballah and C. S. Gyuricza (2015) "Impact of bubbler discharge and irrigation water quantity on 1-hydraulic performance evaluation and maize biomass yield" *International Journal of GEOMATE*, Vol. 9, No. 2, pp. 1538–1544.
[4] Food and Agriculture Organization (FAO) (2024) *Sprinkler Irrigation* Chapter 5, Irrigation Water Management: Irrigation Methods Retrieved from https://www.fao.org/4/s8684e/s8684e06.htm#TopOfPage on 18 May 2024.
[5] Hunter Industries Inc. (2020) Residential sprinkler system design guide Retrieved from https://www.hunterindustries.com/sites/default/files/design_guide_Residential_System_LIT-226-US.pdf on 18 May 2024
[6] General Services Administration Residential Code (2024) Water Supply and Distribution, P2902.5.4.1 Additives or Nonpotable Source.
[7] Theodore G. Santiesteban (2009) *Benefits of Pressure Regulation in Irrigation*. Senninger Irrigation Inc., Clermont, FL Retrieved from https://www.irrigation.org/IA/FileUploads/IA/Resources/TechnicalPapers/2009/BenefitsOfPressureRegulationInIrrigation.pdf on 18 May 2024.
[8] J. Keller and R. D. Bliesner (1990) *Sprinkler and Trickle Irrigation*. Chapman and Hall, New York.
[9] Robert W. Hill and Craig Poulson (2002) "Sprinklers, crop water use, and irrigation time Sanpete county", Utah State University Retrieved from https://digitalcommons.usu.edu/extension_histall/174/ on 19 May, 2024.
[10] A. M. Michael (1978) *Irrigation – Theory and Practice*. Vikas Publishing House, New Delhi: 585–685.
[11] Stephanie Tam and Andrew Petersen (2014) *B.C. Sprinkler Irrigation Manual* Chapter 8, Ministry of Agriculture, British Columbia: Victoria, BC, 126.
[12] M. Cyril Harris (1991) *Practical Plumbing Engineering*, McGraw-Hill Company, New York.
[13] Vahan Bagdasarian (2013) "Pump sizing for efficient irrigation" Retrieved from https://www.pumpsandsystems.com/pump-sizing-efficient-irrigation on 19 May 2024.
[14] SlidePlayer.com Inc. "Vegetable production introduction" Slide no 50, Published by Junior King Retrieved from https://slideplayer.com/slide/5808602/ on 19 May 2024.
[15] The Engineering ToolBox (2003) "Hot water systems - equivalent length vs. fittings resistance" Retrieved from https://www.engineeringtoolbox.com/resistance-equivalent-length-d_192.html on 19 May 2024.

8 Steam supply system

Introduction

In a building development, steam supply may be required for its different purposes of use, such as space heating, steam bathing, sterilizing, etc., which require the production of steam from water and then supplying it to the points of use through pipes. So, steam being a fluid, its supply system falls under the subject of plumbing and can be termed as specialized plumbing, as steam supply development requires special knowledge to apply for its sustainability, which is required in buildings of special requirements. In this chapter, all basic knowledge of steam generation and its supply systems in buildings will be discussed concisely.

Steam

In liquid cold water, H_2O molecules remain constantly joining together and separating. As the water is heated, molecular bonds connecting the molecules start breaking more rapidly than they can form. Eventually, when sufficient heat is applied, molecular bonds break and become free. These free molecules of water constitute the transparent gaseous steam, or more specifically, dry steam. Steam releases heat energy, more heat energy per unit mass than that by hot water. At a pressure of 207 kPa, one kilogram of steam can give up approximately 2154 kilojoules of heat.

When water is continually heated and reach a boiling point of about 100°C (at sea level), steam is generated. When steam is generated it expands tremendously. Each liter of water that is boiled will expand to 1600 liters of steam.

Temperature and pressure are two important variables in steam generation in a pressure or sealed vessel. As pressure is increased in a pressure vessel, the boiling point of water also rises, allowing water to remain in liquid form at higher temperature. This relationship enables the production of steam at much higher temperatures and pressures than would be possible in an open vessel.

Steam condensation

Steam condensation is a process by which steam changes into liquid water again due to loss of heat. It occurs when the steam temperature drops to or below its saturation temperature at a given pressure. During the process of condensation, steam releases its latent heat of vaporization to the surrounding surfaces. Saturated steam is a condition of steam at a particular temperature where any further heat loss will cause it to condense.

DOI: 10.1201/9781003724056-8

Steam dryness

Steam dryness has a direct effect on the total amount of transferable heat energy contained within the steam, which affects heating efficiency and quality. The more the dryness of steam, the more efficient the heating system is. The dryness fraction of steam is considered to quantify the amount of water within the steam. If the steam contains 5 percent water by mass, it is said to be 95 percent dry, or have a dryness fraction of 0.95.

So, in a steam supply system, two commonly referred to types of steam are in use; those are termed as follows:

1 Dry steam, also called "saturated steam"
2 Wet steam

Dry steam

A steam that, when slightly superheated above the saturation point, does not contain any water particles in suspension, is termed dry steam. In dry steam, all its water molecules remain in the gaseous state and become a clear, colorless, transparent vapor with no entrained moisture. In heating applications, dry steam is preferable for having better energy exchange capacity and does not cause corrosion. Pure dry steam generally does not carry any impurities.

Wet steam

When a portion of its water molecules have given up their energy (latent heat) and condense to form tiny water droplets, then it is termed wet steam.

Moisture in steam is susceptible to corrosion. Moisture moving at high steam velocities causes erosion in valves and fittings, a phenomenon known as wiredrawing. It also causes erratic operation of valves. Moisture, i.e., water molecules in steam may have impurities, which can increase scaling on surfaces of pipes and heating elements.

There are multiple effects of wet steam in a steam supply system, as mentioned below:

1 Corrosion and calcification
2 Ram strike
3 Reduction in heat transfer
4 Increase in condensate load
5 Deformation in used fixtures

Vacuum steam

Vacuum steam is the general term used for saturated steam at temperatures below 100°C.

Use of steam in building

In the building sector, steam is mostly and popularly used for various purposes, for its heating applications, as a source of direct heating and indirect heating as well. More specifically, steam is used for the following purposes:

1 Space heating
2 Hot water production

3 Bathing
4 Humidification
5 Sterilizing etc.

Steam distribution method

The steam is generated in a boiler, which is conveyed through pipework to the points where its heat energy is required. At the start of a steam supply pipe network, there is generally one main pipe, called "steam mains", which conveys steam from the boiler to the smaller branch pipes to carry the steam to the individual points of usage. Steam flows in response to the pressure drop along the pipe.

On the boiler main, an isolating valve, also known as a "crown" valve, is installed. When this valve opened, steam flows at high pressure from the boiler into and along the main to the branch pipes and then laterals to reach points of usage at lower pressure. The pipework usually remains initially cooler than the steam, causing heat transfer from the steam to the pipe. The air surrounding the pipes being cooler than the slightly heated pipe, heated piping begins to transfer heat to the surrounding air.

Steam, coming in contact with the cooler pipes, begins to condense immediately. At the start of the steam flow, the condensing rate will be at its maximum for having a maximum temperature difference between the steam and the pipework. This starting condensing rate is termed as "starting load". When the piping has warmed up to its maximum level, i.e., the temperature difference between the steam and piping is minimal, some condensation will occur as the piping transfers heat to the air. This minimal condensing rate is termed the "running load".

When the valve on the steam pipe serving an item (or plant) requiring steam is opened, steam flowing from the feeder piping enters the plant and again comes into contact with cooler surfaces. The steam starts transferring its energy in warming up the item or a plant (starting load), and, till it reaches the desired temperature, it continues to transfer heat to the process (running load). So, there is a need for continuous supply of steam from the boiler to satisfy the connected running loads and to maintain the supply of sufficient steam that is to be generated. As a result, more water and fuel to heat this water are to be supplied to the boiler to make up for that water that has previously been evaporated into steam.

The condensate produced in the piping falls to the bottom of the pipe by gravity due to the gradient to be maintained in the horizontal pipes. The condensate shall therefore be drained out from various strategic points, through drain cocks in the steam piping. The condensate produced in both the steam distribution piping and in the process equipment should not be allowed to run out as wastewater, which is effectively used as boiler feed-water. Returning all condensate to the boiler feed-tank makes the steam loop efficient and should be practiced wherever possible. A typical steam supply system in a building has been shown in Figure 8.1.

Steam supply pressure

Steam being a fluid needs pressure to drive it to flow through a pipe. As steam flows through a pipe, it experiences pressure losses similar to those in water flow, resulting in reduced pressure at the points of use. For these reasons, sufficient pressure is to be maintained, from the start to the end, in the steam supply system. The distribution pressure of steam is influenced by a number of factors, as mentioned below:

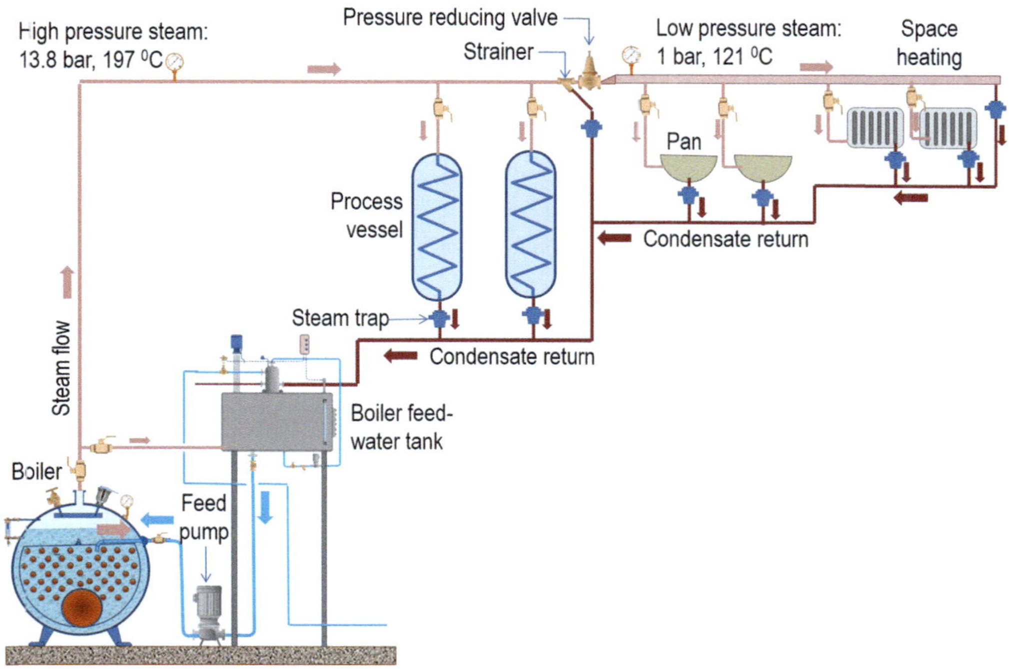

Figure 8.1 A typical steam supply system in a building.

1 The maximum allowable working pressure of the boiler
2 The minimum pressure required at the point of usage
3 Frictional loss within the pipework
4 Condensation rate within the pipework

Considering all these factors, some allowance for this pressure loss should be made while determining the initial distribution pressure.

If steam is generated in a boiler at a high pressure and also distributed at a high pressure, the size of the distribution mains will be smaller than that for a low-pressure system for the same heat load, which is illustrated by Figure 8.2, because at a higher pressure steam occupies less volume than at a lower pressure.

Generating steam and also distributing it at higher pressure offers some advantages, as mentioned below:

1 The thermal storage capacity of the boiler can be increased, working more efficiently with fluctuating loads, minimizing the risk of producing wet and dirty steam.
2 Smaller-sized steam distribution pipes are required, resulting in lower investment costs.
3 Local pressure reduction to suit individual plants or elements will also result in drier steam at the point of use.

Steam distribution at a high pressure has disadvantages as well; it might be necessary to reduce the steam pressure to a particular zone or point of use at a lower pressure in the system in order to correspond with the maximum pressure required for a particular application.

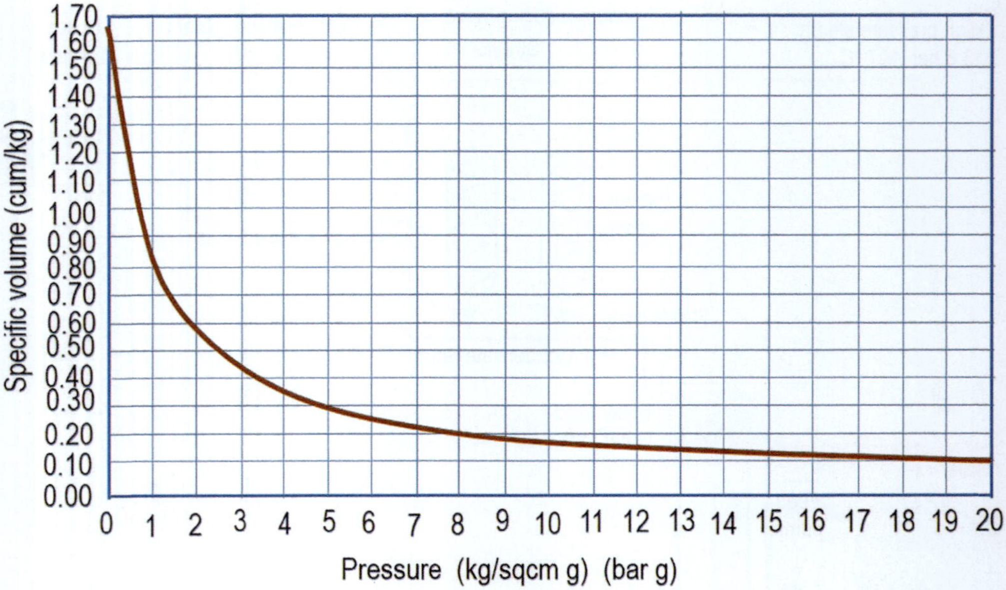

Figure 8.2 Pressure and specific volume relationship for dry saturated steam [1].

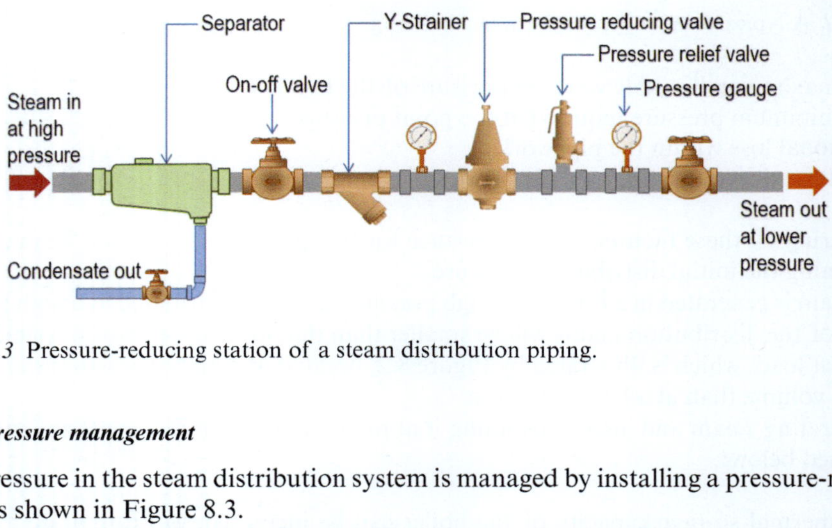

Figure 8.3 Pressure-reducing station of a steam distribution piping.

Steam pressure management

High pressure in the steam distribution system is managed by installing a pressure-reducing valve, as shown in Figure 8.3.

A separator is installed upstream of the pressure reducing valve to remove entrained water from incoming wet steam, which causes high-quality steam to pass through the valve.

Steam distribution system components

The major components of a steam distribution system are as follows:

1 Steam boiler
2 Pipes
3 Drain point
4 Branch line

5 Strainer
6 Filter
7 Separator
8 Steam trap
9 Air vent
10 Condensate recovery system
11 Pressure management system

Steam boiler

A steam boiler is basically a closed vessel into which water is heated by means of a burner that converts fuel into heat and a heat exchanger that transfers the heat into the water as it passes through, until the water is transformed to steam at certain pressure. It requires makeup water to feed in to replace water lost through evaporation and the production of steam during operation. In comparatively smaller steam heating systems, steam boilers that operate at 13.8 kPa pressure or slightly more are preferred.

Basic working principle

In a closed vessel of a boiler, water is filled partially, about three-fourths full of water. Fuel (coal, gas, electricity, etc.) is burned in a furnace, and hot gases are produced, which are brought into contact with a water vessel or tube where the heat of these hot gases transfers heat to the water to raise its temperature; consequently, steam is produced from water in the boiler. The produced steam is then discharged through an outlet and then conveyed to the points of usage through a pipe.

Types of boilers

There are many different types of boilers with respect to methods of heating, pressure range, and type of steam generation.

In terms of ways of heating, there are two types of boilers:

1 Water-tube boiler
2 Fire-tube boiler

In terms of the nature of the fuel used:

1 Gas fired
2 Liquid fuel fired
3 Coal fired
4 Electrically fired
5 Nuclear fired

In terms of type of production, there are two types of boilers:

1 Hot water boiler
2 Steam boiler

In terms of pressure range, there are two types of boilers:

1 Low-pressure boiler: Maximum 1 bar pressure. Generally used in buildings.
2 Medium-pressure boiler
3 High-pressure boiler: 80 bar and more, used mostly in industrial processes

In terms of number of tubes:

1 Single-tube boiler
2 Multi-tube boiler

In terms of the position of the axis of the boiler shell:

1 Vertical boiler
2 Horizontal boiler

In terms of circulation system:

1 Natural circulation boiler
2 Forced circulation boiler

Fire-tube boiler

In a fire tube boiler, there are numbers of tubes through which hot gases are passed. These hot-gas-flowing tubes are kept immersed in water to be boiled in a closed vessel or shell. These fire tubes or hot gas tubes heat up the water and convert the water into steam, which is contained in the vessel. A fire tube boiler cannot produce steam at very high pressure as both the water and steam remain in the vessel. Generally, it can produce maximum 9 metric tons of steam per hour at a pressure of about 17.5 kg/sqcm [2].

Types of fire-tube boilers

A fire tube boiler might have an external furnace and an internal furnace. Fire tube boilers with an external furnace are designed in the following three different categories:

1 Horizontal return tubular boiler
2 Short firebox boiler
3 Compact boiler

On the other hand, an internal furnace fire tube boiler has two categories, as named below:

1 Horizontal tubular
2 Vertical tubular

Normally, horizontal return fire tube boilers are found to be of low capacity. In a horizontal drum-type vessel, there are numbers of horizontal tubes kept submerged in water to be boiled, as shown in Figure 8.4. Normally coal as fuel is burned below these horizontal

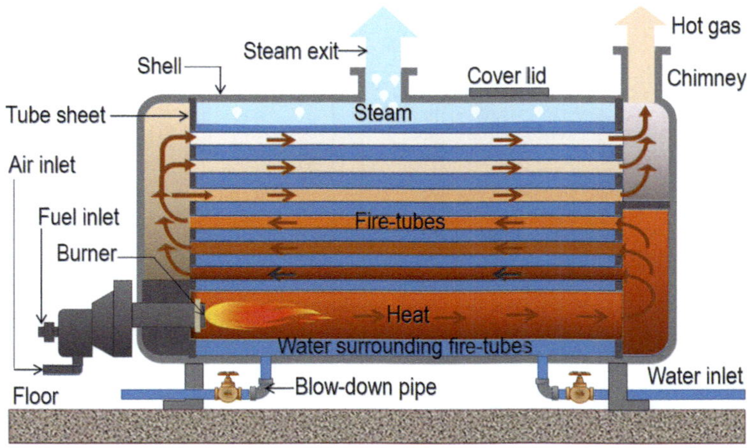

Figure 8.4 A horizontal fire-tube boiler.

drums, and the combustible gases move to the rear from where they enter into fire tubes and travel toward the front into the smoke box. During the travel of hot gases in tubes, heat is transferred into the water, and steam is produced to come up. As steam production continues, pressure in the closed vessel of the boiler develops.

Fire-tube boilers have many advantages and disadvantages as well; those are mentioned below:

Advantages of a fire-tube boiler

1 It operates quietly
2 Compact in construction; so, smaller foot print
3 Fluctuating steam-demand is met easily
4 Economic in price

Disadvantages of a fire tube boiler

1 Longer time is required for rising temperature at desired pressure, as the larger volume of water is required for operation.
2 Development of very high pressure steam is impracticable, as the water and steam remain in the same vessel.
3 It's difficult to generate dry steam.

Water-tube boiler

A water tube boiler is a kind of boiler in which the water is heated inside multiple tubes to produce steam and the hot gases surround the tubes containing water to be heated, as shown in Figure 8.5. Virtually, a water tube boiler is to some extent the opposite of a fire tube boiler.

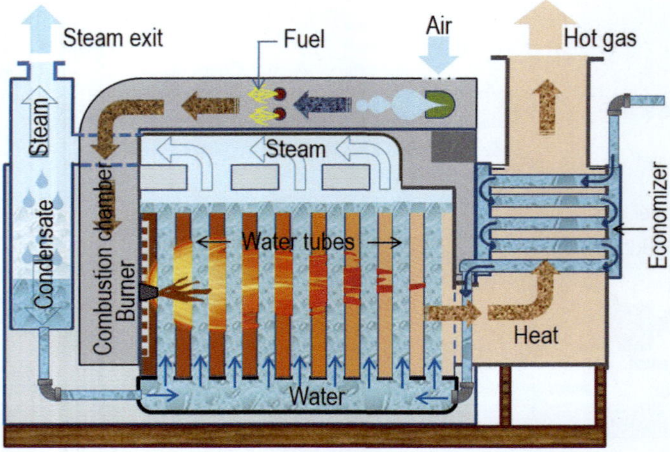

Figure 8.5 A horizontal water-tube boiler.

Types of water-tube boilers

There are also various types of water-tube boilers, as mentioned below:

1 Horizontal straight tube boiler
2 Bent tube boiler
3 Cyclone-fired boiler

Advantages of a water-tube boiler

There are many advantages of a water tube boiler, which make the boiler popular for large volume of steam generation:

1 A larger heating surface can be developed by making more numbers of heating tubes.
2 Due to convectional flow, comparatively faster movement of water is achieved to cause a higher rate of heat transfer, resulting in higher efficiency.
3 Very high pressure in order of 140 kg/sqcm can be developed [2].

Disadvantages of a water tube boiler

1 This type of boiler is not compact in construction.
2 Cost is relatively high.
3 Size is relatively bigger and causes difficulty in transportation and needs more floor area to install.

Comparison between a fire and water tube boiler based on various important decision-making factors is very essential in the selection of a boiler. So, the comparative advantages and disadvantages of fire-tube and water-tube boilers have been furnished in Table 8.1.

In a forced circulation boiler, a pump is used to circulate water inside the boiler, whereas in a natural circulation boiler, current density causes it to circulate water inside the boiler. In some forced circulation boilers, the water is circulated at twenty times the rate of evaporation.

Table 8.1 Comparative advantages and disadvantages of fire-tube and water-tube boilers [3]

Sr. no.	*Parameters*	*Fire-tube boilers*	*Water-tube boilers*
1	Rate of steam generation	Slow	Quick
2	Suitability for power plants	Unsuitable	Suitable
3	Operating steam pressure	Limited to 24.5 × 105 N/sqm	Limited to 200 × 105 N/sqm
4	Chances of explosion	Less	More
5	Risk of being damaged due to explosion	Much high	Much low
6	Water treatment	Not very urgent, as a little scaling would not cause fast overheating and tube-bursting.	Required, as rapid scaling may lead to tube-bursting.
7	Floor space required	Much	Less
8	Cost and construction problem	Much high	Much low
9	Transportation	Inconvenient, as the shell is often of large size.	Comparatively, more convenient
10	Skill required for efficient operation	Less	More
11	Operating cost	Less	High
12	Overall efficiency	Up to 75 percent	Up to 90 percent

Major components of a steam boiler

Boilers of different types might have different designs, but the common five parts remain the same throughout various designs. They are the combustion chamber, the heat exchanger, the burner, the shell, and the exhaust stack.

Combustion chamber

The combustion chamber is usually made from cast iron, which is incorporated for burning the fuel inside. The heat generated in this chamber will be transferred to the heat exchanger part of the boiler. The temperatures inside the combustion chamber can be increased up to several hundred degrees in a short period of time.

Heat exchanger

A heat exchanger is a component of a boiler that allows heat to be exchanged between two fluids, water and gas-flue in a boiler, without letting the substances mix together. The primary function of a heat exchanger in a boiler is to carry the heat from the burner to the water without having direct contact with the water. The elements are made of very sturdy material, able to withstand high temperatures and transfer heat efficiently. Heat exchangers are generally made as bundles of steel, copper, or cast iron tubes.

Burner

The burner initiates the combustion reaction inside the boiler. An outside fuel tank provides fuel to the burner after passing through a filter mechanism, with the help of a pump. The mixing of the air with the fuel is done herein. The fuel then turns into a fine spray with the help of the burner's nozzle and is ignited, getting the electronic signal of the thermostats to the burner, which results in a reaction in the combustion chamber.

Shell

The boiler body is typically a hollow cylindrical shell made of steel plates. The welding seams are carried out with an electric arc. The shell is thermally insulated with a 100 mm rock wool cladding bounded with high-density, thick thermosetting resins, suitably supported, and covered externally with thick enameled aluminum. The heat transfer and steam generation occur in this shell. The two ends of the shell are closed by fixing endplates. About two-thirds of the shell is water space occupied by water, and on top is steam space, where steam got collected inside the shell of the boiler.

Exhaust stack

The exhaust stack, also known as the chimney or flue, is a medium to safely expel flue gas away from the boiler. An exhaust stack may be a metal pipe that expels hot flue gases away from the human reach, near windows or doors. Flue gas may contain water vapor (H_2O), carbon dioxide (CO_2), carbon monoxide (CO), nitrogen (N_2), and complete or incomplete products of fuel combustion. The stack terminal shall be installed in such a way that it does not allow downdrafts of strong winds.

Feed-water tank

In a boiler feed-water tank, as shown in Figure 8.6, the boiler feed-water and make-up water are stored, and condensate is returned into it. Make-up water is the cold water from the water treatment plant that makes up any losses in the system.

It is a very essential supporting element for a boiler. The temperature of water in the feed-water tank should be kept at a high level, with a view to minimizing the content level of dissolved oxygen and other gases that might be present. Maintaining high temperature also reduces energy cost. As a rule of thumb, every 6°C rise in feed-water temperature reduces the

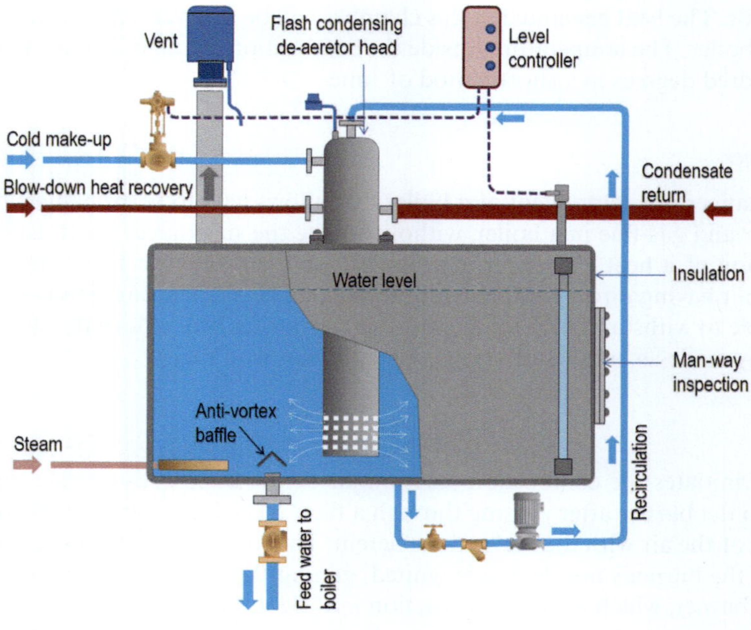

Figure 8.6 A feed-water tank with major accessories.

fuel bill by 1percent. The feed-water tank should be adequately insulated to prevent heat losses. There must be a temperature gauge that can be a local or remote reading device.

The feed-water tank shall be vented to prevent any pressure build-up. As a guide, vent size range from 80 mm on a 2,000-liter tank to 250 mm on a 30,000-liter tank. The vent is fitted with a vent head. A drain connection is fitted in the bottom of the feed-water tank to facilitate its emptying when necessary.

Water level controller are installed on feed-water tank to maintain water level in the tank A level indicator or water level gauge glass is installed on the tank to facilitate viewing the water level in the tank from outside. An adequately sized cover should be kept to enable internal inspection and the fitting of ancillaries, etc.

Flash steam from heat recovery system

A heat recovery system may, for example, recover flash steam from the boiler blowdown. It is another opportunity to use recovered heat to raise the feed-tank temperature, and so, save fuel.

As with pressurized condensate, the flash steam needs to be condensed. Traditionally, this was achieved using spurge pipes, but a modern and much more effective method is the incorporation of the flash condensing deaerator head.

Many water treatment plants need a substantial flow through them in order to achieve optimum performance. A "trickle" flow as a result of a modulating control into the feed-tank can, for example, have an adverse effect on the performance of a softener. For this reason, a small plastic or galvanized steel cold make-up tank is often fitted. The flow from the softener is controlled "on / off" into the make-up tank. From there a modulating valve controls its flow into the feed-tank.

This type of installation leads to "smoother" operation of the boiler plant. To avoid the relatively cold make-up water sinking directly to the bottom of the tank (where it will be drawn directly into the boiler feed-water line), and to ensure uniform temperature distribution, it is common practice to spurge the make-up water into the feed-tank at a higher level.

Steam injection

As previously mentioned, there are significant advantages to maintaining the feed-tank contents at a high temperature. One of the most convenient ways of achieving this higher temperature is by injecting steam into the feed-tank.

Overflow

The overflow of a feed-water tank for a steam boiler is a safety and control feature designed to prevent overfilling and potential damage to the system. Overflow pipe outlet is positioned slightly above the normal operating water level inside the tank. This should be fitted with a "U" tube water seal to prevent flash steam loss.

Feed pump take-off

If the take-off is from the base of the feed-tank there should be a 50 mm internal stub to prevent any dirt in the bottom of the tank from entering the pipeline. It should be generously sized so that friction losses are minimized, and the net positive suction head (NPSH) to the feed-pump is maximized.

Inspection opening

An adequately sized inspection opening should be fitted to enable internal inspection and the fitting of ancillaries, as appropriate.

Level probes can be arranged to signal high water level, the normal working (or control) water level, and low water level. The signals from the probe can be linked to a control valve on the cold water make-up supply. The probe is fitted with a protection tube inside the feed-tank to protect it from turbulence, which can result in false readings.

Mountings and accessories of a boiler

The boiler mountings are those fittings that are found mounted on the boiler for its efficient and safe functioning. Some of the mountings are for the safety of the boiler. Various mountings installed on a boiler are as follows:

1 Water level indicator
2 Pressure gauge
3 Safety valve
4 Steam stop valve
5 Blow off
6 Feed check valve
7 Fusible plug

The boiler accessories are those devices, which though perform as an integral part of the boiler but are not directly mounted on its body. The accessories actually help in controlling and running the boiler safely and efficiently. The accessories used in a boiler are as follows:

1 Economizer
2 Air pre-heater
3 Super-heater
4 Feed pump
5 Steam trap
6 Steam separator

Water level indicator

A water level indicator on a steam boiler is used to facilitate reading the water level inside the boiler. It is generally installed in front of the boiler. There are various types of water level indicators, as mentioned below:

1 Gage glass
2 Indirect sensing method
3 Independent remote

Among these, the gage glass is the most common form of level indicator found on steam boilers, as shown in Figure 8.7. It consists of three cocks: a steam cock, a water cock, a drain cock, and a glass tube.

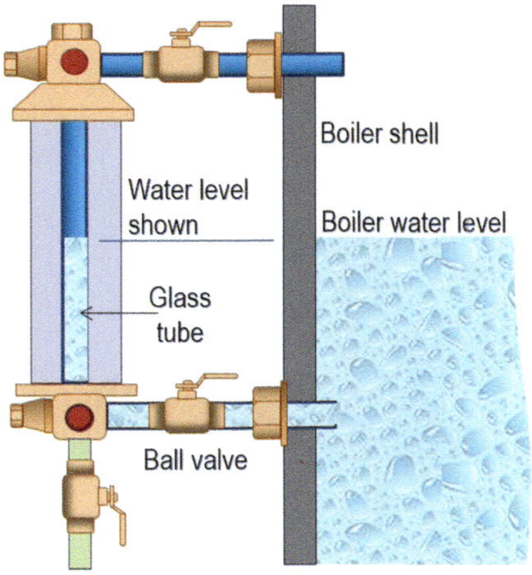

Figure 8.7 A water level indicator.

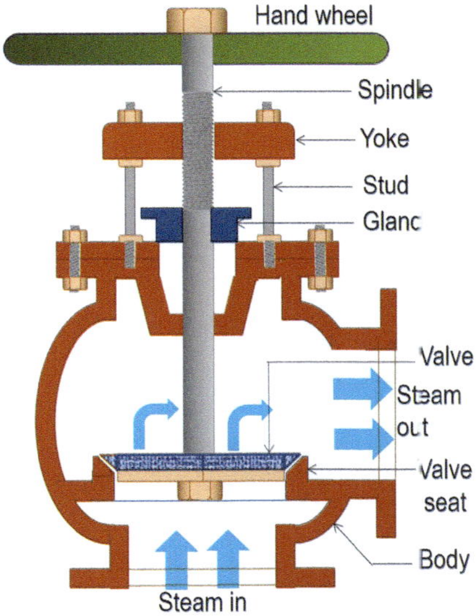

Figure 8.8 A steam stop valve in closed condition.

Steam stop valve

The steam stop valve or junction valve is basically an angular gate valve used to stop or regulate the flow of steam from the boiler to the steam main pipe, or from one main steam pipe to another branch steam pipe. Figure 8.8 represents a steam stop valve in closed condition.

Blow-off valve

The blow-off valve, shown in Figure 8.9, is installed under a boiler, which allows the boiler to drain out the water periodically or when required and to blow out sediments, precipitated sludge, loose scale, various impurities, etc., accumulated during the operation of the boiler.

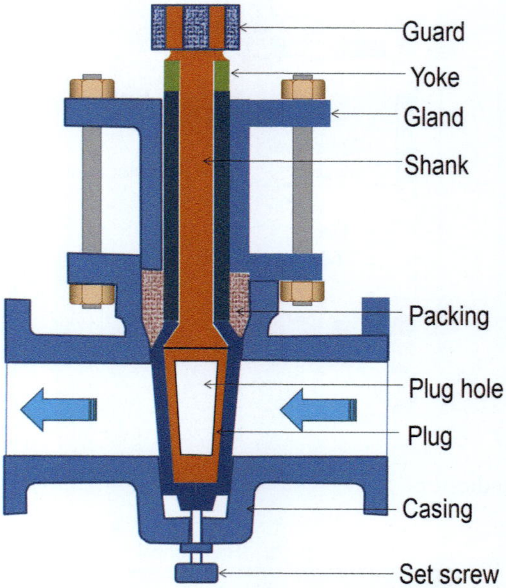

Figure 8.9 A blow-off valve.

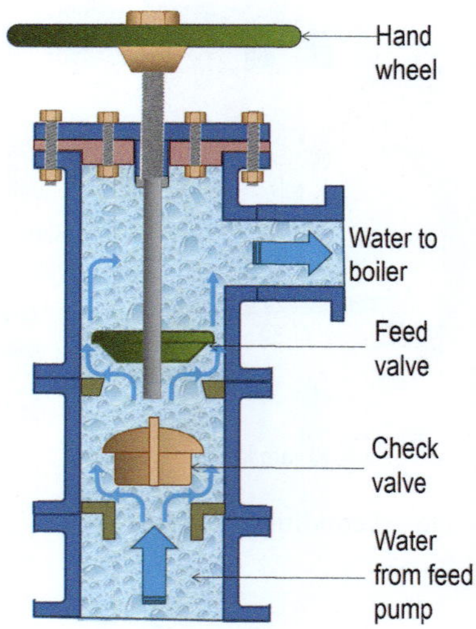

Figure 8.10 A feed check valve.

Feed check valve

The feed check valve, shown in Figure 8.10, is installed to prevent flowing of water back from the boiler, which may be due to malfunctioning of the feed pump or the pump pressure falling below the boiler pressure. It is fitted on the boiler shell, slightly below the normal water level of the boiler.

Steam strainer

In steam supply systems, damages might be caused by various matters, such as scale, rust, jointing compound, weld metal, and other solids, which may find their way into the piping of the steam supply system. To arrest these matters in flowing steam and protect equipment from their harmful effects, strainers are used, which are fitted upstream of every steam trap, flow meter, and control valve.

In steam supply systems, a Y-type strainer is mostly and almost universally used. Y-type strainers, as shown in Figure 8.11, should be installed in horizontal steam lines, keeping the pocket in the horizontal plane. Y-type strainers have comparatively a lower dirt-holding capacity, so it requires more frequent cleaning.

Steam filter

In a steam supply system where "clean steam" is highly demanded, in that case steam-strainers might not be effective, where steam-filters, as shown in Figure 8.12, would be the best choice. A typical steam filter consists of a sintered stainless steel filter element inside it. The sintering process produces a fine porous filtering medium in the stainless steel, which

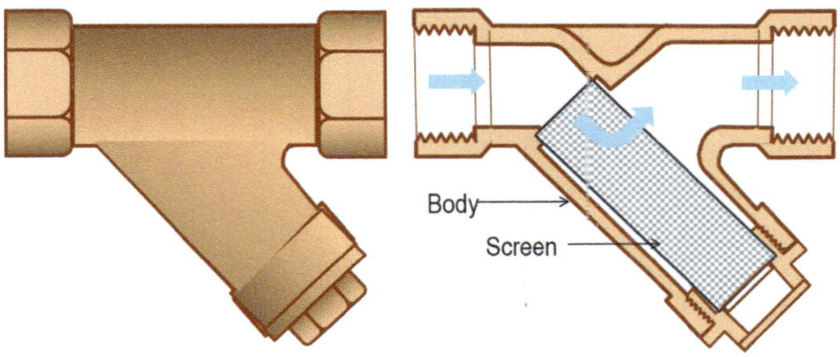

Figure 8.11 Y-type steam strainer.

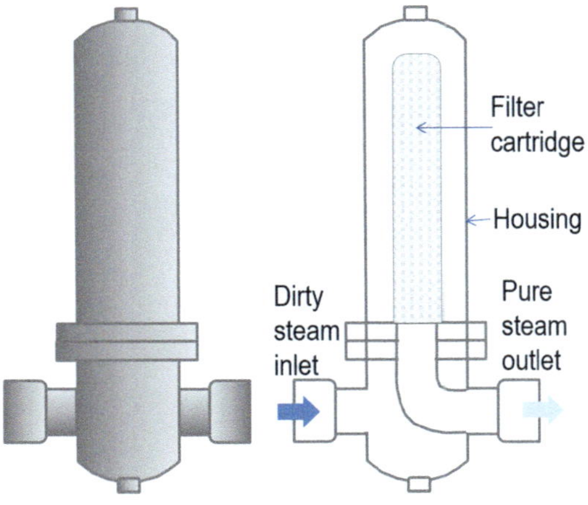

Figure 8.12 A steam filter.

removes the finest particles from steam passing through it. Steam filters have the capability of removing particles as small as 1 μm in size.

The fine, porous nature of the filter element creates a larger pressure drop across the filter. In addition, filters are subject to damage due to excessive flow rates. When the filter is used in steam or gas applications, a separator should be fitted upstream of the filter to remove any droplets of condensate held in suspension. In addition to improving the quality of the steam, this will prolong the life of the filter. A Y-type strainer should also be fitted upstream of the filter to remove all larger particles that would otherwise rapidly block the filter, increase the amount of cleaning required, and reduce the life of the filter element.

With the help of pressure gauges, installed at either side of the filter, the pressure drop across the filter can be measured; the extent of the pressure difference indicates when the filter requires cleaning. Alternatively, installing a pressure switch on the downstream side of the filter, which activates an alarm light when downstream pressure decreases below a set level; thereby can alert an operator to clean the filter.

Steam trap

A steam trap is an automatic valve that holds the steam for the supply of pure steam to the system after removing condensate (water) released from the steam and venting out air from the steam, if any. Based on the applications, there are various types of steam traps, as mentioned below:

1 Thermostatic
2 Float and thermostatic
3 Thermodynamic
4 Inverted bucket

Among various types of steam traps, inverted bucket traps are commonly used, as shown in Figure 8.13.

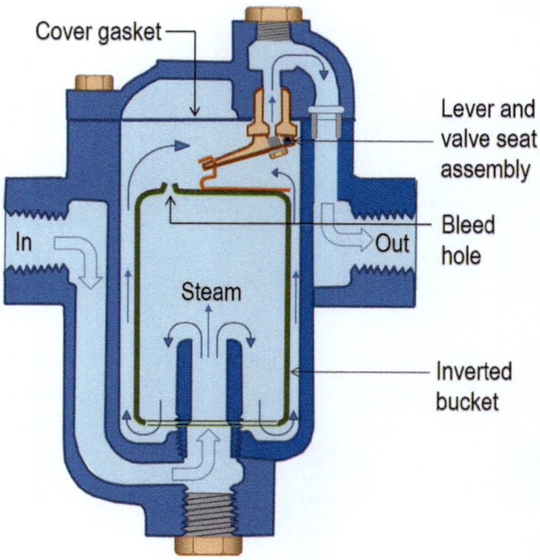

Figure 8.13 An inverted bucket steam trap.

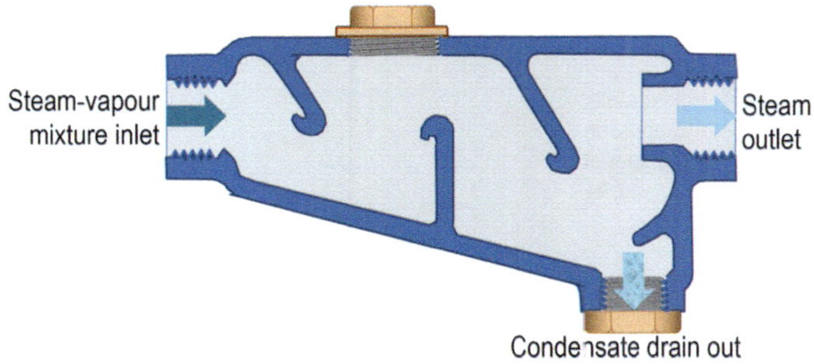

Figure 8.14 A steam separator.

Steam traps remove condensate flowing along with steam at the bottom of the pipes but are unable to remove moisture entrained in the flowing steam; as a result, dry steam cannot be obtained using steam traps alone.

At flow velocities less than 10 mps, changes in flow direction and pressure differentials can cause separated condensate in the piping to be re-entrained in the steam flow. Removing entrained moisture from steam flowing at a velocity of 30 mps or more can be done only by a steam separator.

Steam separator

It is almost impossible to obtain dry steam from steam boilers and during the supply of steam. To remove entrapped moisture in the steam, a steam separator, as shown in Figure 8.14, is used.

Various types of steam separators are available, as mentioned below:

1 Divider plate and (wing type) or (baffle) steam separator
2 Cyclone (cyclonic or centrifugal) type steam separator
3 Vortex type steam separator

Economizer

A boiler economizer or flue gas heat recovery unit is basically a heat exchanger that is installed in the flue gas stack of a boiler with the objective of recovering some heat from hot flue gas that would otherwise be wasted. In Figure 8.5, an economizer is shown installed under the flue gas stack.

Air pre-heater

Air is required in the combustion chamber of a steam boiler. Hot air causes efficient burning. So, using an air preheater, some portion of the waste heat from the flue or chimney-gas is recovered, which is then utilized for preheating the air supplied to the combustion chamber of the boiler. An air preheater is usually placed after the economizer, and before the gases enters the chimney. Figure 8.15 represents a tubular-type preheater.

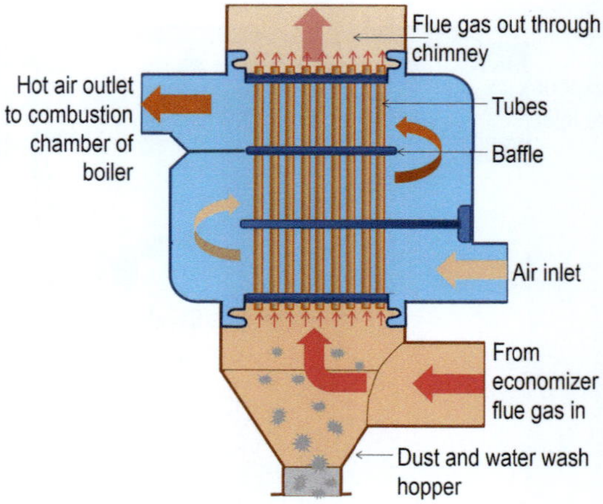

Figure 8.15 Tube type air preheater.

Three types of preheaters are in use, named as follows:

1 Tube type
2 Plate type
3 Regenerative type

Tube type air preheaters are often considered more economical for smaller to medium-sized boilers, while regenerative type air preheaters are typically more suitable for high-capacity boilers, say, of the capacity of 600 tons per hour of steam or higher. Figure 8.15 represents a tube-type air preheater.

Super-heater

By installing super-heaters along the path of flue gases, the wet or saturated dry steam is superheated, thereby increasing steam temperature above its saturation temperature. The super-heater is installed after the furnace, as shown in Figure 8.16. In larger boilers, the super-heater may be placed in an independent fire furnace.

Safety measures for a boiler

The explosion of and the secondary combustion in a boiler are the prime concerns for taking appropriate safety measures for a boiler. Different types of safety equipment and devices are therefore installed on boilers to avert the development of high pressure or temperature inside the boiler, as mentioned below:

1 Safety valves
2 Low-water cutoff switch
3 Pressure switches
4 Fusible plug
5 Expansion tank

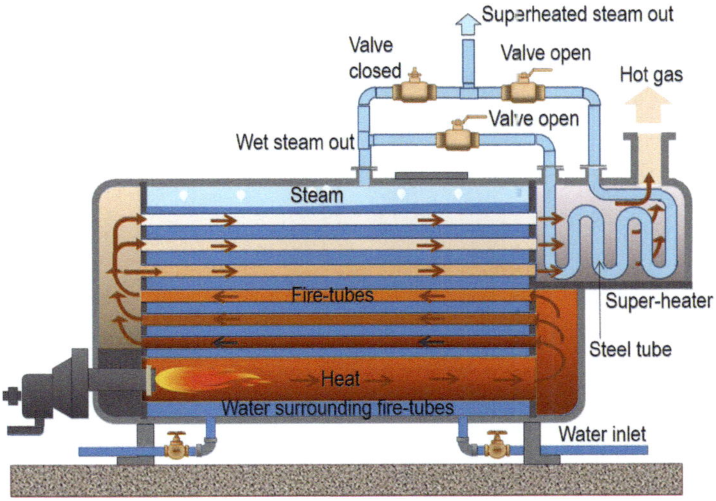

Figure 8.16 Tubular steam super-heater.

Safety valve

It is used to avoid the explosion of a boiler, due to the development of excessive internal pressure of steam. When the excess pressure of steam inside the boiler exceeds the maximum withstanding pressure, the boiler explodes by blowing off steam.

For steam boilers, a small overpressure is allowed, which is usually 3–5 percent above the boiler's working pressure. In some cases, 10 percent overpressure is allowed. Generally, two safety valves are installed on a boiler to ensure release of excess pressure. There are various types of safety valves, as mentioned below:

1 Pilot-operated safety valve
2 Dead weight safety valve
3 Spring-loaded safety valve

Among all types of safety valves, spring-loaded safety valves, also known as pressure relief valves, as shown in Figure 8.17, are the most commonly used safety valves for the protection of steam boilers. It is designed in a way to compel the load of the steam to press the disc against the inlet pressure.

Low-water cutoff switch

Steam boilers shall be equipped with a low-water cutoff switch to protect the appliance from burning out, as it might run out of water or too low water. Water in a boiler must be maintained at a normal operating water level (NOWL). Low water cutoff (LWCO) switches are generally placed 50–150 mm below the NOWL. The LWCO uses an electric conductivity probe as shown in Figure 8.18, or a float, to detect the water level inside the boiler. When

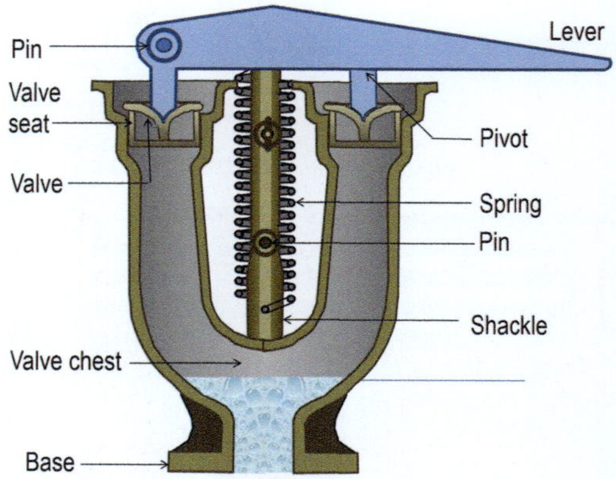

Figure 8.17 A spring-loaded safety valve.

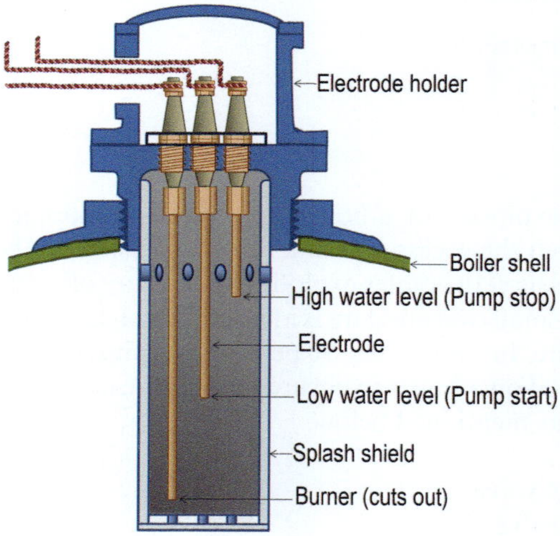

Figure 8.18 Probe type low water level electrode assembly.

water drops below this level, LWCOs prevent the burner from firing. There are two types of low-water cutoff switches; those are as follows:

1 Float type
2 Probe type

Float type low-water cutoff switch

A float type low-water cutoff switch consists of a cutoff switch that operates in conjunction with a ball that floats on the boiler water. The ball is connected through a linkage to an electrical switch that operates a feed-water valve. As the water level falls, the ball drops with it until it reaches a point at which the feed-water switch is actuated to feed-water into the boiler. If the water level still continues to fall, a second auxiliary low-water cutoff switch is connected to the ball by a linkage to another electrical switch that is actuated to shut off automatically the fuel-burning equipment. A float-type low-water cutoff switch is widely used in steam boilers.

Probe low-water cutoff switch

In the probe type LWCO device there are conductivity probes to detect the water level, which cuts off the fuel supply if the water level falls below the minimum water level to be maintained in the boiler. To keep the relay energized the electrical current flows from the probe through the water. When the water level falls, the probe loses direct contact with the water, so the flow of the electrical current is stopped, which ultimately causes shutting off the fuel-burning equipment. Probe-type LWCO devices are housed within a chamber, which requires minimal maintenance.

Pressure switch

Pressure-limit switches are operated by steam pressure. A typical pressure switch is shown in Figure 8.19. The pressure switch controls pressure to allow burner operation within a fixed range of boiler pressures. Basically, they consist of a bellows assembly linked with a snap switch through a pressure-adjusting mechanism. The bellows assembly expands or contracts with any increase or decrease in boiler pressure and cause the snap switch to open or close its contacts. The pressure-adjusting mechanism is set so that the contacts open and close at specified values of cutout and cut-in boiler pressures. The differential between the cutout and cut-in pressures can be made either fixed or adjustable.

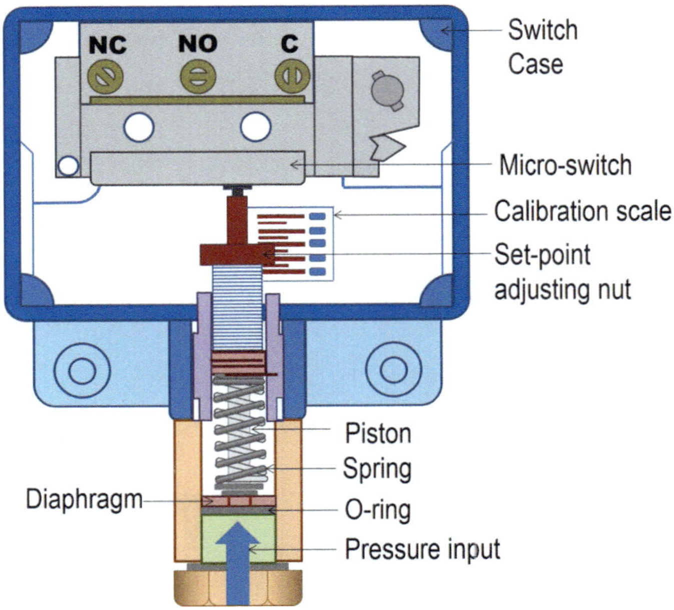

Figure 8.19 A pressure switch.

Pressure switches may be of two types, as mentioned below:

1 Mechanical pressure switch
2 Electronic pressure switch

Fusible plug

A fusible plug shown in Figure 8.20 operates as a safety valve for the boiler when temperature inside reaches above the set danger level in a closed chamber. It extinguishes the fire in the event of the boiler shell failing below a certain specified limit. The fusible plug is installed by screwing it onto the crown of the top plate of the firebox of steam boilers, which is typically kept extending about 25 mm into the water space above it.

Expansion tank

An expansion tank is incorporated for protecting the boiler from excessive pressure and ensuring its safety during operation. An expansion tank is installed on all closed-loop heating systems incorporating a boiler. As the boiler heats the water, the water expands in volume and exerts pressure as water is incompressible. The expansion tank uses compressible air or an inert gas like nitrogen that acts like a cushion against the pressure to get absorbed, as shown in Figure 8.21. It is intended that the excess pressure in the system does not approach the boiler relief valve setting or the relief valve may open. Absorbing initial excess pressure is the main function of an expansion tank.

Flash tank

Flash steam is also steam formed from hot condensate when it is released at reduced pressure. Flash steam occurs when water undergoes from a higher pressure to a lower pressure and flashes off the excess temperature to lower the water temperatures to match the system pressure. The process of flashing condensate can be used to both cool condensates for

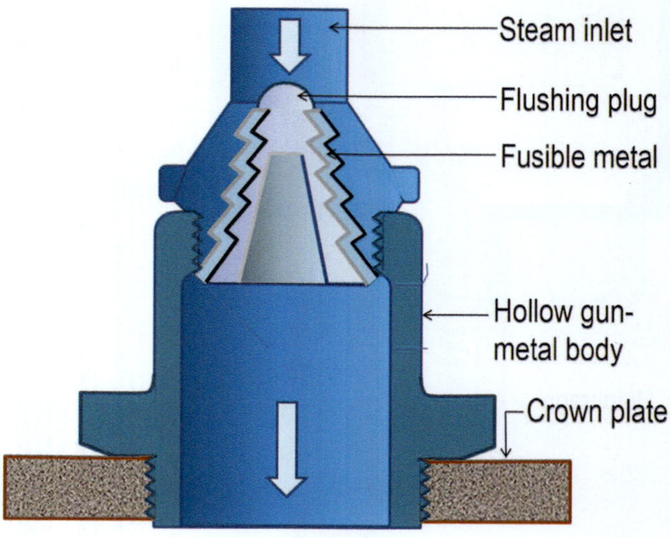

Figure 8.20 A fusible plug.

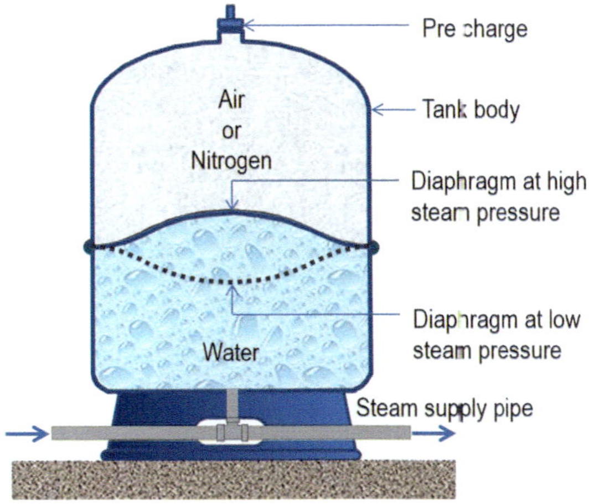

Figure 8.21 An expansion tank.

pumping and also generate opportunities to recover flash steam at lower pressures, which can be used elsewhere. Flash steam works best for heating water or air in a process or application that is continuous. A great example in the boiler plant is to use flash steam to preheat boiler feed-water.

Flash steam can be recovered by two basic devices, as mentioned below:

1 A horizontal flash tank
2 A vertical flash receiver

Figure 8.22 represents a vertical steam flash tank along with its piping supported by valves.

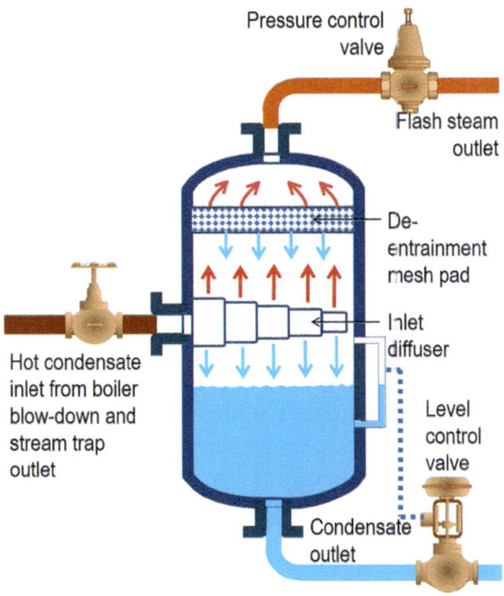

Figure 8.22 A vertical steam flash tank.

Pressure range of boilers

In the early 1900s, as boiler codes and rules were being established, it was decided that a high-pressure boiler would be operating at 1.03 bar and above, with anything lower being defined as a low-pressure boiler. Now boilers operating at 1.03 bar and above are grouped in two categories: medium and high. All these boilers must be able to operate within a maximum allowable working pressure limit. There should be a difference of at least 10 percent between the operating pressure and maximum allowable working pressure.

Based on the pressure range, the boilers can be identified differently as below:

1 Low-pressure boiler
2 Medium-pressure boiler
3 High-pressure boiler

Low-pressure boiler

This type of boiler produces steam below 1.03 bar pressure. These boilers are generally used for heating and some industrial low-pressure steam applications, such as space heating, small process heating, humidification, laundries, bakeries, etc.

Medium-pressure boiler

It has a working pressure of steam in the range between 1.03 and 17 bar pressure. This pressure range is generally used for small industrial steam applications, and may be used in large hospitals.

High-pressure boiler

The boiler in which the pressure is generated above 17 bar is called a high-pressure boiler. These boilers are generally built with thicker steel plates including heavy duty parts to withstand high stress. These boilers are generally used in electric power generation and industrial processes. High pressure boilers are also classified as follows:

1 Ultra-high pressure boiler
2 Subcritical boiler
3 Supercritical and ultra-supercritical boiler

These boilers require comparatively more complex maintenance protocol and skilled operation.

Steam boiler capacity

The output of a steam boiler can be expressed in boiler horsepower (BHP) or in kilograms of steam delivered per hour (kg/h) under specified steam conditions.

The amount of steam delivered by a boiler varies with temperature and pressure. So, a common expression of the boiler capacity is the heat transferred over time, expressed as Watts. Generally, a boiler's capacity is expressed as kilowatts per hour (kilowatts/hour) and can be calculated by the following formula.

$$W = \left(h_g - h_f\right)m \tag{8.1}$$

where,

W = boiler capacity in kW
h_g = enthalpy steam in kJ/kg
h_f = enthalpy condensate in kJ/kg
m = steam evaporated in kg/s

One boiler horsepower is the amount of energy required to produce 15.65 kg of steam per hour at a pressure and temperature of 0 bar and 100°C using feed-water at pressure 0 bar and temperature of 100°C.

One boiler's horsepower is about 9811 watts (8430 kcal/hr).

A BHP is equivalent to 8430 kcal/hr. It is to be noted that a boiler horsepower is 13.1547 times a normal horsepower.

Velocity of steam flow

A reasonable velocity for a particular fluid flowing through pipes is considered the practical sizing factor for the concern pipe. So, in steam supply system, a velocity of 25–40 mps is considered for sizing saturated steam supply pipes. Above this limit, there might be a chance of generating noise and developing erosion, particularly if the steam is wet. Generally, a velocity of 25–40 mps is used when saturated steam is supplied; though some National codes allow velocities up to 76 mps for saturated steam, which might be feasible if the steam is very dry, the pipe is well insulated, relatively short, straight, and horizontal, and the required pressure at the point of use is well managed [4].

Pipes used for steam supply shall have to be sized properly. Use of both undersized and oversized pipes has negative implications for the system, as mentioned below.

Effects of oversized pipework:

1 Investment, operating, and maintenance costs of pipes, fittings, accessories, etc. will be more than necessary.
2 Installation costs will be higher, including insulation, maintenance, etc.
3 A greater volume of condensate will be produced due to the greater heat loss, so more steam trapping will be required, or more wet steam will be delivered to the point of use.

Effects of undersized pipework:

1 Availability of comparatively lower pressure at the point of use; inadequacy of pressure may not be desirable.
2 Excessive pressure drop might cause a risk of steam starvation.
3 An inherent increase in steam velocity might increase the risk of erosion, water hammer, and noise.

Pipes for steam supply

Pipes from carbon steel (ASTM A106) are generally used for steam supply systems; the pipes of the same material are also used for condensate lines. As an alternative, copper pipe is preferred in some cases.

For superheated steam of high-temperature pipe materials, additional alloying elements, such as chromium and molybdenum are preferred to improve the tensile strength and creep resistance of pipe handling fluid at high temperatures.

The piping lengths available are around 5–7 meters for single random length pipes and around 11–13 meters for double random length pipes.

Pipe shall be installed on a slope of 100:1; i.e., for every 100 meter run there shall be a fall of 1 m.

Pumps in steam supply

In a steam supply system, pumps are used at different locations of the supply system for different purposes. The pumps for particular use are as follows:

1 Feed-water pump
2 Condensate pump
3 Steam circulating pump

Feed-water pump

A feed-water pump is used to provide the boiler with the appropriate amount of water needed to maintain safe operation of the boiler. These pumps must generate enough pressure to overcome the steam pressure generated by the boiler, overcoming all the pressure drops in various accessories of the boiler. So, boiler feed pumps are found to be mostly electrically driven multistage pumps, which can be of the centrifugal pump type or positive displacement type. Pump materials are commonly stainless steel or ductile iron.

Boiler feed-water pump control may be of two types depending on the type of feeding water to the boiler: (1) Modulating continuous feed-water or (2) On-off feed-water control.

Modulating continuous feed-water control

Modulating continuous feed control and feeding is based on the water level occurring in the boiler as it gradually lowers with respect to the evaporation rate of water to steam leaving the boiler. In the modulating continuous feed-water system, the pump is run constantly at full rating with an individual actuator-operated modulating feed-water valve controlling the amount of flow to the boiler based on its evaporation rate. Modulating feed-water valves allows for constant incremental adjustments to feed-water into the boiler. In this control system, the water level in the boiler is found to be steady to some extent.

On-off feed-water control

The required water level in the boiler can be maintained by switching the feed pump on and off between two levels of water, typically differing by 50 mm. With the switch off of the pump, the burner will trigger firing to cause the increased steam load and regain the boiler pressure. When the water level drops to the "pump on" level, the pump delivers water at a high flow rate. The low water cutoff and pump control mounted on the broiler, which is electrically connected to the pump circuit, actuates with respect to the specific level reached in the boiler. In this control system, the water level in the boiler is found to be fluctuating.

Boiler feed control methods are as follows:

1 On-off control and fixed speed pumps
2 Modulated feed control valve and fixed speed pumps
3 Modulated feed control valve and variable speed pumps
4 Variable speed pumps only (no control valve)

The base flow rate of the boiler feed pump can be determined from the following formula:

$$Q = \text{Boiler horsepower} \times 0.069 \times C \tag{8.2}$$

where, C = a constant, 2.0 for on-off intermittent operation and 1.5 for modulated continuous feed operation

In SI unit the flow rate is in cum/h. So, the formula is as below:

$$Q = \text{Boiler horsepower} \times 0.069 \times 0.004088 \times C$$

Example: If a boiler has a power of 100 BHP, then the required flow rate for the feed pump would be:

$$Q = 0.069 \times 100 \times 0.004088 \times 1.5 = 0.0423108 \, \text{cum / hr.}$$

Condensate pump

Condensate from boilers cannot be easily drained away under gravity. Therefore, installing a pump can help dispose of condensate effectively. Condensate pumps, basically centrifugal pump types, are used to collect and disperse the hot water and liquid produced due to condensation of steam during its flow in the supply system.

Condensate pumps are designed specially to run intermittently. When condensate accumulates in a collection tank and the liquid in the tank reaches a designated level, a float switch activates the condensate pump to run until the level of liquid reaches the lowest designated level. If the pump fails, discharge is blocked, or the pump is shut off, an alarm is triggered. The system is made fully automatic.

Condensate pumps are generally used for returning the hot condensate back to the feed-water holding tank in the boiler room. Pumps are required to overcome the pressure under gravity, pressure drops in piping, accessories, etc., and backpressures in return lines. Condensate return pumps are made as either electrically driven centrifugal pumps or pressure motive pumps (PMPs) that use steam pressure as the motive force to drive the pump, thereby delivering the condensate.

Circulation pump

The steam generator circulation pumps (CP) circulate feed-water from the hot water drum back to the boiler to temperate the steam temperature.

Boiler circulator pumps are practically rated in accordance with the size of the boiler it is to work for. Most importantly, the flow rate of the pump should be matched to the size of the boiler and the heating system. Another important consideration might be the high suction pressure to raise water from greater depths. In such case, high pressure end suction type circulation pump would be required.

Water heating by steam

Water is generally heated to a maximum of 82.22°C, but must not be more than 93.33°C; therefore, 6.9 kPa steam pressures are sufficient, which can be generated by a low-pressure steam boiler. The steam pressure of 6.9 kPa is not generally used for heating water but is commonly used in vent condensers.

Steam flow requirement for water heating

It is known that temperatures on heated surfaces depend on steam pressure.

Heat transfer

The mean heat transfer can be expressed by the following formula:

$$P = c_p \Delta T m / t \tag{8.3}$$

where,

P = mean heat transfer rate (kW, kJ/s)
m / t = product's mass flow rate (kg/s)
c_p = specific heat of the product (kJ/kg°C)
t = change in fluid temperature (°C)

Amount of steam

If we know the heat transfer rate, the amount of steam can be calculated by the following formula:

$$m_s = P / h_e \tag{8.4}$$

where,

m_s = mass of steam (kg/s)
P = calculated heat transfer (kW)
h_e = steam evaporation energy (kJ/kg)

The evaporation energy at different steam pressures can be found in the steam table, a part of which is furnished in Table 8.2.

Example of continuously heating water by steam

Let us consider that water from a hot water storage tank is flowing at a constant rate of 3 lps, which is to be heated from 10°C to 60°C with steam at 8 bar (9 bar abs).

Table 8.2 Steam table (part). Properties of steam at varying pressures and temperatures [5]

Absolute pressure (kPa, kN/sqm)	*Evaporation temperature (°C)*	*Specific volume (cum/kg)*	*Density: ρ (kg/cum)*	*Specific enthalpy of*			*Specific entropy of steam: s (kJ/kgK)*
				Liquid: h_l *(kJ/kg)*	*Evaporation:* h_e *(kJ/kg)*	*Steam:* h_s *(kJ/kg)*	
0.8	3.8	160	0.00626	15.8	2493	2509	9.058
2.0	17.5	67.0	0.0149	73.5	2460	2534	8.725
5.0	32.9	28.2	0.0354	137.8	2424	2562	8.396
10.0	45.8	14.7	0.0682	191.8	2393	2585	8.151
20.0	60.1	7.65	0.131	251.5	2358	2610	7.909
28	67.5	5.58	0.179	282.7	2340	2623	7.793
35	72.7	4.53	0.221	304.3	2327	2632	7.717
45	78.7	3.58	0.279	329.6	2312	2642	7.631
55	83.7	2.96	0.338	350.6	2299	2650	7.562
65	88.0	2.53	0.395	368.6	2288	2657	7.506
75	91.8	2.22	0.450	384.5	2279	2663	7.457
85	95.2	1.97	0.507	398.6	2270	2668	7.415
95	98.2	1.78	0.563	411.5	2262	2673	7.377
100	99.6	1.69	0.590	417.5	2258	2675	7.360
101.331	100	1.67	0.598	419.1	2257	2676	7.355
110	102.3	1.55	0.646	428.8	2251	2680	7.328
130	107.1	1.33	0.755	449.2	2238	2687	7.271
150	111.4	1.16	0.863	467.1	2226	2698	7.223
170	115.2	1.03	0.970	483.2	2216	2699	7.181
190	118.6	0.929	1.08	497.8	2206	2704	7.144
220	123.3	0.810	1.23	517.6	2193	2711	7.095
260	128.7	0.693	1.44	540.9	2177	2718	7.039
280	131.2	0.646	1.55	551.4	2170	2722	7.014
320	135.8	0.570	1.75	570.9	2157	2728	6.969

The heat flow rate is as follows:

$$P = 4.19\,\text{kJ/kg.}^\circ\text{C} \times (60 - 10)^\circ\,\text{C} \times 3\,\text{lps} \times 1\,\text{kg/l}$$
$$= 628.5\,\text{kW}$$

The steam flow rate is as follows:

$$m_s = 628.5\,\text{kW} / 2030\,\text{kJ/kg}$$
$$= 0.31\,\text{kg/s} = 186\,\text{kg/h}$$

Space heating by steam

A space heating by steam can be efficiently done by taking advantage of the high latent heat that is given off by steam when it condenses to liquid as water. In this system, a volume of space is equipped with a radiator, which receives flow of steam from a source of low-pressure steam, generally a steam boiler. While steam flows through the radiator, it gives up its latent heat, thereby condensing and eventually returning to liquid water.

The released latent heat of steam from the radiators, which basically provide radiant heat directly in the air of a space, and the condensate water generated are returned back to the boiler either by gravity or with the help of a circulator pump.

Steam for space heating can be efficiently supplied long away, between buildings on a campus or the all the floors of a high-rise building, from an efficient central steam boiler using low-cost fuel, taking advantage of the low density of steam.

Advantages of steam heating:

1 The principle is simple, and there are involvements of few moving parts.
2 A well-maintained steam heating system provides better comfort while running efficiently.
3 Generally, with few exceptions, there is no need of a fan or pump associated with moving the heat through the distribution system.

Disadvantages of steam heating:

1 The working environment of a steam boiler is harsh because of the high temperatures and the changes of state that take place.
2 The interior components are variously exposed to air, water, and steam.
3 Steam heat does not support air conditioning, which would be expensive to retrofit.
4 Steam radiators can produce annoying sound.

Comfortable room temperature

The recommended room temperature comfortable for the occupants in a building is general, which may vary slightly subject to a person's concern. The comfortable room temperatures in different rooms of particular use have been furnished in Table 8.3.

Table 8.3 The comfortable room temperatures in different rooms of particular use and the corresponding wattage required [6]

Season	*Room type*	*Recommended temperature (°C)*	*Room type*	*Recommended temperature (°C)*
Winter	Bathroom	22	Libraries	20
	Lecture rooms	20	Classrooms	20
	Living rooms	21	Corridors	16
	Study	22	Offices	20
	Kitchen	20	Recreation rooms	18
	Bedroom	18	Exhibition halls	18
	Dining rooms	20	Restaurants	18
	Hotel rooms	21	Shops	18
	Stores	15	Laboratories	20
	Wards	18	Warehouses	16
Summer	All rooms 20–22°C			

Note: Per degree temperature energy consumed is about 3.864 watts per cum.

Major components of steam supply for space heating

The steam supply system for space heating requires the following major components:

1 Steam boiler
2 Steam supply pipe
3 Radiator
4 Condensate return pipe
5 Circulator pump
6 Vent valves
7 Pressure-reducing valve

Steam boiler for space heating

In domestic and small commercial buildings, the steam is generated at relatively low pressure, less than 200 kPa steam boilers. Steam boilers utilize the natural pressure of steam to distribute heat, eliminating the need for pumps and regulators. Boilers must provide consistent and high-quality steam.

Radiator

Radiators, also known as convectors, are basically heat exchangers designed to transfer thermal heat energy from one medium to another, which is also used for the purpose of space heating, transferring heat from steam to air. Steam radiators are those in which steam is used for heat transfer and are generally installed on walls. There are also freestanding stand-up radiators, generally with a visible set of vertical fins flanked by pipes, mounted on the floor, as shown in Figure 8.23.

Radiators are generally constructed as a set of vertical fins flanked by pipes. The fins become hot and distribute the heat into the air of a space. Radiators virtually cause heat loss by radiating heat to the indoor space, thereby warming the space.

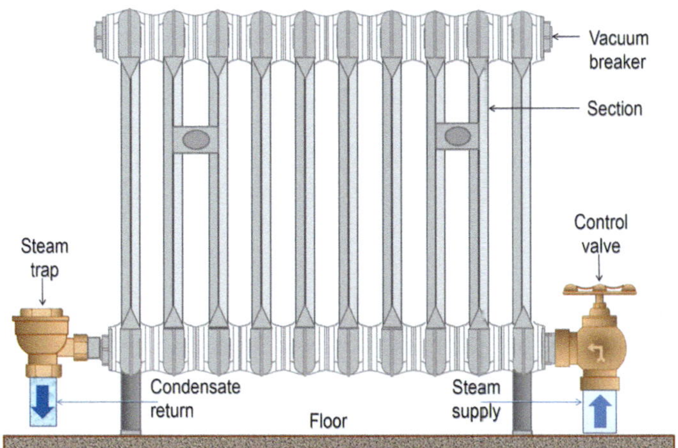

Figure 8.23 A floor-mounted steam heat radiator.

To prevent heat loss, a heat reflector is placed behind these radiators. The reflectors are made either the same size or slightly larger than the radiator. The reflectors shall be cleaned periodically to maintain maximum heat reflection.

Radiators emit different amounts of heat based on their size and the radiator's output. The radiator width considered is the length between the two valves at either end, rather than the panel itself.

Based on the material of construction, radiators may be of the following types:

1 Cast iron radiators
2 Steel radiators
3 Ceramic radiators

Among these radiators, cast iron radiators are an old technology and hot to the touch, but ceramic radiators are not hot to the touch.

Based on ways of mounting, radiators may be of the following types:

Wall-hung radiator: Wall-hung radiators are supported on the appropriate heavy-duty brackets or wall stays for a fully secure installation on a wall. Wall stays shall be of brass, chrome, and stainless steel.

Floor-mounting radiator: It is recommended to put a radiator at low level, preferably under the window, for better performance. In this perspective, floor-mounting radiators are preferred, which are mounted on the floor.

Under floor radiator: Though it is technically possible to integrate steam radiators into a floor system, it is not a common or standard practice. But it is possible to mount steam radiators below the floor subject to fulfillment of the conditions that the steam shall be able to get in, the air shall be able to get out, and the condensate could be drained out easily.

Air vent valve

The radiator air vent, also known as the air valve, steam vent, or steam valve, is installed only on radiators of one-pipe steam boiler systems. It is usually located at the end of the radiator, opposite the supply pipe, about halfway up or toward the top. Mostly air vents are found bullet-shaped, but they can be many different shapes and sizes.

Steam flows from the boiler to the radiators, forcing entrapped air out through the air vent valve. The air vent valve closes automatically when the radiator becomes full of steam.

It is recommended to use a 25 mm valve for radiators up to 5275 kJ or fewer and 32 mm valves above that [7].

Steam distribution systems for space heating

For space heating, steam supply from the boiler to the radiators and returning condensate back to the boiler can be performed by the following piping layout systems:

1 One-pipe system
2 Two-pipe system

One-pipe system

In one pipe radiator, the steam comes from a boiler through a pipe; as shown in Figure 8.24, the heat is dispersed through the radiators, the steam eventually condenses in the radiator, and the condensate water returns to the boiler through the same pipe. In one pipe system, a radiator with an air vent on its side is used, as shown in Figure 8.25.

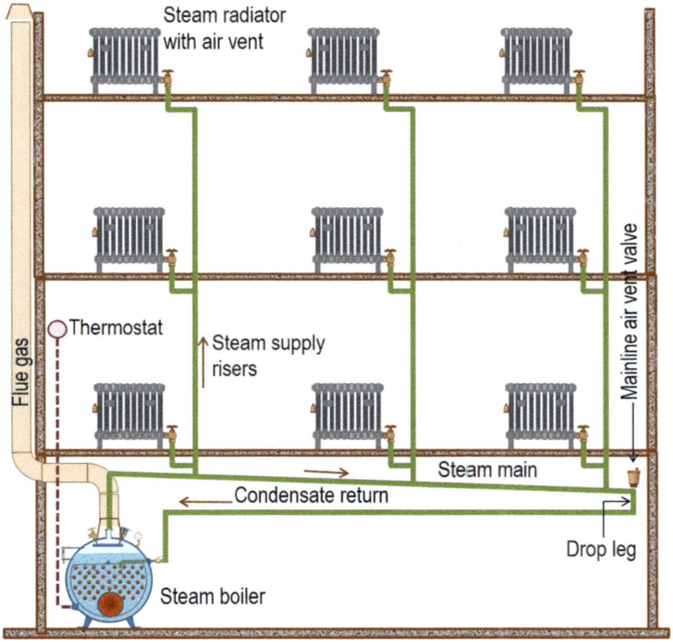

Figure 8.24 Schematic of a one-pipe gravity return steam supp-y system for space heating.

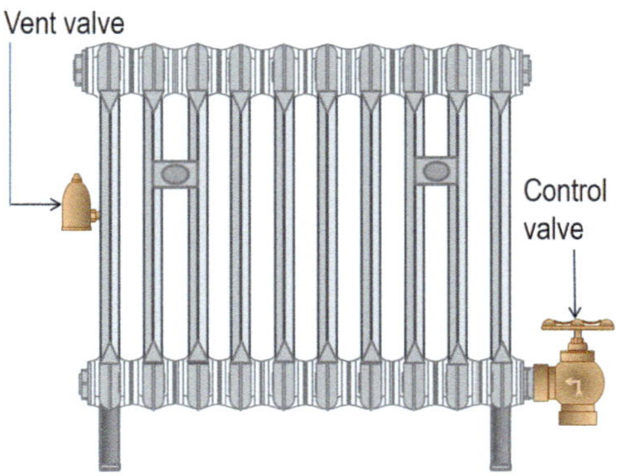

Figure 8.25 A one-pipe radiator with a vent valve on the side.

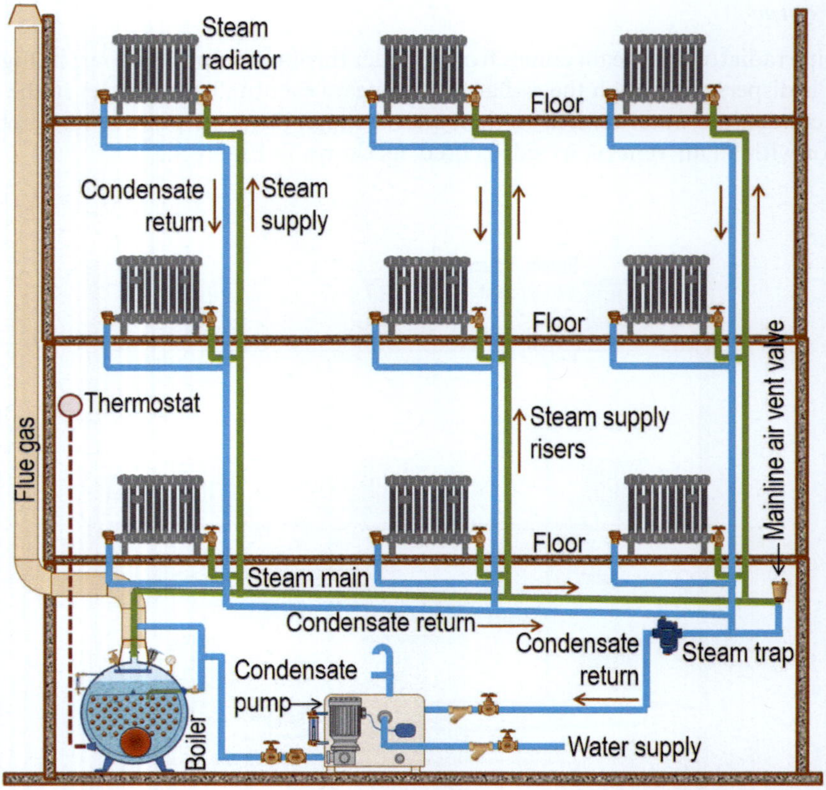

Figure 8.26 Schematic of a two-pipe steam supply system for space heating.

Two-pipe system

In a two-pipe system, the steam comes from a boiler through a piping system, the heat is dispersed through the radiators, the steam eventually condenses, and the condensing water is directed toward the outlet through another pipe to return condensate water to the boiler, as shown in Figure 8.26. In a two-pipe system, the radiator does not have an air vent valve on its side. The radiator shown in Figure 8.23 is a two-pipe radiator.

Steam supply for steam bathing

A steam shower is a type of bathing, also termed steam bathing, which is done under water vapor that is dispersed around a person's body inside a self-contained enclosure. A steam shower is believed to offer various health benefits. Some physiotherapists recommend steam as a rehabilitation practice. Steam is often cited as a reliable way to temporarily alleviate the symptoms of the common cold. Some physicians recommend steam baths as a therapy to relieve the pain and stiffness associated with ankylosing, spondylitis, and rheumato-arthritis, etc.

The self-contained enclosure for steam bathing is essentially a steam room that offers the typical features of a bathroom shower, which prevent the water vapor from escaping into the surrounding rooms, which might cause damage to the drywall, paint, wallpaper, etc. of those rooms due to over humidification by steam. Steam showers combine the functionality of a standard steam room with many additional features, including a shower and hydrotherapy. Steam shower heads are the apparatus in a shower from which the steam comes out, and they are available in a variety of materials, such as tile, acrylic, fiberglass, glass, stone, and wood.

Steam bathing system development requires various compulsory elements and a few optional. The major components are as follows and are shown in Figure 8.27:

1 Bathing enclosure or stall
2 Steam bath generator
3 Steam distribution piping
4 Steam shower head
5 Fragrance pump
6 Temperature sensor and control

Steam bathing stall

Steam bathing stall units shown in Figure 8.27, are generally sized based on the stall's volume in cubic meters, its shape, and the finishing material of the wall. A typical 1.2 m by 1.5 m stall of 2.4 m high, i.e., a stall of 4.32 cum steam bathing stall may be the smallest size of a steam bathing stall.

The finishing of stall walls influences the steam-generating capacity. Walls being finished with ceramic tile will require a generator that would be half the capacity of a steam generator for a stall of the same volume, but the stall walls are finished with stone.

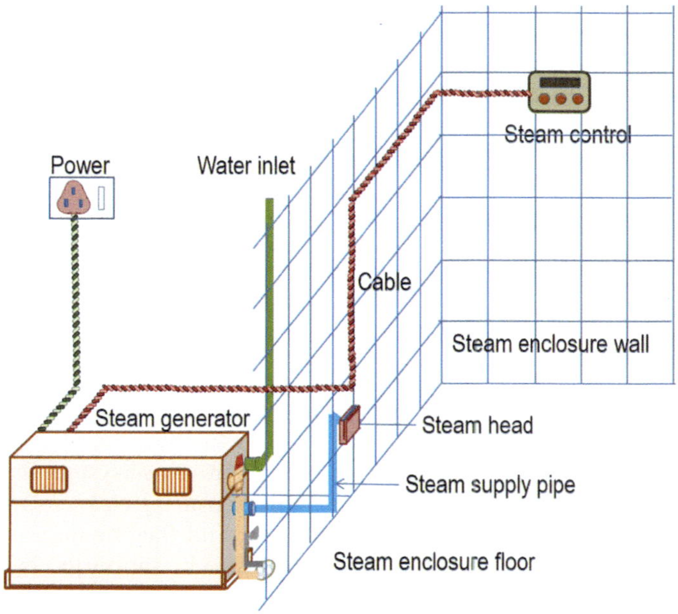

Figure 8.27 Steam bath enclosure and the system components.

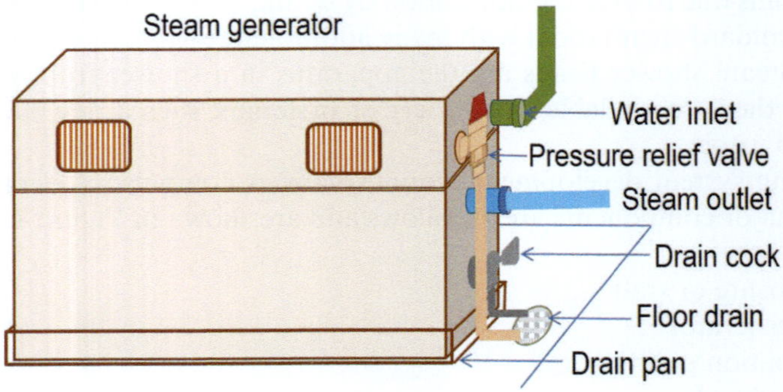

Figure 8.28 Bathing steam generator.

A wall finished with ceramic tile requires a generator of at least 7 kW; however, stalls finished with natural marble, stone, slate, glass block, or concrete require approximately twice that steam-generating capacity [8].

Steam generator for bathing

A steam generator, shown in Figure 8.28, is commonly a rectangular boiler used to heat water up to a specific temperature under required pressure that creates steam. The steam generated is conveyed through a pipe to the shower enclosure to emit steam through the steam head installed inside the enclosure, to instantly fill the room with warm steam.

Steam bath generators typically produce steam for bathing at a temperature range of 100 to 120°C and under atmospheric pressure of 101.3 kPa, and the steam almost at this temperature is emitted from the steam-head, which reaches the bather at a comfortable and effective range of temperature between 38 and 43°C.

Steam generators of various capacities ranging between 5 kW and 30 kW are used subject to the type or size of bathing enclosure, as mentioned in Table 8.4, made with a wide variety of control options and accessories, as mentioned below:

1 Stainless steel water reservoir inside
2 Serviceable heating element
3 Modular "snap-in" connections
4 Electronic water-level control system
5 Full-port drain valve
6 Built-in, low-voltage 24-volt control
7 60-minute electronic countdown with 75-minute limiting safety backup safety valve

Steam generators might be supported by an optional auto-flush system that automatically and electronically drains the generator after each use, ensuring clean steam all the time. It allows the introduction of fresh water prior to using, and thereby improves performance and lengthens system life as well. Generator should also be supported by a condensation pan with a 20 mm diameter drain cock fitting, which might be required by building code. A stainless steel pan is preferable for avoiding corrosion. Drain cock shall discharge into a floor drain supported by a water seal trap and drain pipe withstanding rated temperature.

Table 8.4 Steam generator rating for various volume of steam bathing enclosure

Bathing room volume (cum)	*kW rating*	*Volts*	*Amps*
2.83	5.0	240	21
4.25	6.0	240	25
6.37	7.5	240	32
10.19	9.0	240	38
13.45	10.0	240	42
16.28	12.0	240	50
19.11	15.0	240	63
24.78	20.0	240	84
30.44	24.0	240	100
36.10	30.0	240	126

Conversion in SI unit from source [9] Provided with the Permission of Almost Heaven Group

Note:

- Add 15 percent to capacity for each wall that's exterior to the outside.
- Add 15 percent to capacity for each wall that's tile over concrete or concrete block.
- Add 15 percent to capacity for each wall (other than door) that's glass.
- Double capacity for steam-room enclosures made entirely of natural marble, stone, slate, glass block, or concrete.

To ensure perfect functioning and a long service life of a steam generator, the operating conditions must be supportive. During operation, an ambient temperature of 5–25°C in the shower enclosure is required for optimum mist formation as steam generators operate technically up to 40°C. The steam connection pipe should have a maximum length of 10 meters with a maximum of four bends. The ambient humidity of the device should be 75 percent of relative humidity (RH). Water quality and temperature play an important role for optimum evaporation. Untreated or de-mineralized (de-ionized) water with an inlet temperature of 1–40°C performs the best. The water connection should be provided with 12 mm diameter pipe supported by an angle valve and having a pressure of 1–10 bar.

Steam shower-head

The steam showerhead virtually disperses steam into the shower enclosure, which is generally installed into a wall of the steam shower stall. These are available in a variety of styles and finishes to complement other fixtures in the stall. It may be installed in conjunction with an aromatic steam system as an optimal requirement.

The steam shower-head should include the following features that are highly preferred:

1 The slot area being three times the cross-sectional area of the supply pipe
2 Lower velocity of flow producing less noise
3 Partially "pre-distributed" steam supply
4 Eliminates the pooling of condensate inside
5 Highly corrosion-resistant construction and finish (preferably anodized aluminum)

Controller

Control acts as a secondary power control. The controller provides simple on and off actuation of steam flow for the steam head. There are controllers that allow seamless control of various sorts of steam room functionality over using remote control devices.

Steam for indoor humidification

When the indoor air is heated, the relative humidity of the indoor air drops. Too dry indoor air may cause electrostatic shocks that cause a nuisance, which occur when relative humidity falls below a threshold limit of 40 percent. So, in order to ensure a comfortable, healthy, and productive indoor environment, preservation of books and records, and infection control, at least 40 percent and maximum 60 percent relative humidity is to be maintained by employing a suitable method of humidification. In large buildings, especially in colder climates, use low-pressure saturated steam as the predominant heat source for indoor heating and ventilation systems, often combined with steam humidifiers, for conditioning the indoor air.

Steam radiators installed for space heating may emit steam through leakages, which generate humidity into the rooms. The humidity thus caused by the steam makes an indoor feel more comfortable, particularly in dry winter months. It is to be observed that the indoor humidity should never exceed the maximum allowable percentage of humidity, as according to the Environmental Protection Agency, the best indoor relative humidity should be below 50 percent.

Steam for sterilization

Steam can be used for sterilization using an appliance called a steam sterilizer, also known as an autoclave. A typical standard for steam sterilization is achieved once all surfaces of an element to be sterilized have reached the desired temperature. Autoclaves operating at higher temperatures and sterilizing in a short period are preferred over those operating at lower temperatures but running longer to sterilize. Available standard temperatures to pressures employed in autoclaves are 115°C under 67 kPa, 121°C under 103 kPa, and 134°C under 186 kPa. Operating of an autoclave at 121°C requires about 15 minutes or 134°C requires about 3 minutes of sterilization.

The pressure chamber is the main component of a steam autoclave, consisting of an inner chamber ranging from 100 to 3000 liters, in which the materials are sterilized, which is made up of stainless steel or gunmetal and an outer jacket. Generally, an electrical steam generator or boiler is put underneath the chamber that uses an electric heating system to heat the water and generate steam in the inner and outer chambers. There is an airtight lid to seal off the outside atmosphere and create a sterilized condition on the inside. Autoclaves are supported by a pressure gauge, a pressure relief valve, and a safety valve, as shown in Figure 8.29.

Autoclave effluents are too hot. Some autoclaves are provided with a system to cool the effluent before it is discharged into the draining pipes.

Piping for steam supply

Metal pipes manufactured from carbon steel to ASME B 16.9 A106 are generally preferred for steam supply; the same material may be used for condensate lines as well. Alternatively, copper pipes are also preferred in some cases. For conveying high-temperature superheated steam, additional alloying elements, such as chromium and molybdenum, are included in pipe material to improve tensile strength and creep resistance at high temperatures. Typically, these pipes are manufactured in 6-meter lengths.

Any surface over 49°C should be insulated, including steam and condensate return piping, fittings, valves, etc. The most common types of insulation materials are calcium silicate and mineral fibers. Fiber or cellular glass is also preferred for steam pipe insulation.

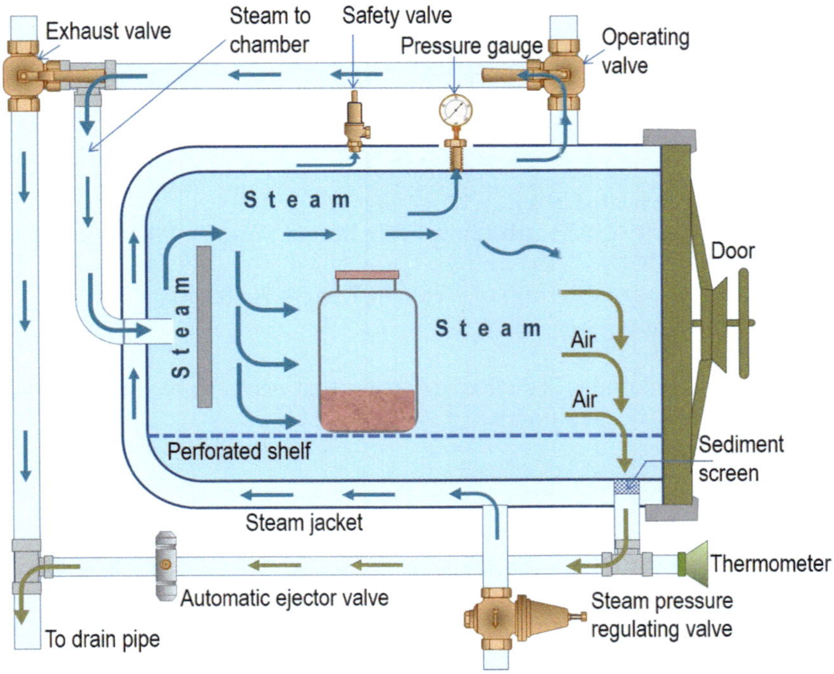

Figure 8.29 Gravity displacement type autoclave.

In steam supply systems, pipes must be carefully sloped to collect condensate properly and prevent trapped condensate blockage in the piping.

Design of major components of a steam supply system

Sizing a boiler

The right selection of a steam boiler output is very important for its smooth operation at its maximum efficiency and longer performance. To find the size of a boiler, it is to be known the flow rate of steam to be generated and the pressure needed to reach the steam at the desired location or area.

The basic equation to calculate the kJ required to heat any substance is as follows:

$$Q = mc_p \Delta T \tag{8.5}$$

where,

Q = amount of heat or energy (kJ)
m = mass of substance (kg)
c_p = specific heat of substance (kJ/kg °C)
ΔT = temperature rise of substance (°C).

The steam flow rate is a function of the total kJ required over the length of time to achieve the temperature change.

The mass flow rate of steam can be expressed as follows:

$$Q = \frac{W \times Sp \times t}{L \times h} \text{kg/h} \tag{8.6}$$

where,

W = Mass of substance heated in kg
S_p = Specific heat capacity of the substance being heated (Water) = 4.186 kJ/kg°C
t = Temperature rise in °C
L = Latent Heat of evaporation (hfg) of steam at 7 bar = 2048 kJ/kg
h = Recovery time in hours

Example: A tank containing 2500 liters of water that needs to be raised in temperature from ambient 10°C to 70°C in 2 hours.

Given that:

W = Mass of substance to be heated = 2500 liter
S_p = Specific heat capacity of the substance being heated (Water) = 4.186 kJ/kg°C
t = Temperature rise = 70−10 = 60°C
L = Latent Heat of evaporation (hfg) of steam at 7 bar = 2048 kJ/kg
h = Recovery time = 2 hours

So,

$$\begin{aligned}\text{mass flow rate of steam required} &= 2500 \times 4.186 \times 60 / (2048 \times 2) \\ &= 154\,\text{kg/h of steam.}\end{aligned}$$

Steam boiler efficiency

The percentage of total heat exported by outlet steam in the total heat supplied by the fuel (coal) is called steam boiler efficiency.

$$\text{Steam boiler efficiency}(\%) = \frac{\text{Heat exported by outlet steam}}{\text{Heat supplied by the fuel}} \times 100 \tag{8.7}$$

The efficiency includes thermal efficiency, combustion efficiency, and fuel-to-steam efficiency. Steam boiler efficiency depends on the size of the boiler used. Typically, the maximum efficiency of a steam boiler running on coal, oil, natural gas, and biomass can reach a limit of 85 percent, 80 percent, 75 percent, and 70 percent, respectively [10]. The inefficiency of the boilers is due to some occurring losses, such as incomplete combustion, radiating losses occurring from the steam boiler surrounding the wall, defective combustion gas, etc.

Sizing feed-water pump

While going for sizing a feed-water pump for a boiler, it is important to know the set pressure at which the boiler operates, as the pump should fill the boiler in the event of an activated safety valve. This requires that the pump be able to reach the safety valve pressure set point.

For a complete pump selection, the total head calculations are required by taking into account the pressure losses in the pipes upstream and downstream in addition to considering the water level heights. Since the feed-water temperature is high and close to the saturated temperature, it is also very important to calculate the available NPSH of the system.

In order to properly supply adequate water to the boiler at its maximum operating pressure, it is required to provide a flow greater than the maximum rated boiler capacity, so that it can raise the water level even when the boiler is operating at full capacity. The required pump flow will be varying depending on whether the boiler has an interruptible type of control (on/off) or a modulating level control. There are two simple rules regarding the required pump flow rate, as follows:

1 Case with a modulating water flow valve, the pump flow rate = 1.25 × the boiler capacity.
2 Case without a modulating water flow valve, the pump flow rate = 2 × the boiler capacity.

Feed-water tank capacity: In most cases, a minimum of ten minutes of water flow should be readily available in the feed-water tank for the boiler [11].

Example:

Steam flow: 65 tons/hour (Q boiler), maximum pressure: 34 bar (P boiler operation), temperature (max): 105°C, Height of feed tank: 4 m, Available pressure in tank: 2 bar, suction pipe: 150 mm.

From the boiler required, we have the following information:

$$Q(\text{Boiler}) = 65\,\text{tons / hour}, P(\text{Boiler max}) = 34\,\text{bar}, \text{Temp}(\text{max}) = 105^{\circ}\text{C}$$

From the steam/vapor table, the following data on water with a temperature of 105°C is found: Density (ρ) = 954.465 kg/cum; Vapor pressure = 1.2 bar.

$$\text{Boiler steam flow}\, Q = 65 \times 103 / 954465 = 70.14\,\text{cum / h}$$

$$\begin{aligned}\text{Pump operation flow} &= 1.25 \times \text{boiler flow}\\ &= 1.25 \times 70.14 = 87.68\,\text{cum / h; Say, } 88\,\text{cum / h}\end{aligned}$$

$$\begin{aligned}\text{Feed pump head}\, H &= 1.1 \times \text{boiler head}\\ &= 1.1 \times 34 \times 10.19 = 381.11\,\text{m water; Say, } 382\,\text{m water}\end{aligned}$$

Sizing steam radiator

In the efforts of finding the proper size of steam radiators for achieving the ideal level of heat in a space or room, the following information is required:

1 The Watts required
2 The volume of space to be warmed in cum
3 Selection of types of radiators
4 Number of radiators of particular size for an even heat

To figure out the size of a radiator, it is to be known how much heat in Watts the space in cum requires. After knowing how many Watts are required, divide it by the number of radiators to work out which size each radiator should be. It is recommended that a radiator size should be within a 10 percent range of the Watts requirement for heating a space.

An example of the energy (in watts) required to raise the temperature of an average room by 10°C is given below:

Dry air-specific heat C_P = 1.00 J/g K
Dry air density ρ = 1275 g/cum
Room area A = 30 sqm
Room height h = 3 m
Mass of air m = ρAh
Change in temperature from 15°C to 22°C; so, ΔT = 22−15 = 7 K.

So, energy required to heat the room:

$$U = C_P m \Delta T = 1 \times 1275 \times 30 \times 3 \times 7 = 803\,\text{kJ}$$

Time required in raising the temperature with a 1000 Watt space heater:

$$P = U / \Delta t$$

So, $\Delta t = U / P$ = 803,000/1000 = 803 seconds ≈13.5 minutes.

Determining steam flow for radiator

Heating indoor air with steam heating by steam radiators needs sufficient steam to flow through.

The formula for heating air with steam is as follows:

$$Q = q \Delta T c_p \rho / h_e \tag{8.8}$$

where,

Q = Steam consumption (rate of condensation), (kg/s)
q = volumetric air flow rate (cum/s)
ΔT = Temperature difference in air flow (°C)
C_p = Specific heat of air (kJ/kg °C)
ρ = Density of air (kg/cum)
h_e = Specific enthalpy of evaporating (condensation) steam (kJ/kg)

Example: 1 cum/s of air is heated from 10°C to 30°C with steam at atmospheric pressure. The steam flow rate (consumption) can be calculated as below:

$$\begin{aligned} Q &= 1 \times (30 - 10) \times 1.0 \times 1.204 / 2257 \\ &= 0.011\,\text{kg/s} \end{aligned}$$

Sizing feed-water tank

It is to be kept in mind that sufficient water should be available for a boiler at any point in time. The feed-water tank should be correctly sized. To design a feed-water tank, following information is needed:

1 Boiler's power (hp): Amount of steam the boiler is capable of generating.
2 Residence time of water: Usually 10 minutes to de-aerate if a surge condensate tank exists or 15–20 minutes if not.

3 Average converting factor: A boiler on average can generate 15.6 kg of steam from each hp.
4 Water density at operating temperature: The recommended temperature to operate is −2°C from saturation temperature.

There is a rule of thumb that the feed-water tank should be sized to 1.5 times of the peak steam demand. So, if the peak steam demand is 3 tons per hour, then the feed-water tank should be 4.5 kl.

There is another practice which instructs ten minutes of water flow should be kept readily available for a boiler.

$$\text{Feed-water requirement}(\text{tons / hr}) = \text{Total steam generation rate}(\text{tons / hr}) \\ / \text{Boiler efficiency}(\text{percent})$$

$$\text{Feed-water tank capacity}(\text{tons}) = \text{Feed-water requirement}(\text{tons / hr}) \\ \times \text{Operating time}(\text{hour}), \text{ the tank is to supply water without needing a refill,} \\ 1 \text{ to } 4 \text{ hours, depending on reliability and operational needs.}$$

$$\text{Feed-water tank capacity}(\text{liters}) = 1000 \times \text{Feed-water tank capacity}(\text{tons})$$

To calculate the storage tank needed the following formula is used:

$$V = \frac{W}{\rho} \times \frac{t}{60} \tag{8.9}$$

where,

V = Volume of tank (cum)
W = Total steam generation rate (tons/hr) = Feed-water requirement (tons/hr) × Boiler efficiency (percent)
ρ = Specific gravity of water at 98°C = 959.49 kg/cum
t = Time (hour)

Example: For a 500 HP boiler of 85 percent efficiency, the tank capacity –

$$W = 500 \times 15.6 = 7800\,\text{kg} = 7.8\,\text{ton of steam}$$
$$\text{Feed water requirement}(\text{tons / hr}) = W / E = 7.8 / .85 = 9176\,\text{kg} \\ = 9176 / 959.49 = 9.563\,\text{cum}$$

For 10-minute run, volume required = 9.563 × 10/60 = 1.594 cum, 1594 liter
Add 50 percent for safety. So, volume required = 1.5 ×1594 = 2390 liter;
So, per hp of a boiler feed-water tank size would be about = 2390 /500 = 4.78 liter.

The typical practice is to use about 4 liters per boiler horsepower of capacity, when the condensate return is less than 50 percent. If the condensate returns exceed 50 percent of the total unit flow, the capacity should be increased to 6 liters or 8 liters per boiler horsepower [12].

Sizing pipes for steam supply

Selection of the appropriate pipe size, for a particular application of steam supply, is based on accurately determining available pressure and flow rate. So, the pipe size may be selected on the basis of the following two criteria:

1 Velocity
2 Pressure drop; as a general rule, the pressure drop should not normally exceed 0.1 bar per 50 m length of pipe

Two principal factors with respect to pressure, as mentioned below, implicates in determining pipe sizes for a steam supply system:

1 The pressure developed by the boiler and the allowable pressure drop in the supply system: The total pressure drop in the supply system should not exceed 20 percent of the total maximum pressure generated at the boiler. The total pressure drop in the supply system includes pressure drop in the pipe, fittings, valves, etc.
2 Steam velocity: The allowable velocity range for process steam is between 30 and 60 mps, but lower pressure heating systems normally have lower velocities.

Pipe size determination should address future expansion of the system where there is possibility for expansion. So, pipes shall be sized based on the foreseeable future expansion. There may be less trouble with slightly oversized pipes considering future expansion than with ones that are designed marginally.

Table 8.5 Sizing steam pipes based on velocity [13]

Steam flow capacity (kg/h)

Gauge pressure (bar)	Steam velocity (mps)	Nominal pipe size (mm)										
		15	20	25	32	40	50	65	80	100	125	150
		Inside diameter (mm)										
		15.8	20.9	26.6	35.0	40.9	52.5	62.7	77.9	102.3	128.2	154.1
0.4	15	9	15	24	42	57	95	135	208	359	564	815
	25	14	25	41	70	96	158	225	347	598	940	1358
	40	23	40	65	112	153	252	360	555	958	1504	2173
0.5	15	9	16	26	45	61	101	144	222	383	602	869
	25	15	27	43	75	102	168	240	370	638	1003	1449
	40	24	43	69	120	163	269	384	592	1021	1604	2318
0.7	15	10	18	29	51	69	113	162	250	431	676	977
	25	17	30	49	84	115	189	270	416	718	1127	1628
	40	27	48	78	135	184	302	431	666	1148	1803	2605
0.9	15	11	20	32	56	76	126	179	277	478	750	1084
	25	19	33	54	93	127	210	299	462	796	1250	1806
	40	30	53	86	149	204	335	478	738	1274	2000	2890
1	15	12	21	34	59	80	132	188	291	501	787	1137
	25	20	35	57	98	133	220	314	484	835	1312	1895
	40	32	56	91	157	214	352	502	775	1336	2099	3032
2	15	17	31	50	86	117	193	275	425	733	1151	1663
	25	29	51	83	143	195	322	459	708	1221	1918	2771
	40	47	82	133	229	312	515	734	1133	1954	3069	4434

Sizing based on velocity

Pipe sizes for a steam distribution can be effectively done based on velocity limitation of steam flow when the pipe length is found to be less than 50 m. In general, steam velocity of 25 mps is considered to be sufficient for most saturated steam applications [13].

Sizing based on pressure drop

As steam is a very compressible gas, the mass flow capacity of a pipe supplying steam depends on steam pressure. Steam pipes can be sized with the Table 8.5 and Figure 8.30, where pressure in bar, velocity in mps and capacity in kg/h.

Example: Finding capacity of 25 mm steam pipe

From Figure 8.29, the capacity of a 25 mm steam pipe with 6.7 bar pressure and steam velocity 25 mps is 200 kg/h.

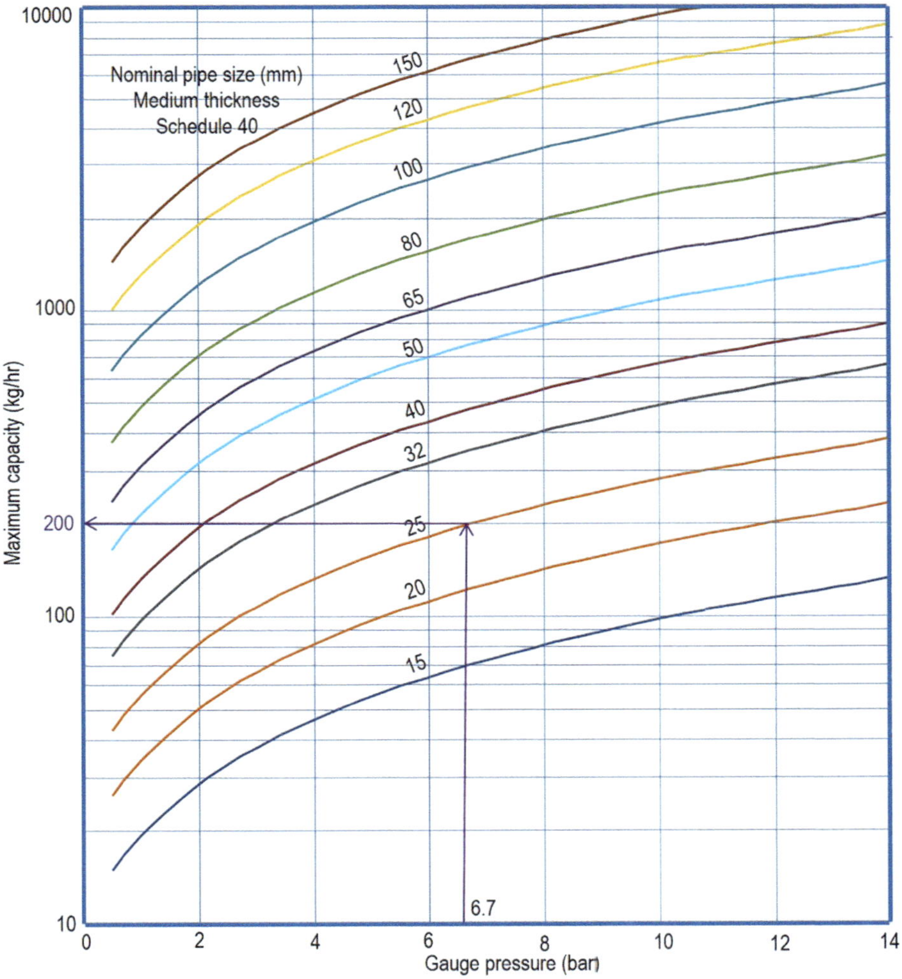

Figure 8.30 Chart for sizing steam supply pipes for flow velocity 25 mps [13].

References

[1] ARI-Armaturen Albert Richter GmbH & Co. (2024) "Steam Paathshaala" Maths Retrieved from https://www.ari-armaturen.com/in/steam-paathshaala/maths on 30 September, 2024.
[2] Electrical 4 U (2021) "Steam boiler: working principle and types of boiler" Retrieved from https://www.electrical4u.com/steam-boiler-working-principle-and-types-of-boiler/ on 21 May 2024.
[3] e-Krishi Shiksha (2013) "Thermodynamics and heat Engine4" Retrieved from http://ecourses online.iasri.res.in/mod/page/view.php?id=2422 on 21 May, 2024
[4] Spirax Sarco Limited (2024) "Steam distribution" Retrieved from https://www.spiraxsarco.com/learn-about-steam/steam-distribution/pipes-and-pipe-sizing?sc_lang=en-GB on 25 May, 2024.
[5] The Engineering ToolBox (2003) "Properties of saturated steam - SI units" Retrieved from https://www.engineeringtoolbox.com/saturated-steam-properties-d_101.html on 2 October, 2024.
[6] US EIA. Residential Energy Consumption Survey (RECS) (2015) Retrieved from https://www.eia.gov/consumption/residential/ Found in article "Article Quantifying the Building Energy Dynamics of Manhattan, New York City, Using an Urban Building Energy Model and Localized Weather Data" by Wenliang Li, 11.
[7] Castrads North America Inc. (2024) "One-pipe steam heating" Retrieved from https://www.castrads.com/us/resources/how-it-works/one-pipe-steam-radiators/ on 1 July 2025. Retrieved from https://www.castrads.com/us/resources/how-it-works/one-pipe-steam-radiators/ on 16 October 2024.
[8] This Old House Ventures, LLC. (2024) "Steam showers: How to create a luxury home spa experience" by Joseph D'Agnese Retrieved from https://www.thisoldhouse.com/bathrooms/21014870/steam-shower-how-it-works on 13 October, 2024.
[9] Almost Heaven Group (2021) "MS series residential steam generators" Retrieved from https://www.almostheaven.net/aho/mrstmms.htm on 04 July. Provided with the Permission of Almost Heaven Group.
[10] European Union (2012) "Study on the state of play of energy efficiency of heat and electricity production technologies" Joint Research Centre Institute for Energy and Transport, p58, Petten.
[11] Nationwide Boiler Incorporated (2024) "Proper sizing of steam boiler feedwater systems" Retrieved from https://www.nationwideboiler.com/resource-center/proper-feedsystem-sizing.html on 14 October, 2024.
[12] Atlantic Feedwater Systems, Inc. (2022) "Tank capacity" Retrieved from https://atlanticfeed watersystemsinc.com/tank-capacity/ on 16 October 2024.
[13] The Engineering ToolBox (2003) "Sizing steam pipes (kg/h)" Retrieved from https://www.engineeringtoolbox.com/steam-pipes-sizes-d_258.html on 05 June 2024.

9 Gas supply system

Introduction

Gas is one of the fundamental states of matter. There are more spaces between the gas particles, so the density of gas particles is very low; as a result, they are highly compressible, and they do not have a definite shape or volume. They expand to fill a container wherein they are introduced. They occupy more spaces compared to liquid and solid. Particles in gas vibrate and move freely at high speed. There are various types of gases, of which some are used as fuel to burn. The unique properties of natural gas virtually made it popular for its various uses in a building for which a plumbing system is to be developed especially. In this chapter, the supply system of natural gas in buildings will be discussed.

Natural fuel gas

This natural fuel gas is a natural gas, basically a mixture of simple hydrocarbon compounds, and so it is a combustible gas and a fossil fuel. Natural fuel gas is predominantly methane (CH_4) gas, containing ethane, propane, butane, pentane, and some heavier hydrocarbons, as well as small amounts of nitrogen, carbon dioxide, hydrogen sulfide, and traces of water. Natural gas is found deep beneath the earth's surface, typically in porous sedimentary rocks.

The gas is colorless, odorless, and also tasteless and free from any toxicity. However, to smell its presence in closed space, a chemical called mercaptan is mixed with it in small volume to make the gas detectable by sensing the smell of eggs. This helps to find out any gas leaks. The gas is 40 percent lighter than air, so when it leaks out from a piping system in a closed room, it dissipates into the indoor air. This gas has a low flammability range and a high ignition temperature. No smoke is produced upon burning but produces high calorific value. The by-products after burning this gas are water vapor and carbon dioxide.

Liquefied petroleum gas

Liquefied petroleum gas (LPG or propane) is widely recognized as an environment-friendly and eco-friendly fuel because it emits less carbon dioxide, which causes global warming, compared to other fossil fuels. In this respect it is needless to compare with others, such as firewood and coal. Also, since it contains no nitrogen and almost no sulfur, it is an effective energy as a measure against SO_x (sulfur oxides) and NO_x (nitrogen oxides).

At atmospheric pressure and temperature, propane exists as a colorless and odorless gaseous state. But it liquefies at minus 42°C, and its volume in liquid state reduces by about 250

DOI: 10.1201/9781003724056-9

times from its natural state. As propane gas liquefies at a much lower temperature, it is easier to liquefy the gas for transport in cylinders and store in storage tanks. This nature of propane makes this fuel reachable to the users who don't have access to the gas pipeline network.

Uses of natural gas

The following are the common uses of natural gas in a building:

1 As a source of energy for cooking, heating, and electricity generation
2 As fuel in vehicles in the form of CNG (compressed natural gas)
3 To manufacture a few chemicals and fertilizers
4 As chemical feedstock for manufacturing plastic and other commercially important organic chemicals
5 In the production of fabrics, glass, paint, plastic, steel, etc.
6 In the production of hydrogen
7 In the production of protein-rich animal and fish feed on a commercial scale.

Gas use in building

Buildings that are generally used for dwelling, as well as those that provide health care and food services, have need for cooking, hot water generating, space heating, cleaning, and provide laundry services, use gas energy for these purposes. The gas that is generally used is natural gas. Gas as energy is also used for providing fuel to stoves, ovens, clothes dryers, lighting fixtures, and some other household appliances. In buildings, natural gas can also be used for air conditioning as well. Natural gas has often been considered a more efficient and cleaner fuel source than other energy, in the following purposes.

Space heating

Heating indoor spaces by using gas is common, as gas is more effective than electric heating pumps. Heat from natural gas is delivered from forced-air systems at temperatures between 50 and 60°C. The air from an electric heat pump is typically delivered at 30 to 35°C, which is warm enough to heat a room.

Cooking

Natural gas can be directly used to cook. It often provides better temperature control and uses significantly less energy than an electric oven. So, it is supplied to some cooking appliances like stoves, barbecues, etc.

Water heating

Natural gas heats faster than electricity and costs less. It can also be used to heat water.

Air conditioning

Natural gas can be used for room air conditioning that is also termed propane air conditioning, in which natural gas is used to fuel the heating elements. Though a gas-powered air conditioning system is comparatively costlier in purchase and installation, its operation cost would be less where the price of gas is low. Natural gas air conditioning offers an efficient and economical system by reducing energy use by 30 to 50 percent compared to electric air conditioning [1].

Lighting a fireplace

Fireplaces, made for room heating, can use natural gas instead of wood burning as a very common age-old practice. Using natural gas as the fuel for the fireplaces, there could be better and easier fire control to obtain an efficient, cost-effective, comfortable, and trouble-free maintenance system, as it does not produce sparks or ashes. Avoid construction of chimneys in which cleaning of deposited creosote, dark brown oil containing various phenols and other organic compounds, is troublesome.

Gas supply system

Natural gas is extracted from subsurface rock formations via drilling through the earth crusts. Advancements in hydraulic fracturing technologies have enabled access to large volumes of natural gas from deep underlying shale.

After extraction of gas from the shale, natural gas processing is carried out, which is very complex and usually involves several steps to remove oil, water, hydrocarbon gas liquids (HGLS), and other impurities such as sulfur, helium, nitrogen, hydrogen sulfide, and carbon dioxide, present in the gas. The impurities potential in extracted natural gas determines the required stages of operations to produce usable quality gas. Processed natural gas may also be stored, mostly underground, for later use.

A network of pipelines is developed to convey natural gas safely from the gas processing facilities to end users, as shown in Figure 9.1. The largest of these pipelines, which may even cross provincial boundaries, are called transmission pipelines. At the final destination,

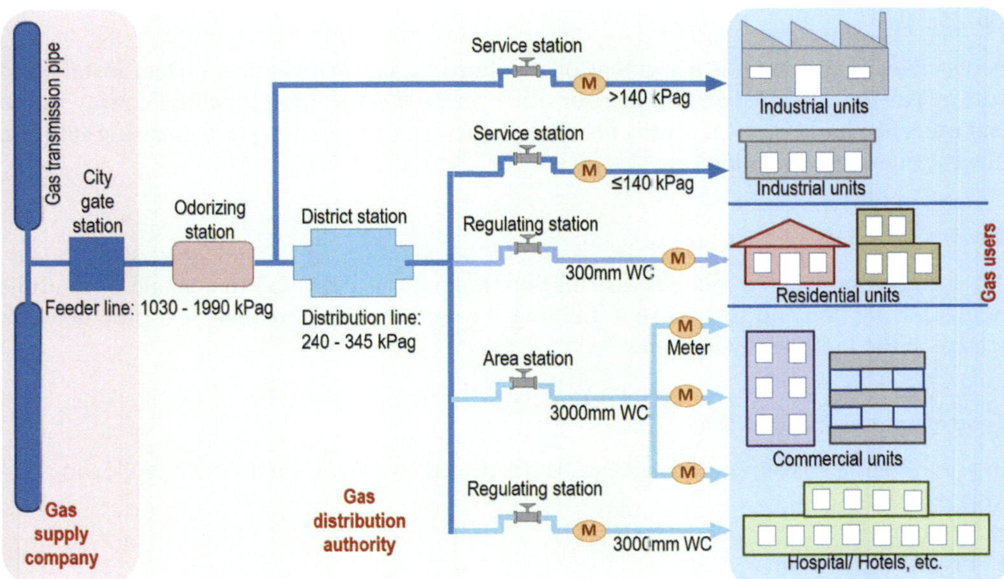

Figure 9.1 Block diagram of a gas distribution pipe line.

local distribution companies or gas utility service providers deliver natural gas to domestic users and businesses.

Gas supply main

According to ASME B31.8 Section 803.3, a gas main or distribution main is a segment of pipeline in a gas distribution system installed to convey gas to individual service lines or other mains. Distribution pipelines convey gas to end-users' locations where gas is needed, such as buildings – mostly residential, factories, industries, gas dispensing stations, power generating facilities, etc.

In a country, a national grid is developed to supply gas, in which gas is supplied through a gas transmission system, from which gas enters the distribution networks at high pressure, where it is then transported through various reducing pressure tiers until it is delivered to the end user, the concerned buildings needing gas.

The major elements involved in the gas supply mains of the national grid are as follows:

Production or importation

Natural gas is either conveyed from a remote gas field after being extracted or imported from other countries as liquefied natural gas (LNG).

Transmission

A transmission line is a segment of gas pipeline installed in a transmission system in which reception terminals receive gas from the producers at gas fields, which is then supplied, after quality checks, to a high- or low-pressure distribution system, a large-volume customer, or another storage field.

Distribution

A distribution gas main is a segment of pipeline in a gas distribution system installed to convey gas to individual service lines or other mains. Distribution pipelines convey gas to end users like buildings – generally residential, factories, industries, gas dispensing stations, power-generating facilities, etc.

Building gas supply

After getting connected to a gas distribution, main gas is delivered to a single or multiple points having demand for gas in a building through the development of a pipe network system in the building by its owner.

Types of gas supply in a building

In a building, a gas supply system can be developed by taking gas from the following two sources:

1 From the gas main
2 Individual gas sources.

Gas supply system from the main

In this system, a gas service pipe is connected to the gas main remaining underground around the building site, subject to obtaining a permit for such connection. Natural gas then flows from the main into the building gas supply system through the service line. Typically, a natural gas utility service authority is made responsible for maintaining and operating gas pipelines and facilities in the distribution gas pipe network so that sufficient gas at the desired pressure reaches up to the gas meters of buildings connected with them. All equipment, gas service pipe, and the meter are under the administration of the gas supply authority, and all downstream of the meter are under the responsibility of the customer.

Disadvantages of the main gas supply system:

1 The main gas pipeline may not be available near the building premise or site
2 High prices of available gas and service pipe connection costs
3 Difficulties encountered in having a permit for gas connection with the main subject of submitting required legal documents
4 Non-flexibility of fulfillment of future higher demand for gas in terms of flow and pressure
5 Unauthorized connections to the main pipeline and trivial gas theft might cause system break-down among the stakeholders
6 Time loss in obtaining permits for connection
7 Shutdowns in main gas might be more frequent.

Service connection to gas-main

To supply gas in a building, gas is collected from the gas main found under a road outside the building premise. The pipe connected to the gas main is termed the building gas service pipe, as shown in Figure 9.2.

Generally, the gas service piping from the gas main to the gas meter set is typically 12 mm or 20 mm in size.

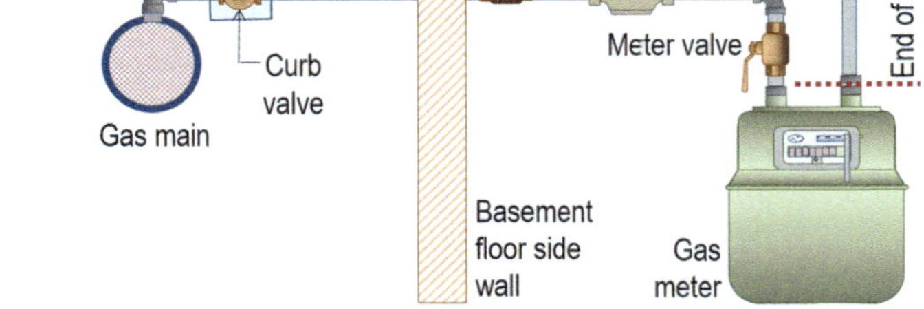

Figure 9.2 Building gas service pipe connection.

The gas service, meter, regulator, meter valve, etc. shall be installed by the gas distribution authority and shall remain their property.

Gas service piping requirements

All piping, either above or below ground, which is located after the outlet connection of the meter is under the jurisdiction of the building owner. So, its repair, maintenance, and periodic inspection for leaks, corrosion, etc., shall have to be carried out by fulfilling the following obligatory requirements:

1. The installation of the gas line to the outlet side of the meter must be made by a qualified plumber who shall adhere to the guidelines set forth in the concern code of practice applicable.
2. Piping connected to the meter set must be black iron. Corrugated Stainless Steel Tubing (CSST) may only be used in a fireplace application, provided it is properly connected and installed, meets the manufacturer's installation guidelines, and is not longer than 2 m. in length. For fireplace applications, a shut-off valve must be installed upstream of the CSST, which must be readily accessible and permanently identified.
3. Joints of piping that will be operated at a pressure of 67 kPa or more must be welded.
4. Installation cannot be considered complete until a pressure test of all piping and appurtenances is conducted.
5. All piping, appliance valves, and vent piping must be in place before conducting a final pressure test, and then a final connection to the meter set is to be made.
6. If polyethylene gas pipe is used, then the following requirements must be fulfilled:

   a. It must be only used in an underground application and shall be installed with #12 copper trace wire. Polyethylene-coated copper wire is sometimes instructed to be buried along with the plastic gas service pipe, terminating at the riser, for allowing electrical-locating of the service pipe [2].
   b. All connections must be made underground and must not terminate aboveground.
   c. Pipe must meet the ASTM D 2513 standard.

Natural gas supply pressure in building

The pressure in the natural gas service line in the street where connection is given to supply gas in a building is generally maintained in the range between 414 kPa the maximum and the minimum of 1.724 kPa or about 175 mm of water column (wc).

Pressure at the inlet of a natural gas meter: Depending on the application, natural gas meters may operate at 175 mm of wc, 3.45–13.79 kPa, or at pressures over 13.79 kPa.

Pressures in the piping between the gas meter and the appliance regulator: The pressure between the building gas meter or regulator and the appliance regulator input is typically about 1.86–2 kPa and needs to be at least 1.724 kPa to meet the appliance regulator's output requirements. This is the pressure delivered to the building's gas piping system by the gas service regulator installed at the gas service meter.

Gas pressure at individual natural gas appliances: Generally the operating pressure needed for natural gas appliances is around 14.9 millibars or 1743 Pa.

Gas supply pressure: Low Pressure: normally 1.724–3.45 kPa. Intermediate Pressure: normally 13.79–68.95 kPa, but may drop to 6.895 kPa during times of high demand.

Medium Pressure: normally 69–414 kPa, but may drop to 13.79 kPa during times of high demand. High Pressure: over 414 kPa and may exceed 6895 kPa.

Pressure testing of gas piping

To ensure the integrity of a gas-supply piping, it must go through a rigorous test, applying some pressure to the system to ascertain that the pipe, along with its fittings, joints, and other essential components, are in a workable condition that averts the risk of any leakage, thereby any explosions. The test procedure involves several steps to follow, as mentioned below:

1 All appliance piping before the appliance regulator shall be capped, if any, and after appliance valves (which must be in place), as appliance regulators may not withstand the higher test pressure.
2 The system should be pressurized to two and one-half (2½) times the system operating pressure or to 34.47 kPa, whichever is greater.
3 Pressurization shall be held for ten (10) minutes, with no drop in pressure. A longer test period may be done if the size of the system is extensive.
4 The pressure gauge shall measure in increments of 3.45 kPa. The gauge shall not be greater than 207 kPa.
5 The installer of piping must be present at the time of the test.
6 The supply system must not be turned on until a successful pressure test is completed.

Independent gas supply system

An independent gas supply system in a building can be developed by supplying gas from a cylinder containing gas. Irrespective of the application, there are basically two types of gas supply to the gas appliances, or points in a building, as follows:

1 Individual cylinders at the point of use from a gas cylinder under pressure, supported by regulators or
2 A central gas distribution system developed by installing multiple gas cylinders as source of gas.

Gas cylinder

The gas cylinder is a cylindrical vessel used to store gas under high pressure. The cylinder is long and narrow, which is generally kept standing upright on a flattened bottom, while gas in use. The pressure cylinders are usually made out of steel, which can store gas at pressure typically as high as 138 bars or greater when full. Gas cylinders are also found made of aluminum, alloys, and composite materials like carbon-fiber composite cylinders. Mechanical strength, corrosion, and impact resistance are the important factors in determining cylinder material to be considered.

Gas cylinders have a stop-angle valve fitted at the top end, used to control the release of the gas. During storage and handling, a cap on the valve is kept screwed over the protruding valve to protect it from damage or breaking off accidentally when the gas is not in use. As an alternative to a cap, there may be a protective collar or neck ring around the service valve assembly. There is a pressure regulator assembly that is used to control the flow of gas at the desired pressure when dispensing gas from the cylinder. When the gas in the cylinder is to be flown at low pressure, the cap is taken off.

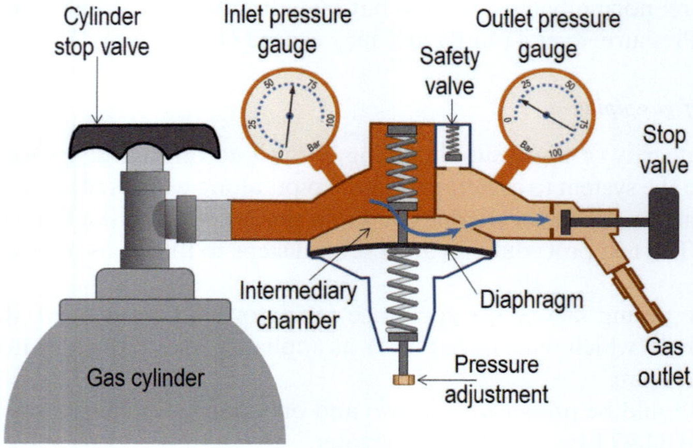

Figure 9.3 Gas cylinder head assembly.

The pressure regulator assembly includes an upstream (inlet) and downstream (outlet) pressure gauges, and at the furthest downstream a needle valve and outlet connection, as shown in Figure 9.3. For gases that remain gaseous under ambient storage conditions, the upstream pressure gauge can be used to estimate how much gas is left in the cylinder with respect to pressure. Depending on the ambient temperature, the storage pressure of the cylinder gas ranges between 2 and 3 bars; however, the operating pressure of the cylinders is designed to be 17.5 bars. Supply pressure starts decreasing while the cylinder starts emptying due to flow out of gas.

Gas cylinders, no matter how big or small they are, contain a higher volume of gas subject to the pressure applied in filling the gas in the cylinder. An empty and open gas cylinder contains the same amount of air when the surrounding atmospheric pressure is 1 bar. When gas is filled under pressure, the pressure inside the cylinder increases, and the cylinder gets filled with more gas according to the following relation.

$$\text{Pressure}(\text{bar}) \times \text{Cylinder volume}(\text{l}) = \text{Gas volume}(\text{l}) \tag{9.1}$$

The amount of gas content in a 20-liter cylinder at various pressures has been shown in Table 9.1.

Depending upon the requirement of gas, steel gas cylinders of various capacities can be used individually or in groups of multiple cylinders connected in series.

Table 9.1 Amount of gas content in a 20-liter cylinder at various pressures

Absolute pressure (bar)	*Gas amount (liter)*
1	20
10	200
100	2,000
200	4,000

Individual cylinder gas supply system

From an independent cylinder, multiple gas appliances can be served by developing a piping system connecting the appliances. The system shall be designed in such a way that overloading of gas supply pipe with the appliances does not occur, as it may make the system unsafe or inefficient due to emerging some pressure issues that can lead to gas leaks or appliance malfunction. To figure out an efficient gas piping system, the following steps are to be carried out:

1 Determining the Watts ratings of all the appliances to be served
2 Calculate the length of gas pipes, from the gas cylinder to the last appliance on the pipe.
3 Consult a pipe sizing chart to determine the main and branch supply pipes serving multiple appliances.

Central gas supply system

When the gas requirement gets higher, expensive appliances are involved, and importantly, efficiency is to be optimized, then a centralized gas supply (CGS) system is the way preferred. Central gas supply system development is based on on-site gas stocking, mostly in multi-cylinder packages in bundles or in special containers or storage tanks.

A centralized gas distribution system is developed through a pipeline networked, starting from centralized cylinder packages in bundles called cylinder banks, manifold systems, or bulk installations (storage tanks), ending with end users like heating appliances, burners in kitchens, etc.

A centralized gas supply system has both advantages and disadvantages, as mentioned below.

Advantages:

1 Reliable, as continuous gas delivery without flow interruptions can be assured
2 More precise adjustment of gas parameters can be done
3 A higher safety level can be maintained by keeping high-pressure gas storage and installation locations in specified and safe places
4 More space can be provided at the working place
5 Usually, lower costs due to the high volume of gas delivery can be done.

Disadvantages:

1 Cannot increase gas volumes by raising pressure in the pipeline
2 Higher installation and maintenance costs.

Components of the central gas supply system

The main components of the central gas supply system are as follows:

1 Cylinder bundle
2 Changeover valve or regulator

3 Bubble tester
4 Gas supply pipe
5 Stop valve
6 Pressure-regulating valve
7 Pressure control panel
8 Tapping points at the various points of use
9 Gas meter

Cylinder bundle

A cylinder bundle holds a number of cylinders permanently interconnected. Bundles are handled by crane or by forklift only and connected to the control panel (unit) through one hose only. This minimizes tremendously the handling time, effort, and safety risks. Depending on the gas consumption, CGS can be designed as multiple cylinder-bundle designs.

In the CGS system, a number of cylinders, minimum 2 to maximum 64 in a bundle, are generally held in a pallet, which is a steel frame structure designed to perfectly hold the cylinders in place. Interconnections of cylinders are made with a copper pipeline. In Figure 9.4, four cylinders are shown, in which a cylinder bundle of two is in operation and the other bundle remains standby.

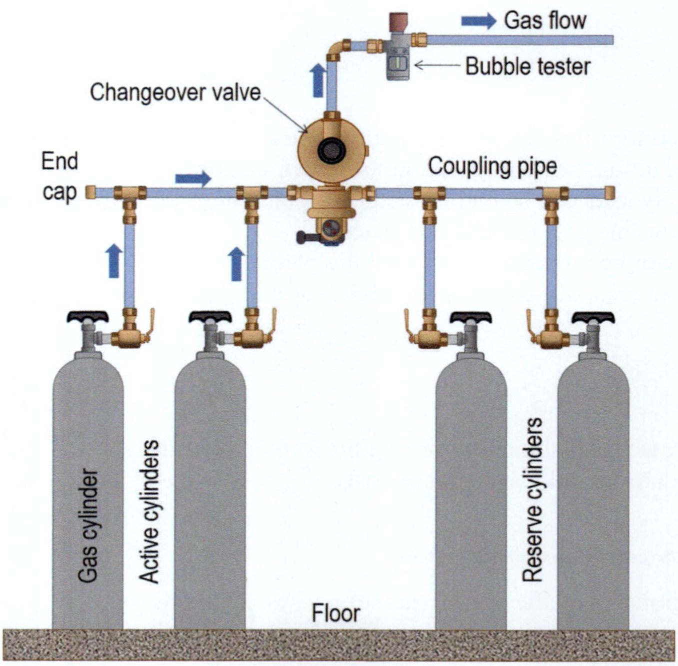

Figure 9.4 A cylinder bundle for supplying gas.

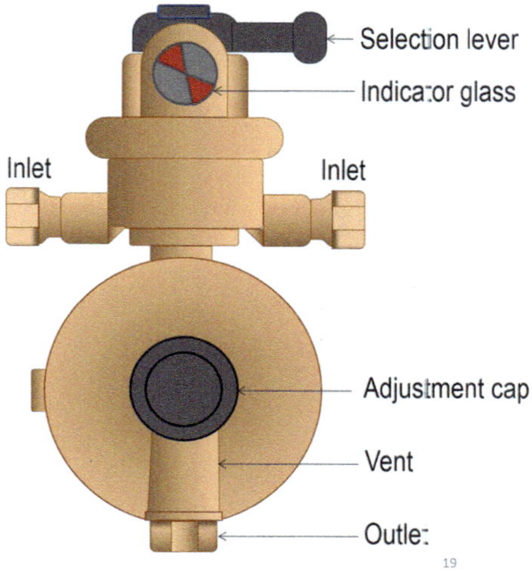

Figure 9.5 A changeover valve regulator.

Changeover valve or regulator

A changeover valve or regulator, as shown in Figure 9.5, is used to allow connection of two sets of gas cylinders to a gas supply main. While one cylinder set is in use, the other is kept on standby. When the first cylinder set is empty, the other cylinder set comes into use by turning the changeover valve. This valve thereby allows the empty cylinder set to be replaced. This process ensures a continuous supply of gas without any interruption.

Bubble tester

A bubble tester is virtually a gas leak detector that is installed inline and connected into the gas system close to the outlet side of the regulator. It helps an instant visual check for gas soundness in a pipe from the outlet side to its end where an appliance is installed. When there is leak, it is positively indicated by the formation of bubbles appearing in the glass-sighting chamber, as shown in Figure 9.6. The chamber contains a propylene glycol solution that produces visible bubbles when exposed to escaping gas through it.

Gas supply pipe

Pipe is used for the development of gas distribution networks of high and medium pressure. The most common type of pipe used for gas supply is black steel. Galvanized steel, copper, brass, or corrugated stainless steel tubing (CSST) are also used in some areas. Some utilities might prohibit the use of copper pipe.

In addition, polyethylene plastic pipes of types PE-80 and PE-100 can be used in gas supply systems, but plastic gas pipes cannot be kept or made exposed above grade.

Gas supply pipe installation

There are important issues to consider when installing a fuel gas supply pipe, based on best practices and codes. All those issues discussed herein shall be well addressed.

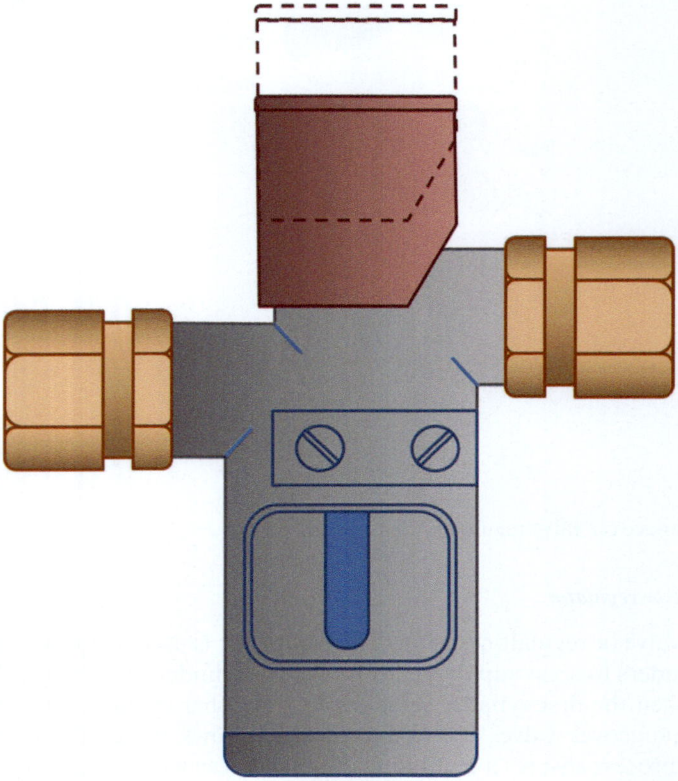

Figure 9.6 A gas bubble tester.

Gas piping-related issues

The following problems are generally encountered with gas piping:

1. Leaks – hardly detectable
2. Improper joints
3. Inadequate support system
4. Piping concealed or installed in inadequate duct systems
5. Rusting
6. No drip leg and drain valve
7. Absence of shut-off valve
8. Inappropriate pipe materials
9. Plastic gas pipe exposed above grade
10. Improperly labeled copper tubing.

Gas pipe installation guidelines

Gas piping installation shall be carried out following the right instructions for its safe and durable functioning, through choosing the right materials, assembling the piping correctly and supporting properly, applying knowledge and skill, and taking all preparation. Following are the guidelines for installation of gas supply piping:

1 The piping system must be designed, drawn and documented well, so that the gas pipes can be properly located by the concerns in the future.
2 The pipe route should be planned, avoiding locations prone to damage or that can damage structural elements of buildings.
3 Gas pipes and their isolation valves should be in locations that are easily accessible.
4 Each floor or section of a building should be provided with an isolation valve to stop flow in case of any emergency.
5 Gas pipes should run, as far as practicable, outside the building.
6 Gas pipes should preferably enter the building aboveground and in a ventilated location.
7 The pipe location must not impair the fire resistance of the building.
8 Pipes shall be kept exposed; in unavoidable cases it may be concealed, keeping provisions for easy access.
9 Pipe riser ducts must have 2-hour fire-rated doors and have one side as an external wall with fixed louvers or ventilated pre-cast block venting naturally to the atmosphere.
10 There should be a gap of at least 50 mm between a gas pipe and any other service pipes. Where electrical services are being used in the same duct, the gas riser should be separated from the electrical services by a gas-tight partition.
11 Pipes passing through a cavity must be sleeved.
12 Pipes are normally run above the ceiling, dropping down to appliances on the floor below it when no other route is available.
13 Pipes must not be laid under or through the foundation of the building.
14 Pipes should never run diagonally across walls or floors. They should run following the plane face of the walls and be installed close to them.
15 The piping system must not be installed in locations exposed to excessive external stress, vibration, or corrosion.
16 The piping system must be installed at a location where inspection and maintenance work can be done easily.
17 To avoid dispersion of leaking gas throughout the building, piping inside any building must not be installed at the following locations:

   a Lift shaft
   b Flues, chimneys, and gas vents
   c Clothes chute
   d Enclosed staircase
   e Air handling room
   f Unventilated void space
   g Under load bearing foundations and walls.

18 In rooms provided with high-voltage power facilities, all piping outlets should be installed to provide sufficient clearance from any obstacles to permit use of appropriate tools. The outlet fitting or piping must not be placed behind doors.
19 Each outlet, not in use, including pipe terminating with a valve, should be securely closed gastight with an approved threaded plug or threaded cap immediately after installation.
20 Pipe shall be made to run straight; where bending is required, it should be done making the inside radius of the bend not less than 6 times the outside diameter of the pipe [3].
21 All external joints shall be protected with chlorinated rubber paint.
22 Gas pipes should be well painted (yellow color) and labeled for quick identification.

Requirements of gas pipe shaft or duct

1 Gas pipes should be installed exposed
2 Areas around the riser pipes should be well ventilated for quick detection of gas leakage and to prevent dangerous gas accumulation.
3 Sufficient clear space and unhindered access shall have to be provided for maintenance.

Gas flow-controlling devices

Stop valve

A stop valve, also known as a shutoff valve, is used to stop the flow of gas, so that pipes can be disconnected or even separated during emergencies or to carry out repair work. A manual stop valve is generally a ball-cock type of gate valve, mostly made of brass or steel. It is opened and closed with the help of a lever.

There are also various automatic gas shutoff valves that can automatically shut off the gas supply in an emergency, mostly when a gas leaks, thereby low flow occurs and pressure drops in the supply system. These valves are mostly combination of electrical and pneumatically control systems to automatically shut off gas flow in response to various signals or fault conditions. Figure 9.7 shows an electro-pneumatically operated gas shutoff valve.

Automatic emergency gas stops or shut-off valves are also found supported by a gas level indicator so that it can be monitored whether the pipe is running low. There are also automatic thermal gas shutoff valves for use on all natural gas-fed appliances. In addition, there are special valves or regulators like Under Pressure Shut Off (UPSO) and Over Pressure Shut Off (OPSO) that are incorporated in the system to ensure safety for the end users.

There are also automatic shut-off valves that close within a few, one to maximum five seconds when subjected to a horizontal and sinusoidal oscillation due to occurrence of an earthquake of certain magnitude.

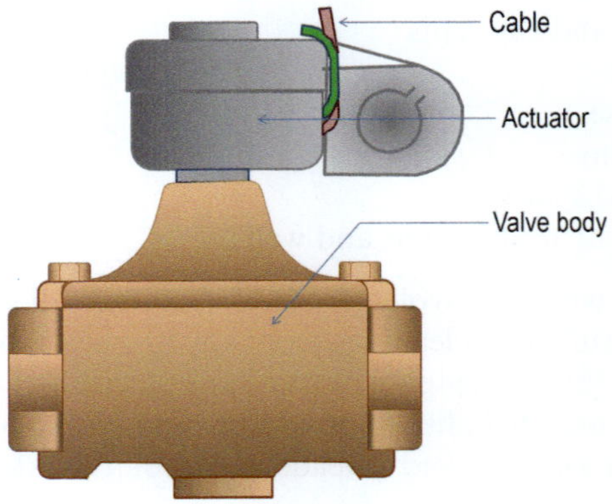

Figure 9.7 An electro-pneumatic gas shutoff valve.

Gas pressure-controlling devices

Pressure regulatory valve

Where the pressure of gas in a low-pressure gas piping system in buildings is more than 3.45 kPa, a gas pressure regulator shall be installed in the service pipe to prevent high pressure from being introduced into a building gas supply piping. The pressure regulator to control the pressure may be located upstream of the gas meter.

Two types of gas pressure regulators are mostly used, depending upon the type of function. Those are as follows:

1 Pressure-reducing regulator
2 Back-pressure regulator

Pressure-reducing valve

It is a control valve that reduces the inlet pressure of a gas to a desired outlet pressure. These valves are installed upstream of pressure-sensitive equipment.

A pressure-reducing regulator is a combination of three functional elements: a restricting element, a loading element, and a measuring element built inside. The restricting element is basically a globe valve that provides a variable restriction to the flow. The loading element, which may be a spring, piston actuator, or diaphragm actuator supported by a spring, is used to apply the required force to the restricting element. The diaphragm also acts as a measuring element to determine the inlet flow as equal to the outlet flow. The principle of pressure regulation is based on the equilibrium of the forces acting on the valve cone. By maneuvering the adjusting screw, the created spring force balances the required outlet pressure. The diaphragm ensures that the gas fills in this equilibrium of forces. The rated inlet pressure of the pressure regulator shall not be less than the gas supply pressure to the inlet.

The reduced outlet pressure by a pressure regulator is dependent on the inlet pressure. When the inlet pressure drops, the outlet pressure increases subject to the adjusting screw remaining unchanged. In many applications, this effect is not acceptable.

Gas pressure regulators are of two types, as mentioned below:

1 Single-stage pressure regulator
2 Two-stage pressure regulator.

Single-stage pressure reducing valve

In a single-stage gas pressure regulator, shown in Figure 9.8, cylinder pressure is reduced to delivery or outlet pressure in one step or stage. Single-stage regulators are mostly used for relatively small reductions in pressure, say 69–346 kPa, where maximum pressure ranges up to 10.34 bar [4]. A single-stage pressure regulator is generally used where inlet pressure does not fluctuate greatly or where periodic readjustment of delivery pressure does not cause considerable problems.

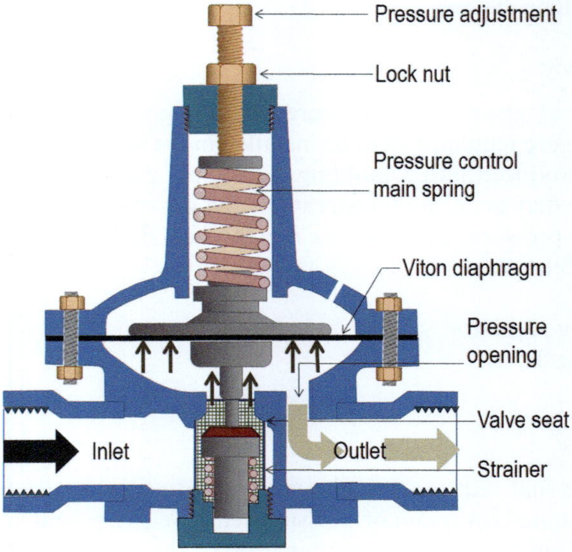

Figure 9.8 Single-stage pressure regulator valve.

Two-(dual-)stage pressure reducing valve

In a two-stage gas pressure regulator, pressure is reduced to delivery or outlet pressure in two steps or stages. These regulators contain an additional second pressure-adjusting spring, as shown in Figure 9.9. The first stage spring is not user adjustable, which is pre-compressed while manufactured.

A two-stage gas pressure regulator provides constant delivery pressure with no periodic readjustment. So, for high pressure stability and/or low outlet pressures to be ensured, a two-stage pressure regulator is recommended. A two-stage regulator provides constant outlet pressure without prior adjustment, despite changes in inlet pressure. The usual maximum delivery pressure for a two-stage pressure regulator is 34.5 bars. So, these are not required when using gases from a low pressure source less than 34.5 bar [4].

Back-pressure regulator

Back-pressure regulators or valves are pressure control devices that maintain the set pressure at its inlet side by opening it, as shown in Figure 9.10, to allow flow when the inlet pressure exceeds the set pressure value. These valves or regulators differ from pressure-reducing regulators in that the pressure-reducing regulator controls downstream pressure and is insensitive to upstream pressure. Back-pressure valves are also termed pressure-sustaining valves or regulators.

When flow is occurring, the sensing diaphragm transmits inlet pressure to the spring, pushing the valve open when pressure exceeds the spring setting. Once the back pressure on the inlet side is reduced, the valve reverts back to its normal, closed position. Pressure is maintained upstream by varying the flow through the valve as inlet pressure changes; opening up of the valve occurs as much as necessary to maintain the desired upstream pressure.

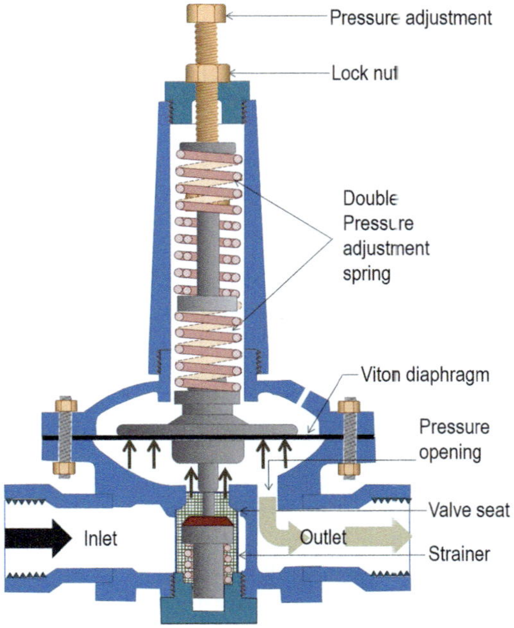

Figure 9.9 Two-stage pressure regulator valve.

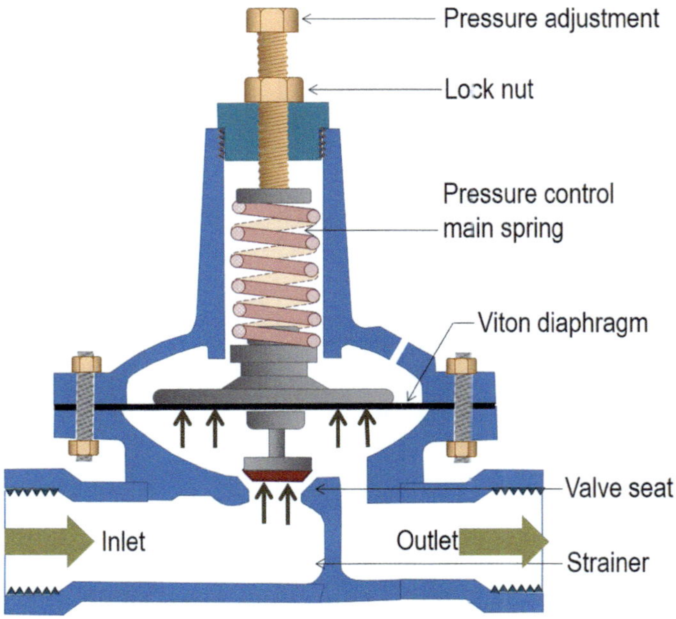

Figure 9.10 A back-pressure regulator.

A back pressure regulator is a normally closed valve used as an obstruction to reduce flow and maintain upstream pressure. Unlike the pressure regulator, a back pressure regulating valve is installed at the end of a piping system.

Installing pressure-reducing valve

With a view to installing a pressure reducing regulator properly various guidelines are to be followed those are mentioned below.

1 Regulators must be horizontally installed, keeping the vent limiter upward.
2 Care should be taken to protect the regulators; they shall be kept protected from damage due to ice and salt accumulation and hail, sharps, vandalism, etc.
3 Gas flow direction must be aligned in the direction as indicated on the body of the regulator.
   If installed indoors, the regulator shall comply with the following:
   a If any of the diaphragms of the regulator rupture, the gas shall be directed to an outlet vent pipe made of brass or plastic in order to vent the gas out of the building. The vent pipe shall be installed about 1 m. above the topmost story of the building, in open air. Necessary measures shall be taken to prevent water from entering the pipe and to prevent blocking it by any means.
   b If the gas pressure at the outlet of the regulator falls below 50 percent of the operating gas pressure or rises above twice the operating pressure, the gas supply to the pressure regulator shall shut off.
   c In the event of malfunctioning of this safety device, a supplementary device shall connect the low-pressure circuit to the vent pipe as soon as the exit pressure reaches 6.90 kPa [3].
4 Outdoor installations can be permitted where the operating temperatures range between 4.44 and 96°C.

Gas meter

A gas meter is a specialized flow meter used to measure the volume of fuel gases. It measures a defined volume of the pressurized quantity or quality of the gas flowing through the meter. Gas meters are installed on service pipes connected to a gas supply system of a building that consumes fuel gas supplied by a gas utility service provider.

Gas meters of various different designs are available, which depend on the volumetric flow rate of gas to be measured, the range of flows anticipated, the type of gas being measured, and various other factors. To get accurate measurement by a gas meter, the desired flow of gas is very essential. Inappropriate selection of meter will deliver unsatisfactory results in measurement.

Types of gas meters

There are eight types of gas meters available in the market; those are as follows:

1 Diaphragm meter
2 Rotary meter
3 Turbine meter

4 Orifice meter
5 Ultrasonic flow meter
6 Thermal mass flow meter
7 Vortex flow meter
8 Coriolis flow meter.

Diaphragm meter

These meters, as shown in Figure 9.11, are the most widespread in use. There is a pump or a pair of pumps, which contracts and expands while gas passes through the pumps. The pump is connected to a counter, which counts the number of times the pump contracts and expands. As the volume of the pump's chamber is known, the counter can calculate the volume of gas that has passed through.

Rotary meter

There are two rotors, also called impellers or pistons, in the meter; those are geared together in such a manner that with each turn, they move a specific quantity of gas through the meter. The rotation of these rotors indicates how much gas is flowing through and thus measured in the counter. These meters, as shown in Figure 9.12, are capable of measuring higher volumes and pressures than diaphragm meters.

Turbine meter

These meters are generally preferred in industrial applications where gas flow continuity is crucial. These meters use a small turbine or helix, as shown in Figure 9.13, in which gas flow velocity is used for measuring the volume of gas passing through the turbine. These meters can handle a wide range of gas flows and pressures. Turbine gas meters can be used for measuring natural gas, propane, air, and other non-corrosive gases.

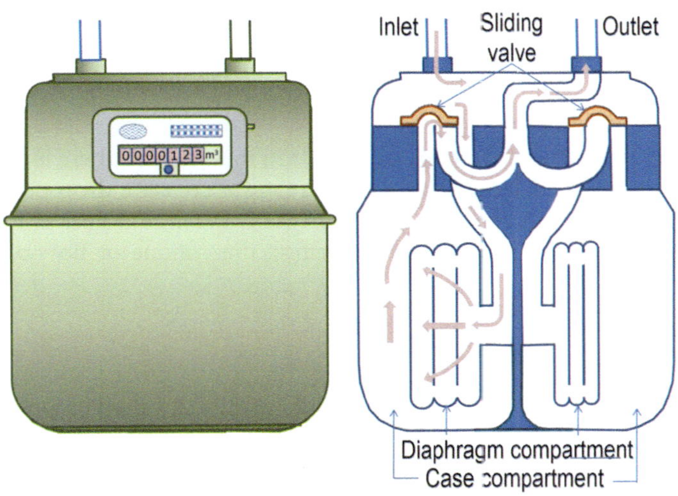

Figure 9.11 A diaphragm gas flow meter.

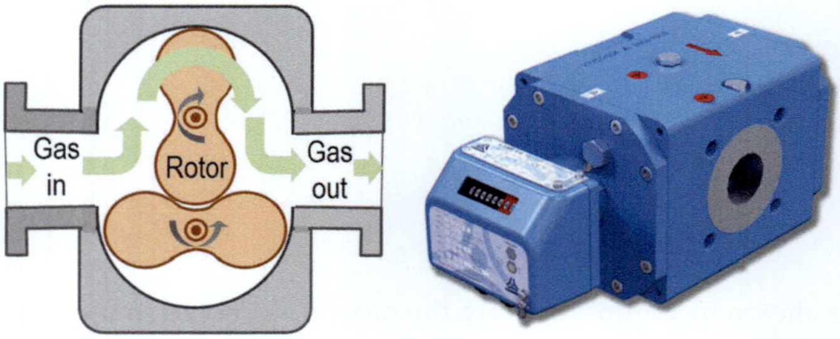

Figure 9.12 A rotary gas flow meter.

Courtesy: www.common.pl [5]

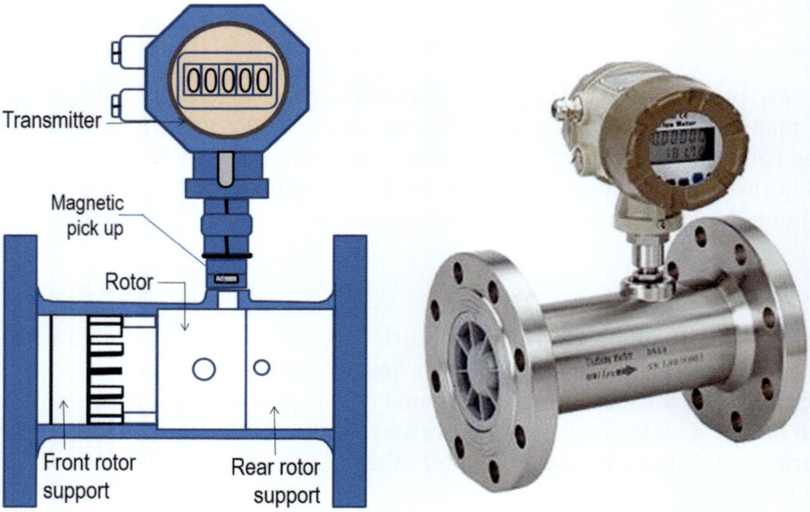

Figure 9.13 A turbine gas flow meter.

Courtesy: www.sino-inst.com [6].

Orifice meter

These meters, as shown in Figure 9.14, are simple in mechanism; there is a straight tube, in the middle of which there is an orifice, which creates a pressure drop proportional to the gas flow. If the flow is high, the pressure decreases, and if the flow is low, the pressure increases. By measuring the pressure at the orifice, the volume of flow of the gas is thus calculated.

Ultrasonic flow meter

Ultrasonic refers to the frequency of the sound waves. There are two types of gas flow meters using ultrasonic technologies: Transit-time and Doppler effect. Transit-time ultrasonic flow meters are very common for inline use as gas flow meters.

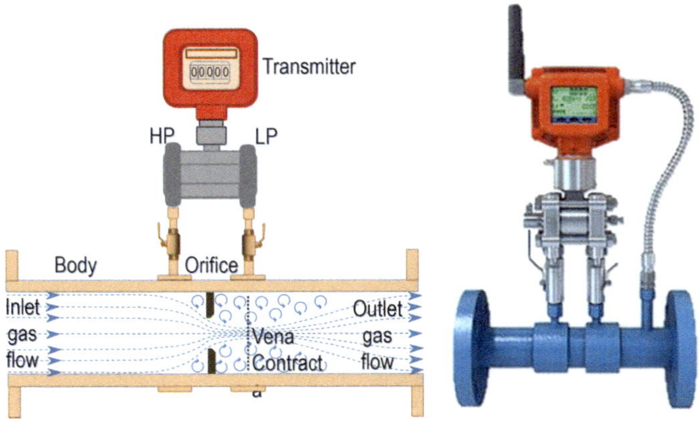

Figure 9.14 An Orifice gas flow meter.

Courtesy: www.cleanpng.com [7]

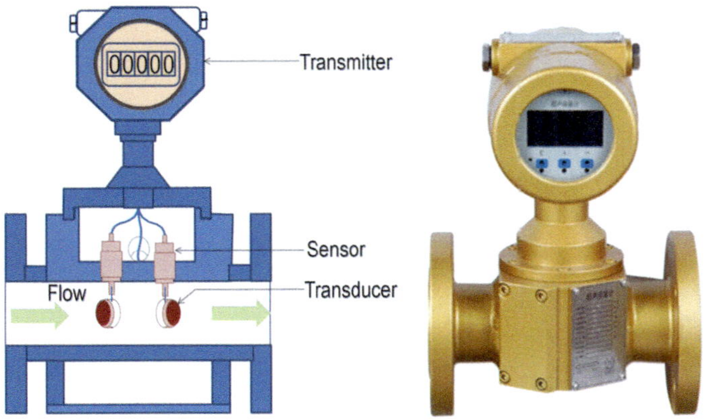

Figure 9.15 An ultrasonic gas flow meter.

Courtesy: Focus Technology Co. Ltd [9].

An ultrasonic flow meter consists of a sensor and transducer pair. Transducers emit ultrasonic waves into the gas flow and subsequently register their echo. Ultrasound is used to measure the rate of flow of gas in a pipe by measuring the speed of the flowing gas inside the pipe. Ultrasonic flow meters have two transmitters that send the sound and two receivers that receive the sound, as shown in Figure 9.15. The transmitters send out ultrasonic signals or pulses through the flowing gas. The transit time of these signals is measured and used to calculate the velocity of the gas. The time it takes to move from transmitter to receiver is dependent on the gas flow velocity. As the volumetric flow rate is equal to the flow velocity times the cross-sectional area of the pipe that is fixed at the flow meter, this calculation yields the gas flow rate.

These meters are comparatively durable and highly reliable and perform with high precision, as precise as ≤ 0.1 percent [8]. These meters are also equipped with a seismic shutoff function.

There are also portable ultrasonic gas flow meters, in which case transducers are attached to the outside of the pipe being measured.

Thermal mass flow meter

This is a gas meter that measures gas mass flow, preferred in many industrial applications of low gas flows. Thermal mass flow meters are designed based on the thermal dispersion principle that the rate of heat absorbed by a fluid flowing in a pipe is directly proportional to its mass flow. The gas molecules create heat transfer; the greater the numbers of gas molecules in contact with the heated surface, the greater the heat transfer. So, when the flow of gas increases, more heat is absorbed by the gas. The amount of heat dissipated from the heat source is proportional to the gas mass flow and its thermal properties.

The meters consist of a transmitter and probe with two platinum resistance temperature sensors located at the bottom of the probe, as shown in Figure 9.16. One temperature sensor measures the process temperature, and the other sensor is heated to a specific temperature above this.

The temperature sensor monitors the actual process temperature while the heated one or heater is maintained at a constant differential temperature above this by varying the power consumed by the sensor. The greater the gas flow velocity, the greater the cooling effect and power consumption to maintain the differential temperature. The measured heater power is therefore a measure of the gas mass flow rate shown by the meter.

Vortex flow meter

Vortex gas flow meters operate under the vortex shedding principle, in which a series of vortices alternating from each side of an obstruction are created when gas flows past a bar

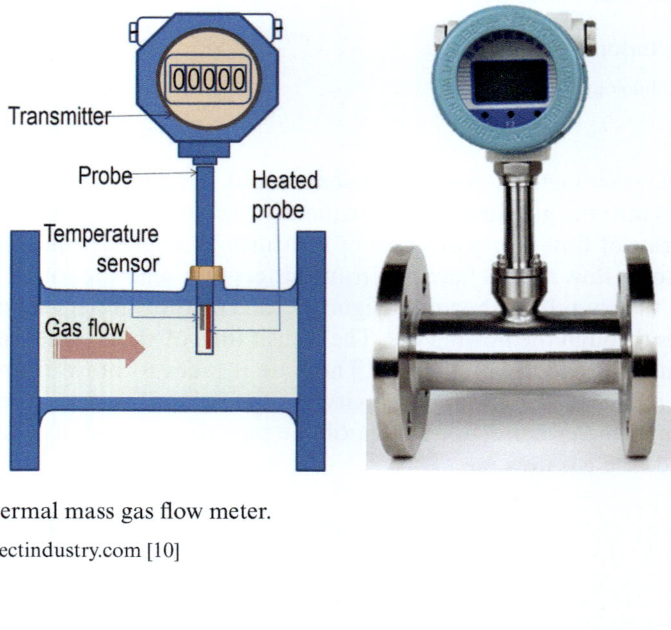

Figure 9.16 A thermal mass gas flow meter.

Courtesy: www.directindustry.com [10]

called a bluff body or shedder bar, positioned in the middle of the flow path, which generates a frequency that can be read by the sensor of the meter. Vortex meters, as shown in Figure 9.17, measure the frequency of vortices generated due to passing a fixed obstruction, as the frequency is directly proportional to gas flow velocity.

By measuring the time interval between successive vortices detected by the sensor, the flow meter calculates the velocity of the gas flowing. Since the frequency of vortex shedding is proportional to the flow rate, the meter can determine the flow rate by analyzing these vortex patterns.

Coriolis flow meter

These meters are designed for mass flow measurement based upon the deflection force of gas moving through a vibrating tube – a double-bend tube structure, as shown in Figure 9.18.

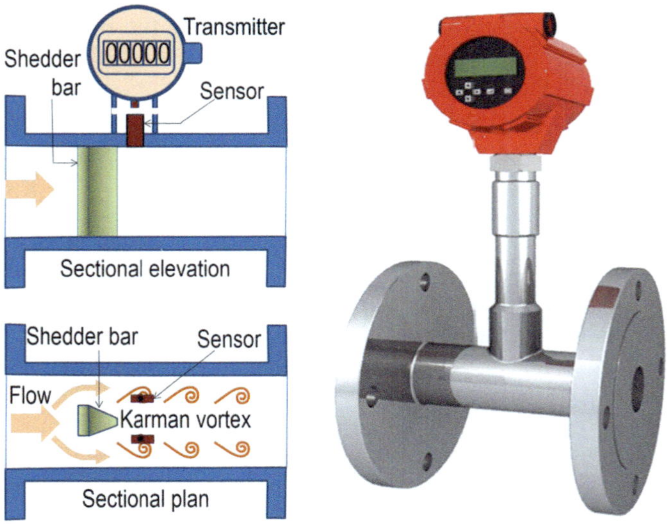

Figure 9.17 A vortex gas flow meter.

Courtesy: Industrial Quick Search [11]

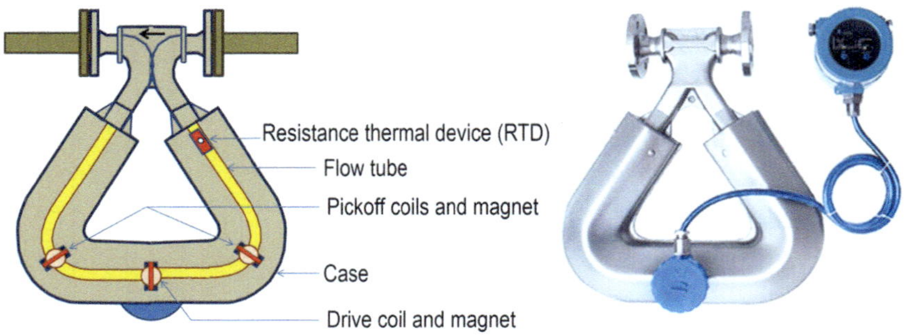

Figure 9.18 A Coriolis gas flow meter.

Courtesy: Zhejiang Sealand Technology Co., Ltd [12].

These meters measure mass through inertia. Gas is flown through the tube, which is vibrated by a small actuator. This process artificially produces Coriolis force (acceleration) into the flowing gas stream. Due to having relative motion in a uniformly rotating reference frame, an inertial force that differs from the usual centrifugal force acts on the object, which is called composite centrifugal force or Coriolis force. This twisting Coriolis force is proportional to the mass; thereby the meter measures mass flow by detecting the resulting angular momentum.

These meters are very accurate and are independent of gas properties but are relatively very expensive, need high skill in installation, and are not suitable for larger pipe sizes. Coriolis gas flow meters offer advantages, especially when measuring supercritical gases and varying or unknown gas mixtures. It is found most suitable for high-pressure gas supply.

Sizing of gas meter

Gas meters shall be sized based on the total gas requirement of a building. Meter capacity shall be greater than the demand. It is recommended that the maximum flow rate is no more than 80 percent capacity of the gas meter [13].

In majority occupancy buildings, diaphragm-type gas meters are used. In Table 9.2 sizes of different diaphragm gas meters for various gas flow requirements have been furnished.

Typical residential meters are as follows:

1 Common gas meter capacity available are 6 cum/h, 16 cum/h, 25 cum/h, and 40 cum/h
2 Normal delivery pressure to the customer is 175 mmWC, or 1.724 kPa.

Typical commercial or industrial meters are as follows:

Commercial or industrial meter sets: Commercial or industrial meter sets may vary in design and configuration because of increased flow and pressure requirements.

1 Flow requirements may vary from 17 scmh to over 2832 scmh.
2 Delivery pressure may vary from 175 mmWC. or 1.724 kPa to distribution pressure 276–414 kPa.
3 Meters may be large diaphragm (600–1000 class), rotary, or turbine meters with or without electronic instruments.

Table 9.2 Sizes of different diaphragm gas meters for various gas flows [13]

Sl.no	*Meter type*	*Flow rate (kW)*	*Capacity SCMH (Standard cubic meter per hour)*
1	U6	0–65	6
2	U16	66–173	16
3	U25	174–271	25
4	U40	272–430	40
5	U65	431–704	65
6	U100	705–1083	100
7	U160	1084–1733	160

Installation of gas meter

Gas meters are to be installed in safe locations, sufficiently away from potential hazards, for the safety of properties and life near around. Applicable code instructions are therefore to be strictly followed. It is to be kept in mind that the gas service pipe, regulator, meter valve, and meter set shall be installed by the concerned department and shall remain their property. Following are some important guidelines for the installation of gas service meters for use in buildings gas supply systems.

Location: Gas meters shall be placed in a position fulfilling the following guidelines:

1. Any potential ignition sources, such as electric meters, air conditioners, etc., may not be located within 1 m in all directions of any regulator or relief vents, regardless of meter set type and design, as shown in Figure 9.19.
2. Must be located in a readily accessible location and in an area that protects the gas meter from damage such as vehicles, water, ice, falling objects, etc. Gas meters cannot be allowed to be installed behind any obstacles or enclosures that may prohibit 24-hour access.

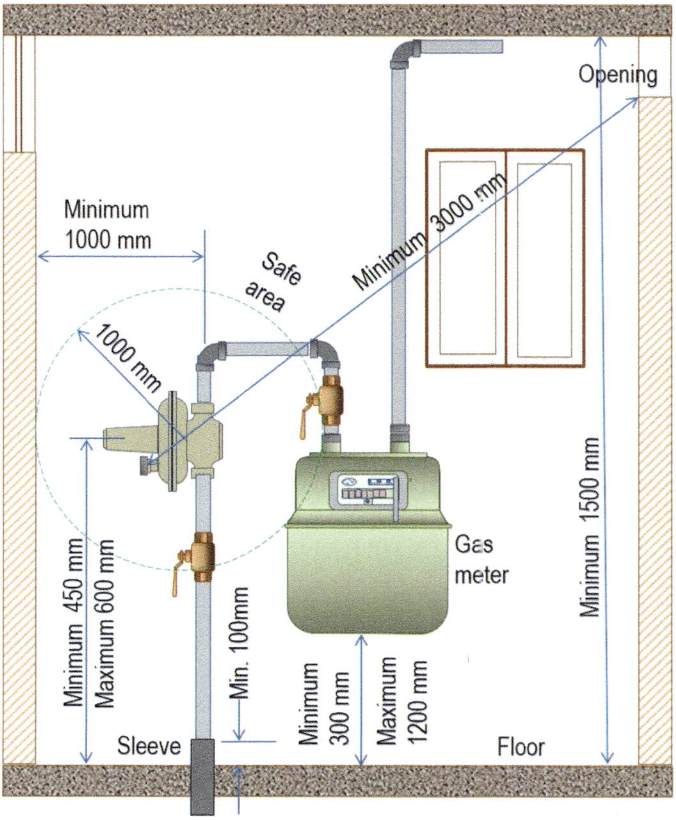

Figure 9.19 Gas meter location and installation indoors.

3 Meter sets will not be allowed to be installed in the following locations:

a Under or in front of operable windows or any other building openings and doors
b Under or in front of buildings and appliance vents or other air intakes.

4 Distance from the centerline of the service riser to the outlet union (customer connection point) is 500 mm for a residential meter set.
5 Meters shall not be installed under decks or porches. If a deck or porch is elevated, a meter can be installed under the deck if a height clearance of 1.5 m or more is available and no sources of ignition are present.

Installation: While installing a gas meter and its piping along with the accessories, utmost care shall have to be taken following the guidelines as mentioned below, for its efficient performance and safety as well.

1 The gas service pipe shall be cleared to prevent dirt, debris, or liquids from being carried into the meter when the gas is flowing
2 Install the gas meter as close to level as possible, and make sure it is not at a low point where liquids could be trapped
3 The meter must be properly supported to avoid excessive stress on the meter housing and connecting pipes
4 The meter's flow direction must correspond to the arrow on the meter body
5 A filter or strainer should be fitted on the inlet side of the meter to protect it from possible contaminants in the gas.
6 Joints shall not be strained when the nuts are tightened. Based on the connection size, connection nuts should be tightened to the appropriate torque without over-tightening
7 The meter should never be in contact with soil, mulch, or concrete walls, as these can cause premature corrosion and potential gas leaks.

Gas service line sizing

In determining the size of service lines for a gas piping system, following factors shall have to be considered:

1 Pipe Material: Plastic or steel pipe
2 Gas supply pressure
3 Allowable pressure loss from the connection on main to the meter
4 Specific gravity and heating-value content of the gas
5 Length of the service pipe, from the connection on main to the meter
6 Demand of gas flow.

In a gas meter, its capacity is stated in cubic meter per hour (cmh) for a given service drop gas pipe size. The sizes to be provided for gas service connection plastic pipe based on meter capacity and length of service pipe from the connection on main to the meter, operating at various pressure levels, allowable pressure drop, have been furnished in Table 9.3.

Gas supply pipe sizing

Correct sizing of gas supply pipes is required to ensure the desired flow rate of gas at system pressure maintained. The size of gas supply pipe depends on the length of supply pipe

Table 9.3 Sizes of gas service connection plastic pipe based on meter capacity and length of service pipe from the connection on main to the meter, operating at various pressure level and allowable pressure drop [14]

Sl. no.	*Size (mm)*	*Main pressure*	*Meter capacity (cum/h) for length (m)*			
			15	*30*	*60*	*90*
1	12	High pressure	27.75	19.54	13.60	10.76
		Medium pressure	6.23	4.25	2.97	2.38
2	20	Medium pressure	18.12	13.31	9.34	7.65
		Intermediate pressure	6.51	4.81	3.40	2.83
3	25	High pressure	174.72	129.12	92.03	74.76
		Medium pressure	39.08	28.60	20.39	16.37
		Intermediate pressure	13.59	10.19	7.65	6.23
		Low pressure	4.73	3.34	2.35	1.93

Note:

- Medium-density polyethylene plastic (MDPE) can be installed for low to medium pressure gas supply. High-density polyethylene plastic (HDPE) shall be installed for high pressure gas supply, but must not be more than 682.58 kPa (99 psig).
- Specific gravity of gas considered 0.6.
- Medium-density polyethylene plastic (MDPE) can be installed for low to medium pressure gas supply.
- High-density polyethylene plastic (HDPE) shall be installed for high pressure gas supply, but must not be more than 682.58 kPa.
- Specific gravity of gas considered 0.6
- Low pressure supply is based on a pressure drop of 0.124 kPa.
- Intermediate pressure supply is based on a pressure drop of 1.24 kPa.
- Medium pressure supply is based on a pressure drop of 3.98 kPa.
- High pressure supply is based on a pressure drop of 13.79 kPa.

to the appliance, the number of appliances being served, and the appliance input kW required. In any case, gas pipes smaller than 12 mm in diameter shall not be used.

The size of gas piping depends upon the following factors:

1 Allowable pressure drop from meter or service regulator, when there is no meter, to an appliance
2 Maximum gas demand or consumption, to be provided
3 Pipe length, number of fittings, valves, and accessories on pipe
4 Specific gravity of the gas
5 Diversity factor.

Maximum gas demand

The volumetric flow rate in cum/h of gas to be provided shall be calculated using the manufacturer's input ratings of the gas using appliances to be served, adjusted for altitude. Where the input rating is not indicated, the gas supplier, appliance manufacturer, or qualified agency shall be contacted, or the rating from Table 9.4 shall be used for estimating the flow rate of gas to be maintained while supplying.

The total connected load operating during an hour shall be used as the basis for pipe sizing, assuming that the appliances are operating simultaneously and in full capacity.

Table 9.4 Approximate natural gas input for typical appliances. SI conversion from source [15]

Appliance	*Input kW (Approx.)*	*Input cum/h*
Space heating units		
Hydronic boiler		
i Single family	29.31	170
ii Multifamily, per unit	17.58	102
Warm air furnace		
i Single family	29.31	170
ii Multifamily, per unit	17.58	102
Space-and water-heating units		
Hydronic boilers:		
i Single family	35.17	204
ii Multifamily, per unit	21.98	127.50
Water-heating appliances		
Water heater, automatic instantaneous:		
i Capacity at 7.57 lpm	41.85	243
ii Capacity at 15.14 lpm	83.53	484.22
iii Capacity at 22.71 lpm	125.55	728
Water heaters, automatic: storage 114–152 liter tank	10.26	59.50
Water heater, automatic: storage 190 liter tank	14.65	85
Water heater, domestic, circulating, or side-arm	10.26	59.50
Cooking appliances		
Built-in oven or broiler unit, domestic	7.33	42.50
Built-in top unit, domestic	11.72	51
Range, freestanding, domestic	19.05	110.50
Other appliances		
Barbecue	11.72	51
Clothes dryer, Type 1 (domestic)	10.26	59.50
Gas fireplace, direct vent	11.72	51
Gaslight	0.73	4.25
Gas log	23.45	136
Refrigerator	0.88	5.10

For SI units: 1 Btu per hour = 0.0293 kW/h, 1000 Btu per hour = 1cfm = 1.699 cum/h

Gas flow

The desired gas flow rate is intended through a pipe that is to be well determined. It is to be noted that the following factors affect significantly the gas rate of flow in a piping system; those are considered in pipe sizing.

1 Length of pipe
2 Pressure drop
3 Relative density of the gas
4 Diameter of pipe.

Length of pipe

Length of pipe supplying gas plays an important role in sizing the pipe. Table 9.5 and Figure 9.20 can be used for sizing pipes supplying natural gas at low pressure, adopting the following considerations:

Table 9.5 Capacity of natural gas supplying pipe (MBH ≈ CFH) for different sizes of pipes of steel pipe of schedule 40 [16]

Pipe size (mm)	*Capacity of pipe (kW)*				
	Pipe length (m)				
	3	*6*	*12*	*24*	*50*
12	41	28	19	13	9
19	86	59	40	28	19
25	161	110	76	52	35
32	331	228	155	108	72
40	495	340	235	161	108
50	956	655	450	310	210
65	1520	1045	720	495	330
75	2690	1850	1270	875	590
100	5490	3775	2595	1780	1200
125	9935	6830	4695	3225	2170
150	16090	11055	7600	5220	3510
200	33050	22715	15615	10730	7210

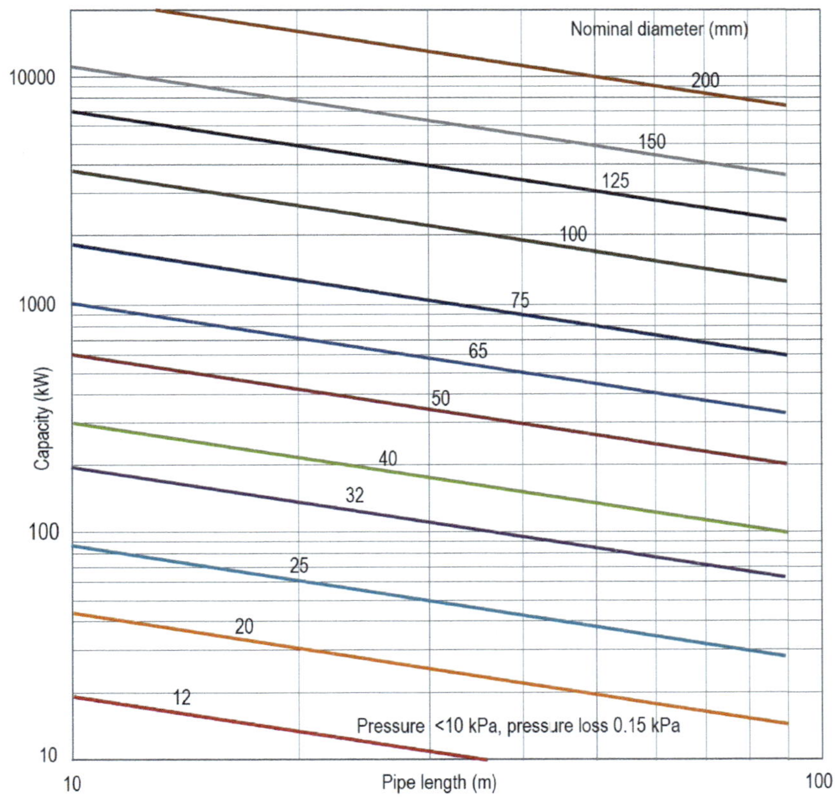

Figure 9.20 Capacity of steel pipe of schedule 40 [16]

1 Pressure less than 10 kPa
2 Fittings factor 1.5−equivalent pipe length = pipe length + 50 percent
3 Pressure drop 125 Pa
4 Energy content in natural gas 37 MJ/cum
5 Steel pipes of schedule 40

Gas supply pressure loss

Pressure in gas supply or distribution pipe is generally measured in two ways:

1 Pascal (Pa) or
2 Millimeters of water column (mmWC).

The pressure in a high-pressure gas system is commonly measured in kilo Pascal (kPa) or bars. This pressure is generally maintained around 13.79 kPa. On the other hand, in low-pressure gas systems, the pressure of gas supply is measured in millimeters of water column (mmWC). Gas pressure of most appliances is indicated in millimeters of water column, which generally ranges between 100 and 150 mmWC. Generally, 1 mmWC is equivalent to 9.81 Pascal.

The pipe size can be determined by the amount of pressure (energy) that can be lost, which is also called the pressure drop, to cause flow of the gas while leaving enough pressure to meet the minimum requirements of the appliance controls. As the pressure goes up, higher allowable pressure drops can be accommodated. The greater the pressure drop, the greater the amount of gas that can be flown through the pipe for a given size of pipe. For a given flow of gas at particular pressure, the pipe size can be decreased as the allowable pressure drop is increased.

A single-phase gas pipe is a pipe in which only gas is flown, which should be sized in such a way that the resulting end pressure is high enough to satisfy the requirements of the next piece of equipment or appliance.

Allowable pressure drop

While flowing any fluid, as in flowing gas through a pipe, frictional loss occurs more or less. This frictional loss in a gas-supplying pipe from a meter to the end at a gas appliance is kept limited. Pressure drop again varies with respect to supply pressure. The allowable pressure drop for various supply pressures has been tabulated in Table 9.6.

Table 9.6 Allowable pressure drop for various supply pressures [17]

Gas supply pressures	*Maximum allowable pressure drop*
Less than 175 mmWC (1.75 kPa)	150 mmWC (125 Pa).
175 mm to wc to 350 mmWC (1.75 kPa to 3.5 kPa)	25 mmWC (250 Pa).
14 kPa natural gas	10 kPa or 7 kPa, depending on the pressure required at the end of the supply.
14 kPa on propane	7 kPa
34 kPa	17 kPa
70 kPa	34 kPa
140 kPa	70 kPa

Diversity factor

In determining the total demand, a diversity factor is considered to reduce the demand, as an understanding that all the gas appliances in a building rarely be used at the same time.

In buildings, residential apartment buildings in particular, there are many apartments or units where many gas ranges, water heaters, and furnaces or boilers might exist; those might be served by a common gas supply piping. It is assumed that all the appliances using gas will not operate simultaneously. It is further observed that with the increased number of such appliances, the number of appliances operating simultaneously also reduced; so, the demand for the supply of gas also reduced. To reduce the load of gas supply based on the number of appliances, a factor is considered termed a load diversity factor. Table 9.7 furnishes the load diversity factor in percent of total load for domestic gas ranges with connected load ratings of 1.78 cum/h or more per range.

Change of gas pressure corresponding to height

In a vertical gas riser, pressure increases. This increase in pressure depends upon the specific gravity of the gas, as shown in Table 9.8.

Specific gravity and other properties of fuel gases have been shown in Table 9.9.

Table 9.7 Multi-family dwellings diversity factor in percent of total load for domestic gas ranges with connected load ratings of 1.78 cum/h or more per range [18]

No. of ranges	*Demand cum/h*	*Percent of ratings*	*No. of ranges*	*Demand cum/h*	*Percent of ratings*
1	1.78	100	13	7.65	33
2	3.03	85	14	8.01	32
3	3.74	70	15	8.30	31
4	4.28	60	16	8.83	31
5	4.90	55	17	–	–
6	5.35	50	18	9.63	30
7	5.60	50	19	9.80	29
8	6.00	42	20	10	28
9	6.37	39	21	10.14	27
10	6.60	38	22	10.59	27
11	6.88	37	23	11.04	27
12	7.28	35	24 and over	11.56	27

1 cum/hr = 35.314 cft/hr

Table 9.8 Increase of pressure with altitude (measured from sea level) [19]

Specific gravity of gas	*Pressure gain per 30 m of vertical riser in Pa*
0.35	239
0.40	221.50
0.45	201.50
0.50	184
0.55	164
0.60	147
0.65	129.50
0.70	109.50

Table 9.9 Specific gravity and other properties of fuel gases [20]

Property	*Units*	*Natural Gas*	*Propane*	*Butane*
Chemical formula		CH_4	C_3H_8	C_4H_{10}
Molecular weight	kg/k/mole	16	44.10	58.12
Specific gravity (air = 1)		0.6	1.53	2.09
Volume occupied/mass of liquid	liter/ton		1957	1743
Calorific value at 20°C	MJ/cum	38.7		
Latent heat of vaporization	kJ/kg		358	372
Amount of air to burn 1 cum of gas	cum	9.78		
Auto ignition temperature	°C		470	410
Octane Number		100	Over 100	92

The density of a gas (LP gas or natural gas) will decrease as pressure decreases, and the pressure of a fixed quantity of gas in a closed vertical cylinder, also in a vertical pipe, will decrease exponentially as height increases.

Let consider the volume of the gas in a vertical pipe constant as any arbitrary volume V per unit h of height.

In the case of natural fuel gas lighter than air, the pressure at the top of a vertical pipe will be less than the pressure of the same gas in the same pipe at the bottom of the pipe, as calculated by this formula:

$$P_h = P_o e^{-\frac{mgh}{kT}} \tag{9.2}$$

Where

P_h = Pressure at the top of the column of height h
P_0 = pressure at the bottom of the column of gas
e = roughly 2.71828. This number, also called Napier's constant, forms the base of natural logarithms.
m = mass of the gas per molecule, kg/mol
g = gravity, 9.81 m/s^2
h = height (m) above the starting point, or 0 represents ground level (or the bottom of vertical gas pipe).
k = universal gas constant (Boltzmann constant) 8.314 J/(mol·K))
T = Temperature of the gas in Kelvin.

Velocity of gas flow

The velocity of gas flow may cause a noise and erosion problem if it is too high. The velocity of gas flow in a pipe should be less than 20 mps. Velocity of natural gas through long-distance pipe should be in the range of 5–10 mps for continuous operation and a maximum up to 20 mps for intermittent operation. For underground installations, gas flow velocity should be 20 mps. The velocity of 20 mps should not be considered as absolute criteria. Higher velocity of flow may be acceptable when appropriate preventive measures like pipe routing, safety valve choice, and installation are well planned to minimize erosion or isolate noise. In such a case, it may be as high as 100 mps.

References

[1] Energy Solution Centers "Natural gas air conditioning" Retrieved from https://gasairconditioning.com/market-segments/residential/advantages-of-natural-gas-air-conditioning/ on 6 June 2024

[2] Village of Morton Zoning & Code Enforcement Department (2022) "Standards for gas meter installation and customer piping requirements" Retrieved from https://www.morton-il.gov/wp-content/uploads/2025/10/Residential-Home-Addition-Building-Permit-Application-with-Natural-Gas-App-Fillable.pdf on 29 October 2024.

[3] Housing and Building Research Institute (HBRI) (2020) *Bangladesh National Building Code, 2020*. Dhaka.

[4] Beswick Engineering Co. Inc. "The basics of pressure regulators" Retrieved from https://www.beswick.com/wp-content/uploads/2018/12/Basics-of-Pressure-Regulators-240417.pdf on 16 June 2024.

[5] www.common.pl "CGR-01 rotary gas meter" Retrieved from https://www.common.pl/produkt/cgr-01-rotary-gas-meter/?lang=en on 16 November 2024

[6] www.sino-inst.com (2024) "SI-3204 flange type turbine flow meter" Retrieved from https://sino-inst.com/si-3204-flange-type-turbine-flow-meter/ on 16 November 2024

[7] www.cleanpng.com (2024) "Industrial flow meter device" Retrieved from https://www.cleanpng.com/png-flow-measurement-gas-meter-orifice-plate-gauge-bri-1511903/ on 16 November 2024

[8] Unictron Technologies Corporation (2024) "Diaphragm gas flow meter vs ultrasonic gas flow meter" Retrieved from https://www.unictron.com/ultrasonic-transducers/piezoelectric-technologies/ultrasonic-gas-flow-meter/ on 21 June 2024.

[9] Focus Technology Co., Ltd. (2024) "Build-in type ultrasonic gas flowmeter" Retrieved from https://shyinuo.en.made-in-china.com/product/SjCQrAlubqWU/China-Build-in-Type-Ultrasonic-Gas-Flowmeter.html on 16 November 2024

[10] "Thermal mass flow meter HLRS" Retrieved from https://www.directindustry.com/prod/kamboda/product-233727-2372156.html on 16 November 2024

[11] www.iqsdirectory.com "Types of flow meters" Retrieved from https://www.iqsdirectory.com/articles/flow-meter/types-of-flow-meters.html on 16 November 2024

[12] Zhejiang Sealand Technology Co., Ltd. (2024) "ATEX approved 25mm coriolis mass flowmeter" Retrieved from https://www.sealandflowmeters.com/quality-mass-flow-meter-for-gas-industry-62316510.html on 26 December 2024

[13] Tiga Ltd t/a Niccolo (2024) "Guide to gas meter sizes" Retrieved from https://niccolo.co.uk/guides/guide-to-gas-meter-sizes on 25 June 2024

[14] NiSource Distribution Operations (2018) "Columbia gas pipe standards for customer service lines, meters, and service regulators" *Form C* 2235 (Plumbers Guide). Revised 2018-04-01

[15] International Code Council (ICC) (2010) "Pipe sizing based on the 2009 IFGC and IRC" *Code Notes* Retrieved from https://iccsafe.org/wp-content/uploads/membership_councils/4531S09_CodeNotes.pdf on 17 December, 2024

[16] The Engineering ToolBox (2005) "Natural gas pipes - low pressure capacities vs. Size" Retrieved from https://www.engineeringtoolbox.com/natural-gas-pipe-sizing-d_826.html on 27 December 2024

[17] Industry Training Authority BC (2023) "Fuel gas systems" *Learning Task 2* Retrieved from https://opentextbc.ca/plumbing3e/chapter/fluid-flow-in-a-piping-system/ 0n 26 June 2024.

[18] Legal Information Institute (2025) "248 CMR, § 5.05 - Modifications to Chapter 5, Gas piping system design, materials, and components" Retrieved from https://www.law.cornell.edu/regulations/massachusetts/248-CMR-5-05 on 13 July 2025.

[19] Svend Plumb (1950) *Plumbing Practice and Design* Volume 2, p 242. John Wiley and Sons, New York.

[20] Air & Gas Holdings Pty Ltd. (2024) "Fuel & gas properties" Retrieved from https://www.airgas.com.au/fuel-gas-properties/ on 27 December 2024

10 Medical gas supply system

Introduction

In a health care establishment, particular types of gases are required for several reasons, primarily to support patient care, surgeries, and medical procedures. These gases serve different functions, from assisting patients in respiratory support to preparing medical devices and instruments for their intended service. So, the medical gases used in a hospital are considered as the life-supporting drugs that give direct influence in maintaining good health of a patient. Therefore, at the points or areas where the medical gases are used or stored, it must be qualitatively pure and quantitatively adequate.

In a large medical gas supply system, a source of gases is setup, from which required medical gases are supplied to operating theatres, ICU, wards, and gas outlets, through piping.

The safe and reliable supply of these gases is essential for patient care and maintaining the functionality of concerned medical equipment. Medical gas supply systems are critically regulated to ensure purity, proper pressure, and delivery to patients and devices. Any interruption or inconsistency in the supply of these gases could potentially compromise patient care and the successful outcome of medical procedures. Therefore, developing, operating, and maintaining a consistent and dependable medical gas supply is critical in healthcare facilities. So, in this chapter the development and maintenance of the medical gas supply system will be discussed comprehensively.

Medical gas

Medical gases are known to be those that are used for treatment, diagnosis, therapy, and to produce anesthesia for a patient, and some for driving medical devices and tools. There are several gases that make up the average medical gas. Those are supplied from a source to where there is need, developing a supply system.

Types of medical gases

Mostly the following gases are used in a healthcare establishment:

1 Oxygen
2 Nitrous oxide
3 Nitrogen
4 Carbon dioxide
5 Helium

DOI: 10.1201/9781003724056-10

6 Argon
7 Medical compressed air
8 Medical vacuum

Oxygen

Oxygen gas (O_2) is used for respiratory therapy and life support for patients who require supplemental oxygen via mask and is additionally used in anesthetic procedures. Oxygen supply in large healthcare establishments is generally accomplished by installing a large storage system of liquid oxygen, which is evaporated into concentrated oxygen to supply to the points of usage through a piping network. The supply pressures are usually maintained between 345 and 380 kPa minimum and 400–500 kPa maximum [1].

Nitrous oxide

Nitrous oxide gas (N_2O) is used, generally in surgical operations, for anesthetic and analgesic purposes. This gas is contained in high-pressure cylinders and then supplied through piping to the points of use. Though there is a bulk system of producing this gas, it is less preferred due to environmental, particularly atmospheric, concerns and overall reduced consumption. Supply pressures are maintained between 345 kPa and 400 kPa [1].

Nitrogen

Nitrogen gas (N_2) is generally used to provide power to pneumatic surgical equipment during various procedures. Nitrogen is stored in, and supplied by, high-pressure cylinders. Pressures range needed around 1.2 mPa to various locations [1].

Carbon dioxide

Carbon dioxide gas (CO_2) is generally used in medical centers to aid laparoscopic surgeries like endoscopies and arthroscopies. Carbon dioxide gas helps insufflation, i.e., inflating the body organs slightly, which simplifies internal visibility, inside the organ, by the surgeons.

Helium

Helium gas helps operate certain medical devices like MRI machines due to its ability to maintain low temperatures. A helium-oxygen mixture called Heliox is used in treating respiratory or upper airway obstruction, asthma exacerbation, bronchiolitis, chronic obstructive pulmonary disease, acute respiratory distress syndrome, pulmonary function testing, etc. It is considered a drug, and its use without a medical practitioner is unsafe.

Medical air

Medical air is basically compressed pure air, used extensively in the ICU, PICU, and NICU areas in medical centers to reduce the risk of excess oxygen in the lungs during surgical procedures. in order to maintain correct dew point levels, this gas is supplied by a specialized air compressor through a dryer, maintaining a supply pressure of 400 kPa (4 bar); so, specified as air-400 [1]. Medical air is basically a specific grade of compressed air intended for patient treatment.

Medical air should be free of dust and moisture, which is performed by oil filters, dust filters, moisture separators, and the dryer at the equipment level. A separate dryer is used for the air compressor system. If compressed air quality is not maintained properly, it may damage equipment and contaminate chemicals, food, and drugs.

Medical compressed air

Medical compressed air is used for both medical and non-medical purposes. Medical use includes laboratory work, inhalation therapy equipment, and powering of surgical tools and non-medical uses are maintaining tools or equipment and in the engineering department. Medical compressed air can include medical air but may also be used for non-respiratory purposes in medical settings.

Non-medical air or surgical air is also known as instrument air, like nitrogen. Its use includes maintaining tools and equipment in the engineering department. Non-medical air is used at a higher pressure to power a variety of surgical tools and other devices. This air is supplied at 700–800 kPa at the point of use, specified as air-700 or air-800 accordingly.

Medical vacuum

In a hospital, a medical vacuum or suction is developed by installing suction equipment by which evacuation procedures are carried out by vacuum pump systems, exhausting to the atmosphere. A medical vacuum system is used for the following applications in a healthcare establishment:

1 Draining wounds, absorbing phlegm, pus, blood, etc.
2 Lung draining
3 Collection and removal of fluids
4 Wound draining
5 Tube cleaning

Generally, in the operating theater, the majority of types of gases, like O_2, N_2O, medical air, instrument air, and vacuum, are required, and in other areas like the cath-lab, intensive care beds, recovery beds, and general beds, mostly O_2, medical air, and vacuum are provided.

Medical gas demand

It is experienced that healthcare activities are of diverse nature, so the demand for medical gas is quite variable in terms of quantification, and to some extent, in assessment of types of medical gas consumption that predominantly depends on the extent of facilities to be provided. To assess the size of their facilities as well as to optimize monitoring and maintenance management, the demand for essential medical gases and others is to be estimated judiciously based on general health and environmental conditions in the area to be served. For these purposes, the following guidelines may be followed.

The annual consumption of a particular type of medical gas per hospital bed may be considered as mentioned below [2]:

1 Oxygen: 350 cum
2 Medicinal air: 325 cum

3 Nitrogen protoxide: 9 cum
4 Carbon dioxide: 3 cum

Medical gas supply system

Medical gas supply systems are developed primarily to sustain patient life. So, medical gases are regulated as drugs, and their supply system is therefore developed following multiple layers of restrictions and instructions for their proper and safe operation, maintaining precision, and hygiene requirements.

Medical gases are to be supplied in airtight, color-coded, well-labeled containers through a network of pipelines running through hospital buildings For more safety and security, an audio-visual monitoring system capable of checking the supply situation and a device to prevent cross-connection at medical gas outlets are provided.

Medical gas mixing up with any unwanted gases, air in particular, might even result in the death of patients. Regular monitoring is thereby needed. The medical gas supply system must be developed and maintained, satisfying the respective regulatory authorities.

Types of medical gas supply systems

Medical gas supply system is developed in the following two forms:

1 Individual cylinder system
2 Centralized piped medical gas supply and vacuum system

Centralized piped medical gas supply and vacuum system

Centralized medical gas supplies are also termed as piped medical gas and vacuum (PMGV) systems, where medical gases are delivered from a central supply location to different outlet points at various sites in the hospital at a controlled pressure of about 400 kPa.

Advantages of a centralized medical gas delivery system

For patients and attendants:

1 No distressing sight (sign) of and pounding on oxygen cylinder at bedside
2 Elimination of noise produced by cylinder movement
3 Protection of sterile areas from contamination due to use and movement of cylinder
4 Uninterrupted and clean gas supply at each work station
5 Safe and relief system

6 Makes effective use of space, economically advantageous and hygienic

For hospital administrator:

1 Easy purchase of gases in bulk quantities at favorable terms
2 Economy on purchase of cylinders

3 Fewer breakages
4 Rationalization in ordering, storing, and transporting a wide variety and size of gas cylinders
5 Minimum damage to building due to handling of cylinders

For hospital staff:

1 Instant availability of gas on taps
2 Clean, safe, and reliable delivery of gases
3 Continuous flow of gases when and where required
4 Minimal accident hazards due to mishandling of cylinders

Components of medical gas supply system

Medical gas supply systems include the following major components:

1 Medical oxygen plant
2 Medical nitrous oxide plant
3 Medical compressed air plant
4 Vacuum plant
5 Medical gas pipeline
6 Terminal unit
7 Pressure regulator
8 Shutoff valve
9 Monitoring and alarm system

Medical oxygen plant

Oxygen for medical use can be produced on-site, generally done in the hospital premises. Pressure Swing Adsorption (PSA) is an economical alternative for onsite production of Oxygen from air, as it contains about 78 percent nitrogen, 21 percent oxygen, less than 1 percent argon, and other gases. In the production process, nitrogen is separated, and the resulting product gas is up to 93 percent oxygen purity.

Oxygen plants mostly used are designed based on the pressure swing adsorption (PSA) technology. PSA technology is used to separate particular gases from a mixture of gases under pressure according to the molecular characteristics and affinity for an adsorbent material like zeolite used as a filter. Afterward the process is caused to swing to low pressure to desorb the adsorbed gas. So, two zeolite beds are used together, as shown in Figure 10.1.

By using an air compressor, air is compressed and purified by being freed from moisture and oil. Compressed air is passed through a molecular sieve made of zeolite. The sieves adsorb nitrogen, allowing 93 percent purified oxygen to pass through being less adsorbed. This is the generation stage of Oxygen. The process of Oxygen generation continues until the filter bed gets saturated with nitrogen. The second zeolite bed begins to discharge nitrogen, while the gas is desorbed due to the release of pressure. This is the regeneration stage. The automatic recycling process of adsorption and desorption between the two zeolite filter beds causes continuous generation of oxygen.

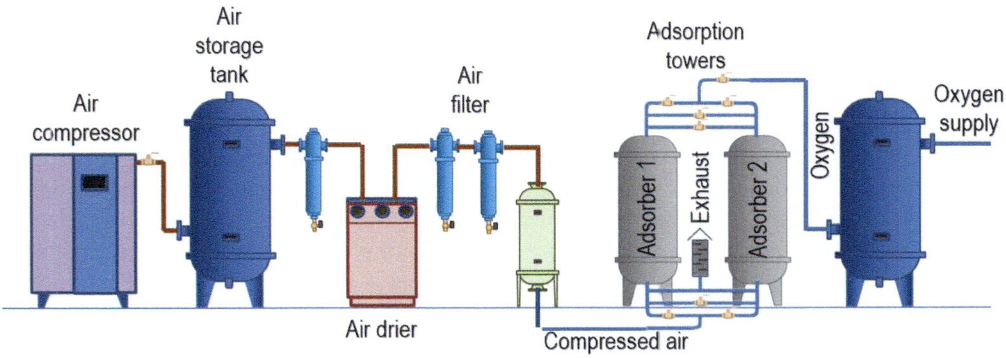

Figure 10.1 Schematic diagram of an oxygen plant.

The oxygen generator can be connected directly to the distribution pipeline through the oxygen buffer tank. The pipeline system is connected in parallel to reserve Oxygen cylinders to maintain Oxygen supply in the event of Oxygen supply pressure falling below 3.5 bar or a fall in oxygen purity.

Nitrous oxide plant

The Nitrous Oxide Plant is designed for producing bulk amounts of Nitrous oxide at high efficiency. The plant production capacity ranges between 5 and 50 kg per hour depending on the process of manufacturing.

The most common process for the manufacture of nitrous oxide is based upon the thermal decomposition of ammonium nitrate. There are a number of other nitrous oxide production processes, like direct oxidation of ammonia or purification of off-gas from adipic acid production (polyamide chain), etc.

Ammonium nitrate is decomposed in the melters, as shown in Figure 10.2, to give a mixture of Nitrous oxide and steam, along with some impurities including ammonium nitrate fumes, nitrogen, and other oxides of nitrogen. The steam and impurities are removed by scrubbing with water, caustic soda, and sulfuric acid in sequence. Nitrogen present in traces is removed by bleeding from the top of storage vessels where Nitrous oxide is stored after compression by a compressor.

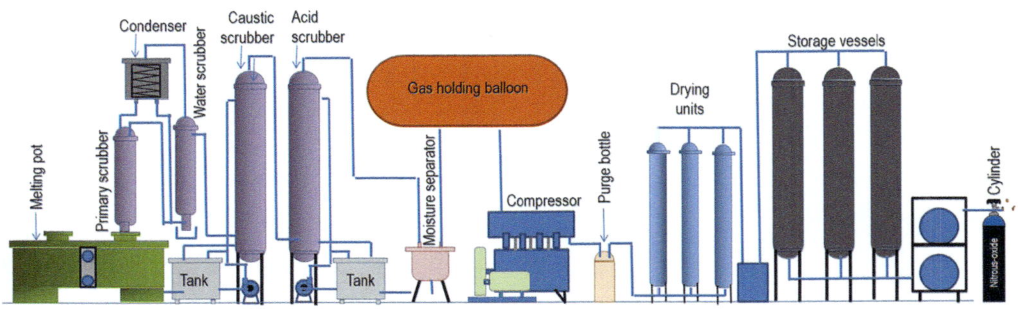

Figure 10.2 Schematic diagram of a nitrous oxide plant.

Vacuum plant

The vacuum generation system is an essential component of the medical gas supply system. A central vacuum plant consists of vacuum pumps, a buffer tank, and a control.

A vacuum pump is virtually used to remove gases and liquids from enclosed spaces. Vacuum pumps are generally single stage rotary vane pumps with ultimate pressure around 1 mba. Among various types of vacuum pumps liquid ring vacuum pumps are popular in medical vacuum supply systems, predominantly due to the fact that these pumps use water or oil as a sealing fluid, which helps ensure that the pump operates without contaminants being released into the surrounding environment. This is especially important in medical settings with a view to preventing contamination of sterile environments or sensitive equipment. Figure 10.3 represents a medical vacuum supply using two liquid ring vacuum pumps.

Minimum of two and in many cases three or four vacuum pumps are used. The pumps are run alternately or started additionally when more pumping speed is required.

The vacuum pump operates quietly and vibration-free. Capacities of a vacuum pump are based on actual cubic centimeter per minute (accm), not standard cubic centimeter per minute (sccm), Pumping flow capacity may vary from 25 to 840 cum/h.

The duplex pumping system should be equipped with a motor starting device and overload protection. Depending on the number of inlets that open or close, the demand for removing gases or liquids rises or falls. The vacuum pumps that are connected to the vacuum tank maintain the desired range of vacuum. Vacuum fluctuation across the drawing pipeline is generally desired to be maintained between −75 kPa and −60 kPa in the UK [1].

A vacuum tank, also termed a vacuum buffer tank, is used to provide a constant "reserve" of vacuum with a view to accommodate sudden or unusually high vacuum demands, reduce fluctuations in vacuum demand, maintain a constant vacuum level, and prevent overloading or excessive cycling of the vacuum pumps. A smaller vacuum pump will be required when a larger vacuum tank is used, as the tank acts as a reservoir for a greater volume of air to evacuate, allowing the smaller pump to work at a more manageable pace to achieve the desired vacuum level.

The buffer tanks are made in volumes of 60, 160, 300, 500, or 1000 liters, operating within a pressure range of 0–5–500 mbar, generally maintaining a final pressure ≤ 0.5 mbar.

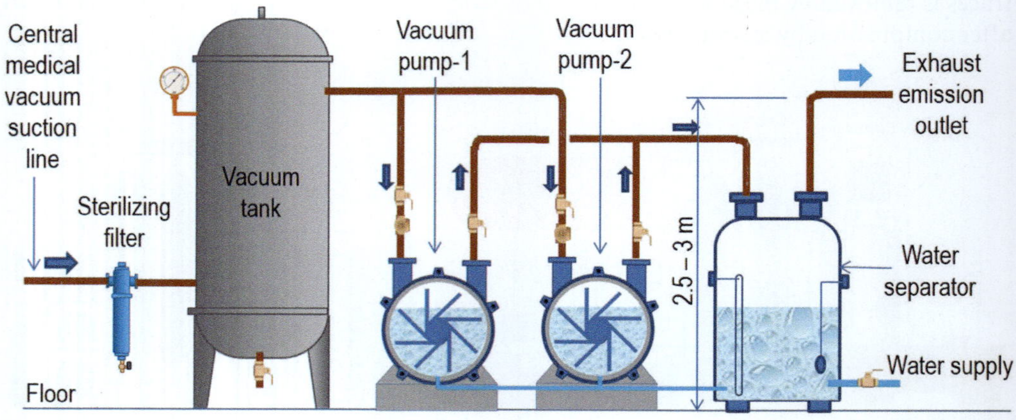

Figure 10.3 Schematic diagram of a central vacuum plant.

The system consumes a power range between 0.8 and 18.9 kW. The buffer tanks are supported by inlet filters and a draining valve to drain out condensed liquids, as it separates moisture from gas and particles. Draining can also be done automatically.

The vacuum control panel has both the manual and automatic control systems. While one pump is running, a reserve pump will start if the running pump cannot meet the required capacity. A pressure transducer controls the cut-in and cut-out of the compressors.

In addition, a central vacuum system typically consists of the following accessories:

1 Manually operated valve (basic control system) or electro-pneumatic operated valve (PLC system)
2 Dust filter on the pump inlet
3 Gauge/sensor
4 Two-stage pressure regulator
5 Switchboard with controller
6 Water separator

A check valve is generally used on the pump side of the tank, and an appropriate on-off valve is used on the downstream side.

In a central medical vacuum system water separator is used to remove water and moisture from the air or gases drawn so that it cannot accumulate in the vacuum lines, thus the system remains efficient, safe, and free from moisture-related problems.

Medical compressed air plant

The plant takes in air, compresses it by a compressor, as shown in Figure 10.4, and supplies the compressed air to the tanks, which serve as intermediate storage elements to ensure the continuous supply of compressed air. Depending on the compressor, a power distribution cabinet or control cabinet is required for the respective control system. The supporting oil

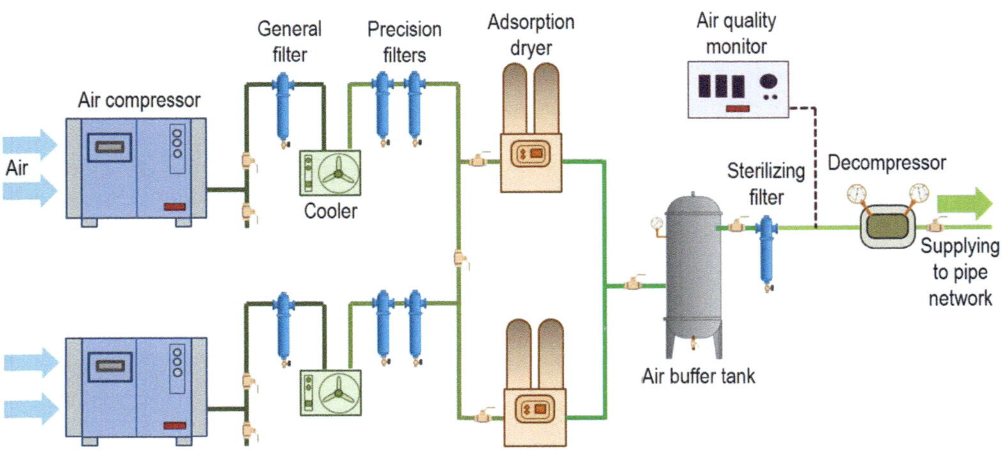

Figure 10.4 Schematic diagram of a central compressed medical air plant.

filters, dust filters, moisture separator, and the dryer condition the compressed air free of oil, dust, and moisture. The pressure-reducing system maintains the desired operating pressures of 4 bar or 5 bar for the ventilation of a patient and 7 bar or 8 bar for the operation of surgical tools. The compressed air is distributed to the points of use. Condensate in the form of water and oil is generated when air is compressed. Condensate drains contain the condensate; afterward, it is drained out to the optional expansion tank through a separate condensate pipe. Medical air compressor capacities are based on actual cubic centimeter per minute (accm), not standard cubic centimeter per minute (sccm).

Gas bank manifold

A gas bank manifold system is a specialized gas supply infrastructure designed to store, distribute, and regulate the supply of particular gas, often in buildings of various occupancies. It consists of a network of interconnected pipes, control valves, monitoring appurtenances, and storage tanks or cylinders that work together to ensure a continuous, controlled, and reliable gas supply to various points of use. In healthcare facilities, a particular type of medical gas is distributed to different locations through pipelines conveying gas from gas banks of that medical gas. Medical gas bank manifolds include oxygen manifolds, carbon-dioxide manifolds, nitrous oxide manifolds, nitrogen manifolds, etc.

A gas bank is a storage unit of a particular type of gas, which is used to store gas in healthcare facilities. It typically consists of a series of cylinders that are connected together, which can be located either inside or outside of a building. The cylinders are refilled by a gas-delivery truck or a pipeline supplying from a gas production plant. Generally, equal numbers of high-pressure cylinders are divided into two banks that alternate. The primary bank will be supplying the gas while the secondary bank will remain ready to supply it when the primary bank is exhausted and refilled. To size the banks, the rule of thumb is that each bank should serve 1–2 weeks.

Medical gas bank manifolds are designed to supply sufficient quantities of gas at correct pressure from gas cylinders or tanks through a pipeline system to terminal units in a healthcare facility. A typical manifold for medical gas supply usually consists of at least two-sided cylinder supply with automatic changeover between the empty and full side. There might be an additional third cylinder as a source for emergency supply.

The high-pressure multiple cylinders are connected to a header bar, which connects to the inlet of the manifold.

There are many kinds of medical gas manifolds, from single and simple gas supplies to complex systems that regulate several gas flows. Manifolds can be designed incorporating at least 2 banks of 2 cylinders each, with a maximum of 20 cylinders each.

Medical gas cylinders are available in a variety of sizes, typically designated by letters A, B, C, etc. with larger letters indicating larger capacities. The most common sizes used in healthcare settings include D, E, and G cylinders, with G being the largest of height 1.52 m, diameter 230 mm.

A medical gas manifold is made up of the following mandatory elements:

1 Inlet and outlet pipe connections
2 Flow control valve
3 Pressure regulator
4 Pressure release valve
5 Safety valve to shut the system down when it detects overly high pressure.
6 Monitor

Optional elements of manifolds are as follows:

1 Alarms that notify when gas is running low
2 Heater where CO_2 and N_2O gases can potentially freeze up when the temperature drops below 24°C

There are various types of manifolds, like single-stage, two-stage, and multi-stage manifolds. Single-stage manifolds are the simplest type; those are used for low-pressure applications ranging from 0 to a maximum of 13 bar, drawing gas from a single source. Two-stage manifolds are used for higher pressure applications and consist of two banks of cylinders—operating and reserve—on each side of the pressure control assembly, as shown in Figure 10.5. It regulates flow pressure in two stages, so it provides better pressure regulation and safety.

Manifold room

The manifold room houses the gas manifolds and a control panel for their operation. The control panel is supported by a warning lamp and is used for primary and secondary pressure regulations.

Cylinder and manifold rooms must fulfill the following criteria while planning, designing, developing, and maintaining:

1 There must be a dedicated room for the purpose.
2 These should be on the ground floor and near the loading dock.
3 It must be well ventilated.
4 If heated, it should be by either steam or hot water.

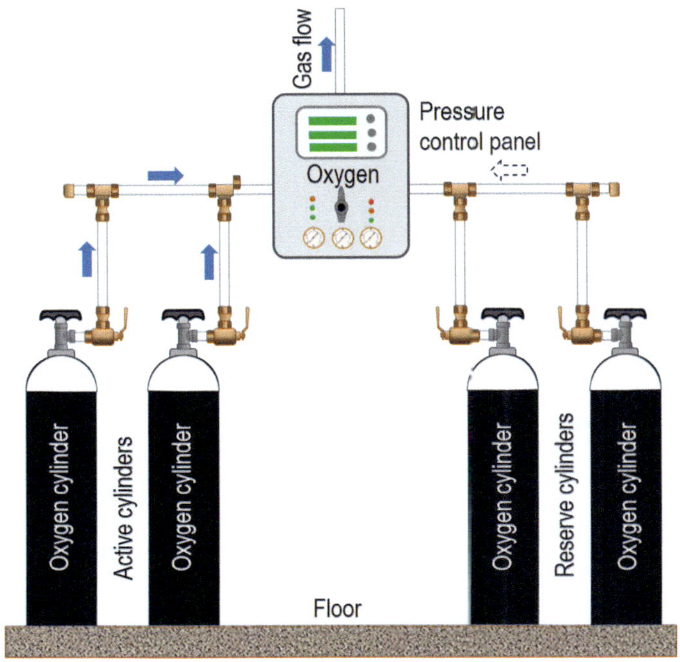

Figure 10.5 Two stage medical oxygen manifolds.

5 Room surface finish must have at least a 1-hour fire resistance rating.
6 Electrical devices shall be located at least 1.5m above floor finish (AFF).
7 All relief valves shall be vented to the outside.
8 Minimum temperature for N_2O and CO_2 shall be maintained at -6.67 °C (20 °F).
9 Each cylinder must be secured individually.

Location and space: A separate enclosure should be made for full cylinders, empty cylinders, compressed air, vacuum supply, etc. These should be located preferably on the ground floor, near the maintenance area. The enclosure or room shall be free from any fire hazard. The compressor unit should be located in a soundproof area, and the liquid oxygen area shall be cordoned off.

Control panel

Where two separate banks are used, gas from one bank, known as the primary side that is utilized at any time while the other bank known as the secondary side remains standing by. A control panel installed between the two banks ensures that both banks are not depleted simultaneously, and the control of the manifold automatically shifts the flow of gas from the primary side to the secondary side when the primary bank pressure falls below the set level. The control panel used should be built in accordance with international standards.

For every cylinder bank, there should be a pressure regulator to initially reduce the cylinder pressure to the two-line pressure regulators that control the final desired pressure. Both the line pressure regulators should be in good service at all times. The changeover from the primary depleted bank to the secondary bank should be immediate without fluctuation of line pressure. Isolation of banks for ensuring on-line servicing is necessary. The manifold should have an intermediate and line pressure relief valve.

There is a fully automatic gas supply control panel incorporating a microprocessor-based digital display that causes a switchover from "Active bank" to "Reserve bank" without fluctuation in delivery line pressure and without the need of external electrical power. Control panel logic will automatically maneuver the electronic control switch to cause the switchover. For manual system there is a valve to changeover the supply. Figure 10.6

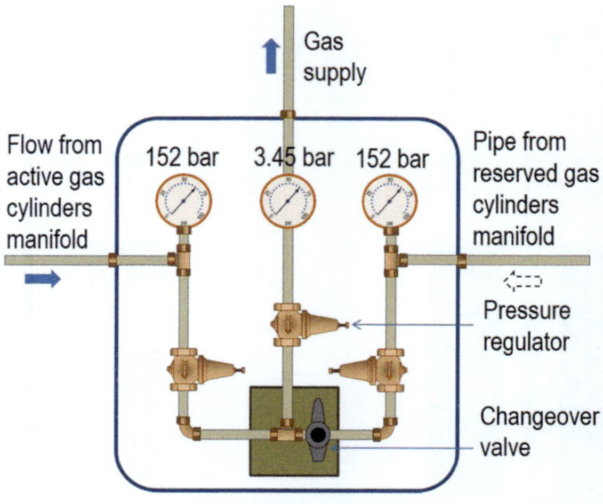

Figure 10.6 Schematic diagram of a manual control panel for cylinder gas cylinders manifold.

represents a schematic diagram of a manual pressure regulator to cause switchover from "Active bank" to "Reserve bank" and maintain the supply pressure.

Alarm system

Alarms are devices that monitor the flow and pressure of medical gases throughout the supply network, subject to sensing reduction in pressure irrespective of causes or variations in gas flow rate. These alarms warn maintenance personnel and help decide when the concerned components of the supply system need to be serviced or maintained, reducing the percentage of supply failures.

Two kinds of alarms: audio and visual, are usually incorporated in the centralized medical gas system. In case of a visual alarm, the control panel incorporates six colored LED lights: green is on when the bank is in operation, amber when the bank is ready and red when the bank is empty. When the light glows red, it indicates that the pressure is low, and alarm is activated. The other alarm is a remote signal lamp It is preferably both visual and audible.

Medical gas supply piping

Medical gas supply is generally developed by pipe-work made of special high-quality copper alloy for its high effectiveness in preventing degradation of the medical gases contained in the pipe and due to possessing good bacteriostatic properties. The fittings used in the piping system are made from brass, and the joints are brazed rather than soldered.

Copper piping shall be degreased, sealed, and steam-cleaned. Piping shall be kept protected against freezing, corrosion, and physical damage. Tubing must be stored and delivered, keeping it capped and cleaned. Medical gas piping ends shall be capped in place until further connection.

Stainless steel pipe and fittings can also be used for medical gas supply, The stainless steel pipe, though, has a big investment cost, but its sealing performance, process adaptability, and service life are comparatively long.

Plastic pipe is not generally recommended for medical gas supply purpose. In some non-critical, low-pressure systems like vacuum or some waste gas lines, certain plastic piping under specific conditions might be allowed.

Pipe installation

The guidelines below for installing medical gas supplying pipes shall be followed.

1 Pipes and electrical services shall either

   i run in separate compartments, or
   ii kept separated by more than 50 mm.

2 If pipelines are installed underground, they shall be protected by installing inside tunnels or ducts.
3 All pipelines for medical gases shall be routed in such a way that they are not exposed to a temperature less than 5 °C above the dew point of the gas at pipeline pressure.
4 Maximum intervals between supports for gas pipes shall be according to the values mentioned in Table 10.1.

Table 10.1 Maximum intervals between supports for gas pipes [3]

Sl. no.	*Pipe outside diameter (mm)*	*Maximum intervals between supports (m)*
1	Up to 15	1.5
2	22–28	2.0
3	35–54	2.5
4	54	3.0

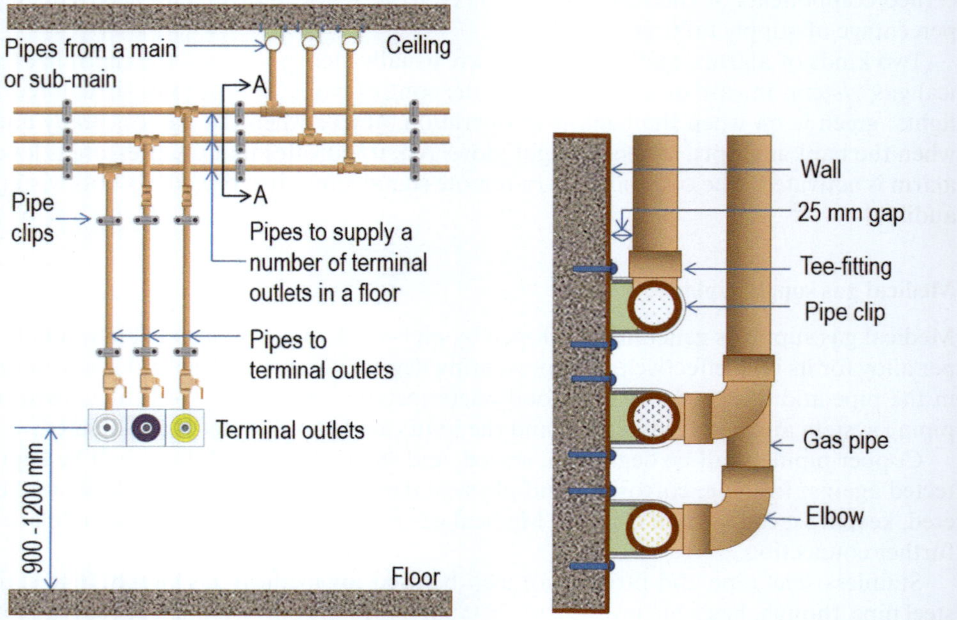

Figure 10.7 Medical gas pipe installation.

Pipes cannot be installed in contact with any masonry surface. There shall be at least 25 mm gap between the piping and supporting surface, as shown in Figure 10.7.

Medical gas supply pressure management

The medical gas supply system is very much concerned with pressure that must be maintained, incorporating all sorts of gas pressure management systems that are applicable. Both high and low levels of pressure than the required pressure level in the system are detrimental for the patients treated by medical gas. For this purpose, the delivery pressure of various medical gases shall have to be well maintained. The delivery pressure of various medical gases has been shown as below [4]:

1 Oxygen—delivery pressure of 345 kPa
2 Nitrous Oxide—delivery pressure of 345 kPa
3 Medical Air—delivery pressure of 345 kPa
4 Carbon Dioxide—delivery pressure of 345 – 690 kPa
5 Nitrogen—delivery pressure of 1241 kPa
6 Instrument Air—delivery pressure of 1241 kPa

Medical gas supply pressure regulator

In case of higher pressure level in the gas supply pipe, it must be regulated to the desired level in the quickest possible time. For this purpose, various pressure-regulating valves or devices are incorporated in the system as discussed below

Medical gas regulators are designed in two categories: single-stage regulators and dual- or two-stage regulators.

Single-stage pressure regulators perform the pressure reduction in a single step or stage. Single-stage regulators should only be used where quick adjustment of delivery pressure is ensured as needed or where there is a nearly constant inlet pressure. With this type of regulator, outlet delivery pressure may not be tightly controlled, as it can be made with a dual-stage regulator.

A dual-stage pressure regulator reduces the inlet pressure down to the desired outlet pressure in two steps or stages. Each stage consists of a separate spring, diaphragm, and control valve to operate. The first stage reduces the inlet pressure to an intermediate pressure level which is about two to three times the desired working pressure. The first stage pressure reduction is not user adjustable; it is factory-set and sealed. The final reduction achieved, to attain delivery pressure level, is accomplished in the second stage. The advantage of a dual-stage regulator is its ability to deliver a constant outlet pressure, even in the event of a decrease in inlet pressure.

The dual-stage regulator is recommended in the following conditions.

1 For gas supply to the instruments, where constant delivery pressure is critical.
2 Where the delivery pressure rise from cylinder pressure decay cannot be accepted at all.
3 Situations where proper pressure is critical and the cylinders in use are in a remote or difficult-to-monitor location.
4 Typically, is not required when using gases from low-pressure sources of less than 34.5 bar [5].

Cylinder pressure regulator

Cylinder pressure regulators are mounted directly on gas cylinders. These pressure regulators are available operating in single- or dual-stage designs. Cylinder pressure regulators are normally found supported by an inlet and delivery pressure gauge.

Line pressure regulator

Line gas pressure regulators are used for installing directly on a pipe of supply systems, such as downstream of a manifold or bulk storage vessel. Because the inlet pressure in piping systems is normally considered to be constant, line pressure regulators are typically single-stage configurations supported by a delivery pressure gauge only.

Terminal units or outlets or inlets

In a medical gas supply system, terminal units are the points at the end of a medical gas pipeline where the operator of medical appliances or equipment makes connections for the supply of specified medical gas and disconnections after use. For vacuum pipeline systems, terminal units are also used that is an inlet for supply vacuum. Medical gas outlets come in different styles and they are generally installed on headwalls, which can also be installed under a ceiling or on floor.

There are two ways of connections; quick connect and twist-on. Outlets are gas-specific, non-interchangeable and color coded as per gas content.

Terminal units are made for making a quick, easy, and tight connection; otherwise, a loose or wrong connection can create a hazard to the patient or operator. Disconnection from the terminal units should also be smooth and easy. It is important that terminal units and their components be designed, manufactured, installed, and maintained in such a way as to meet the requirements specified by the concerned regulatory authorities.

The terminal unit shall be highly wear-resistant, having long shelf lives. Other features include non-interchangeability between probes, socket assemblies, and terminal blocks, as well as quick release of the probe with single-handed operation. Terminals shall be supported by a secondary valve to facilitate ease of maintenance without interruption of the gas supply.

Terminal outlets are mostly provided for the following pressure range:

1 Oxygen: 34–500 kpa
2 Nitrous oxide: 345–500 kpa
3 Medical air: 345–500 kpa
4 Nitrogen or Surgical air: 700–800 kpa
5 Medical vacuum: 40–53 kpa below atmospheric pressure

Terminal units are designed for three types of mounting: trunking, flush, and surface. Various types of terminal units have been shown in Figure 10.8.

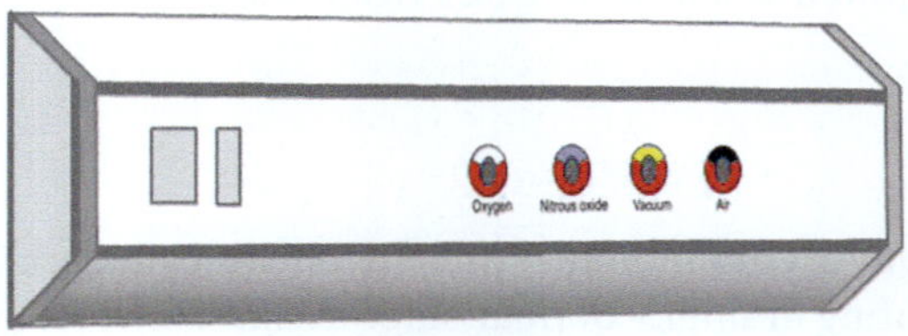

a. Trunking medical gas outlets.

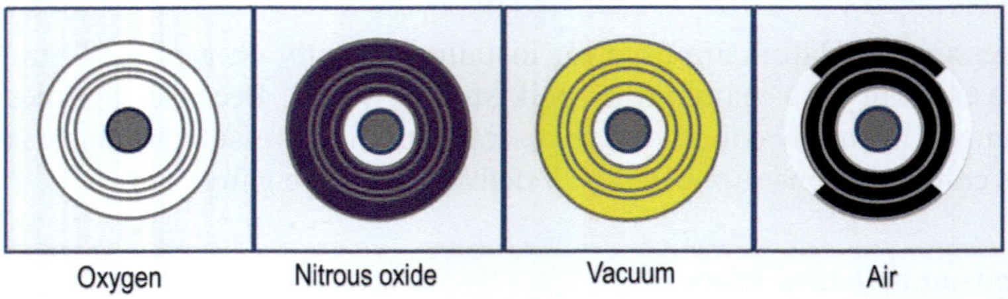

b. Flush-mounted medical gas outlets.

Figure 10.8 Various types of medical gas outlets: (a) Trunking medical gas outlets; (b) Flush-mounted medical gas outlets;

(*Continued*)

c. Surface-mounted medical gas outlet.

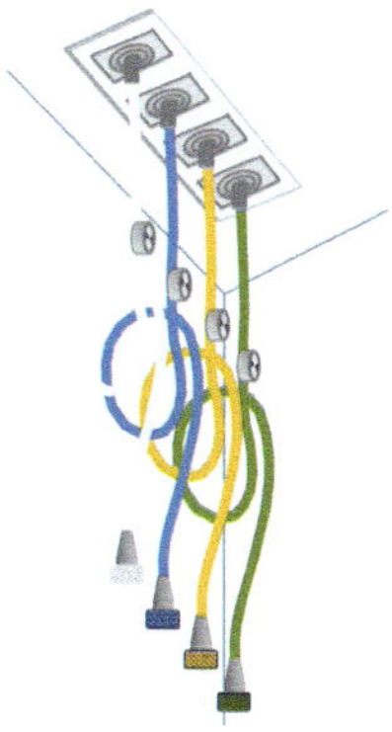

d. Ceiling-mounted hanging medical gas outlets.

Figure 10.8 (Continued) (c) Surface-mounted medical gas outlet; (d) Ceiling-mounted hanging medical gas outlets.

Wall-mounted terminal unis (outlet)

A gas terminal outlet point, kept closed by a self-sealing valve, is kept fixed to the wall near the patient bed and is encased in a small, rectangular-shaped box labeled and colored for instant identification. Its use is activated as soon as a safety key plug connector is inserted into it. It should be at least 20 cm away from any electrical fitments nearby.

Trunking-type terminal unit

Trunking-type medical gas terminal units are found of adjustable length for future expansion and up-gradation. These are found flexible in terms of the positioning of the terminal units.

Individual medical gas terminal units are also installed on walls; they may be surface-mounted or flush-mounted.

There are also medical gas terminal units (outlets) that are kept hanging from the ceiling as shown in Figure 10.8d.

Table 10.2 Color coding of various medical gases [3]

Gas	*USA*	*International*
Oxygen	Green	White
Carbon-di-oxide	Gray	Gray
Nitrous oxide	Blue	Blue
Helium	Brown	Brown
Nitrogen	Black	Black
Air	Yellow	White and black

Vacuum inlet unit

Vacuum inlet unit is connected to source equipment through the distribution system, which is kept covered by a lid. It is located at actual usage point, generally at patient's bedside wall. This is fitted with a locking system and non-return valves. The minimum size of vacuum piping shall be 19 mm, making 12 mm pipe drops to individual inlets permitted.

Color coding usage of medical gases

Color-coding is used predominantly for identifying either the type of gas contained or the hazard associated with contained gas. All the components including cylinders, pipes, fittings and connections are colored. The Table 10.2 furnishes the color-coding chart, which is followed in medical gases supply.

Sizing of medical gas supply components

All the components of a medical gas supply system shall be sized very carefully. Incorrect sizing of the major components can create several issues like insufficient gas flow, fluctuating pressure etc., which can lead to potential life threatening hazards for the patients, equipment malfunction, etc.

For sustainability, piping and equipment should be sized accommodating future growth.

Medical gas pipe sizing

Medical gas piping networks are sized using the following four items:

1 Total flow rate of connected outlets
2 The diversity factor
3 The allowable friction loss
4 The equivalent pipe length

Total flow rate

Total flow rates of medical gases depend on the quantity of outlets, type of outlets, and diversity of use. To determine the required flow rate through a pipe, the cumulative flow rate required in the distributing branch pipes is to be known. To estimate the flow rate required for different types of medical gases, the following outlet flow requirements can be used.

Typical outlet flow requirements

1. O_2 – 28316.847 sccm for infant ventilator and 184059.50 sccm for adult ventilator
2. N_2O – 28316.847 sccm
3. N_2 – 42475.27 sccm per operating room
4. MA – 28316.847 sccm
5. MV – Varies
6. WAGD – 28316.847 sccm (Some anesthesia systems use modified interface valves with larger inflows of up to 56633.70 sccm)

Diversity factor

The diversity factor is a correction factor that is established on the fact that the estimated probable load of a system is part of the sum of loads in the system. The diversity factor for designing a pipe depends on the number and type of outlets to be served. Multiply the total flow rate by the diversity factor specified in Table 10.3 for each section of pipe to determine the required flow rate for the piping.

There is a formula, as shown below, for calculating a diversity factor to be used for the design of pipelines for those locations and applications where no diversity factor is nominated.

$$\text{Diversity factor} \quad F = \frac{1.3}{\log(10n)} \tag{10.1}$$

Where n = the total number of terminal units on the branch or main line.

It is noticed that the diversity factor tends to be asymptotic toward 0.4 or 40 percent.

Equivalent pipe length

The equivalent length of pipe run is calculated by finding the actual measured run and adding 50 percent of the measured run of pipe to account for friction losses in fittings, valves, etc. supposed to be on the piping. The total is the equivalent length of pipe measured.

Allowable friction loss

This is the excess pressure available more than the desired pressure at the outlet of a pipe, which can be lost due to friction that would be developed for flowing any gas through the

Table 10.3 Diversity factor for probable flow of gas in a pipe supplying multiple pipes leading to terminal outlets or inlets [6]

Number of outlets and inlets terminal units per facility	*Diversity percentage of average flow per outlets and inlets terminal units (percent)*
1–10	100
11–25	75
26–50	50
51–100	50

Table 10.4 Maximum permitted pressure loss in medical gas and medical vacuum systems [6]

Type of system	*Maximum allowable system pressure loss (kPa)*
Medical air	34.5
Nitrogen	103.5
Nitrous oxide	34.5
Carbon dioxide	34.5
Oxygen	34.5
Medical vacuum	13.33 (100 mm Hg)

pipe of a particular diameter. Dividing the equivalent pipe length, generally in hundreds of meters, by the allowable system loss, will determine the allowable friction loss per 100 meters of pipe.

Example: If 50 kPa is allowed for total system loss with an equivalent pipe length of 200 meters, then 50 divided by 2 equals 25 kPa friction loss is allowable per 100 meters long pipe.

Maximum permitted pressure loss in medical gas and medical vacuum systems have been furnished in Table 10.4.

The final delivery pressure for all medical gas supply systems, except for instrument air and vacuum, is 345 kPa. Considering allowable friction loss of 10 percent, the gas source must ensure it delivers 379.50 in order for the pressure loss through the piping to be 34.50 kPa.

In the supply of instrument air, the supply pressure is usually around 1379 kPa. So, the delivery pressure of this type of pneumatic instrument must not exceed 1379 kPa because of the standards that are used for these systems. So, a system loss of 138 kPa can be allowed in this case. Updated instruments may operate at lesser pressure than the older instruments.

Branch and drop pipe size

The medical gas supply reaches the using points in a room as a drop pipe or hose. Quick-coupler mechanisms or adaptors connect one end of the hose to the appropriate gas terminals fitted at the end of drop pipes or hoses.

It is a good practice to use the minimum size of the drop pipes leading to the terminal gas appliances or instruments requiring particular gas.

For the compressed gases, 12 mm, sub-mains should be a minimum of 20 mm in size and the main size should be no less than 25 mm. The minimum size of vacuum piping shall be 20 mm, with 12 mm drops to individual inlets permitted.

Runouts to alarm panels and connecting tubing for gauges and alarm devices are of 6 mm.

Particular medical gas supply pipe sizing

For most of the compressed medical gases, except for instrument air, the typical design criteria are the same. As vacuum is considered a medical gas, the same type of criteria is applied.

To determine the particular type of medical gas supply pipe size, all the outlets to be installed in the system shall be added together to calculate a cumulative connected load.

The diversity factor, which is provided in Table 10.2, is used to adjust the total connected load to a level of approximate probable flow rate, as it is expected that all the outlets might not be in simultaneous operation. The maximum estimated flow rate for the supply pipe is calculated by multiplying the diversity factor by the total connected load for the concern pipe.

The equivalent pipe length is determined by multiplying the actual length of pipe that is to be designed by 1.5, as it is assumed that a 50 percent increase of pipe length will cause the same amount of pressure drop as would be done by the fittings and accessories installed on it.

The sizes of various medical gas pipes can then be determined from Table 10.4, which is based on the allowable maximum pressure drop for the respective gases.

Oxygen supply pipe sizing

An example of oxygen supply pipe sizing has been given below:

Oxygen flow required (Total) = 364 lpm (calculated)
Length of pipe from source (cylinder/tank) to end (farthest or highest terminal) = 80 m
Equivalent pipe length = 1.5 × 80 = 120 m

From Table 10.4, for pipe sizing for maximum allowable pressure drop of 250 mm Hg, the required pipe size is 12 mm.

Nitrous oxide supply pipe sizing

An example of nitrous oxide supply pipe sizing has been given below:

Nitrous oxide (N_2O) flow required (Total) = 76 lpm (Calculated)
Length of pipe from source (N_2O cylinder) to end (farthest or highest terminal) = 80 m
Equivalent pipe length = 1.5 × 80 = 120 m

From Table 10.4, for pipe sizing for maximum allowable pressure drop of 250 mm Hg, the required pipe size is 12 mm.

Vacuum pipe sizing

Vacuum pipe sizing depends on the length of the pipe from vacuum pump to end, the flow rate, and the pressure drop.

Anesthesia gas flow required (Total) = 1365 lpm (calculated)
Length of pipe from source (from AGSS vacuum pump at roof) to end (farthest or highest terminal) = 100 m
= 100 × 1.5 = 150 m

From Table 10.5, for pipe sizing for vacuum;
Recommended pipe size is 50 mm, as for 152 m length flow will be 1,586 lpm.

Table 10.5 Size of medical gases and vacuum piping [7]

Maximum delivery capacity[3] *in lpm*						
Length of piping in m[1]						
Medical system	*Gas pipesize mm*[2]	*30*	*76*	*152*	*228*	*304*
Oxygen	12	425	300	209	167	144
	20	1,133	801	555	445	377
	25	1,416	1,416	1,138	912	784
Nitrous oxide	12	425	269	184	150	127
	20	849	699	484	388	331
	25	1,113	1,133	983	7,98	688
Medical air	12	512	314	221	177	151
	20	1,133	847	595	467	399
	25	1,416	1,416	1,192	1,013	826
Vacuum	25	645	388	269	215	184
	32	1,135	694	473	377	317
	40	1,804	1,101	759	600	507
	50	3,758	2,305	1,586	1,274	1,084
Nitrogen	12	708	708	708	674	583
	20	1,699	1,699	1,699	1,699	1,535
	25	3,115	3,115	3,115	3,115	3,115

1 Length of piping includes a 30 percent allowance for fittings.
2 12 mm diameter pipe is the minimum size allowed in medical gas systems.
3 Based on the following maximum pressure drops:
Oxygen, nitrous oxide, and medical air − 250 mm Hg (33.33 kPa).
Vacuum − 100 mm Hg (13.33 kPa).
Nitrogen − 1025 mm Hg (137 kPa).

Medical compressed air supply pipe sizing

Example of medical air supply pipe sizing at various pressure has been given below:

Example of pipe sizing for medical air supply at 4 bar:

Medical air flow required (Total) = 445 lpm (Calculated)
Length of pipe from source (cylinder/tank) to end (farthest or highest terminal) = 80 m
Equivalent pipe length = 1.5 × 80 = 120 m

From Table 10.6, for pipe sizing at 4 bar pressure;

Required pipe size is 22 mm at 14 kPa pressure loss which addresses 513 lpm for 122 m length

Example of pipe sizing for instrumental (medical) air supply at 7 bar

Medical air flow required (Total) = 1793 lpm (calculated)
Length of pipe from source (cylinder/tank) to end (farthest or highest terminal) = 80 m
Equivalent pipe length = 1.5 × 80 = 120 m

From Table 10.7, for pipe sizing at 7 bar pressure;

Pipe size is 35mm at 14 kPa pressure loss, which addresses 2453 lpm for 122 m length

Table 10.6 Medical gas pipe sizing for supplying pressure at 4 bar, based on pipe length and flow rate at various pressure loss [8]

Outsi-de diameter (mm)	*Pressure loss (kPa)*	*Distance from source (m) at 4 bar, for 7, 14 and 21 kPa pressure loss.*																
		8	*15*	*30*	*61*	*91*	*122*	*152*	*183*	*213*	*244*	*274*	*305*	*335*	*366*	*396*	*427*	*457*
		Free air flow rate (lpm)																
12	7	311	209	141	95	75	64	56	50	46	43	40	37	35	34	32	31	30
	14	455	307	207	139	110	94	82	74	68	63	59	55	52	50	47	45	44
	21	564	382	258	174	138	117	103	93	85	78	73	69	65	62	59	57	55
15	7	579	391	263	177	140	119	105	94	86	80	75	70	66	63	60	58	56
	14	845	572	386	260	207	175	154	139	127	118	110	104	98	93	89	85	82
	21	1038	711	481	325	258	219	192	173	159	147	137	129	122	117	111	107	102
22	7	1677	1135	768	518	411	349	307	277	254	235	220	207	196	186	178	170	164
	14	2441	1656	1123	759	604	513	451	407	373	345	323	304	288	274	262	251	241
	21	3023	2053	1395	945	751	638	562	507	465	431	403	379	359	342	326	313	301
28	7	3363	2283	1547	1047	832	706	622	560	514	476	445	419	397	378	361	346	332
	14	4881	3320	2257	1530	1218	1035	912	823	754	699	653	615	583	555	530	508	488
	21	6034	4109	2800	1901	1514	1287	1135	1024	938	870	814	767	726	691	660	633	609
35	7	6023	4096	2783	1886	1500	1275	1124	1013	928	861	805	758	718	683	653	626	602
	14	8720	5943	4051	2752	2192	1865	1644	1483	1360	1261	1180	1111	1053	1002	957	918	883
	21	10758	7344	5018	3415	2723	2317	2044	1845	1692	1569	1468	1383	1310	1248	1192	1143	1099
42	7	10758	7344	5018	3415	2723	2317	2044	1845	1692	1569	1468	1383	1310	1248	1192	1143	1099
	14	14587	9963	6806	4633	3694	3145	2775	2504	2296	2130	1993	1878	1780	1694	1619	1553	1493
	21	17963	12290	8421	5743	4584	3904	3446	3112	2855	2648	2478	2335	2213	2107	2014	1932	1858
54	7	14974	10588	7487	5294	4323	3743	3348	3056	2830	2647	2496	2368	2257	2161	2076	2001	1933
	14	21176	14974	10588	7487	6113	5294	4735	4323	4002	3743	3529	3348	3192	3056	2937	2830	2734
	21	25935	18339	12968	9169	7487	6484	5799	5294	4901	4585	4323	4101	3910	3743	3597	3466	3348
76	7	37754	26696	18877	13348	10899	9438	8442	7706	7135	6674	6292	5969	5692	5449	5236	5045	4874
	14	53392	37754	26696	18877	15413	13348	11939	10899	10090	9438	8899	8442	8049	7706	7404	7135	6893
	21	65392	46239	32696	23119	18877	16348	14622	13348	12358	11560	10899	10339	9858	9438	9068	8738	8442

Table 10.7 Medical gas pipe sizing for supplying pressure at 7 bar, based on pipe length and flow rate at various pressure loss [8]

Outside diameter (mm)	*Pressure loss (kPa)*	*Distance from source (m) at 7 bar, for 7, 14 and 21 kPa pressure loss.*																
		8	*15*	*30*	*61*	*91*	*122*	*152*	*183*	*213*	*244*	*274*	*305*	*335*	*366*	*396*	*427*	*457*
		Free air flow rate (lpm)																
12	7	408	276	186	125	99	84	74	67	61	56	53	50	47	45	43	41	39
	14	599	405	274	185	147	124	109	99	90	84	78	74	70	66	63	61	58
	21	979	664	450	304	242	205	181	163	149	138	129	122	115	110	105	100	96
15	7	759	514	347	234	186	158	139	125	114	106	99	93	88	84	80	77	74
	14	1112	754	510	345	274	232	205	184	169	156	146	138	130	124	118	114	109
	21	1811	1231	836	566	450	383	337	304	279	258	242	227	215	205	196	188	180
22	7	2192	1488	1009	682	542	460	406	366	335	310	290	273	259	246	235	225	217
	14	3198	2175	1478	1001	797	677	597	538	493	457	428	403	381	363	347	332	320
	21	5180	3533	2410	1638	1306	1111	980	884	811	752	704	663	628	598	571	548	527
28	7	4387	2984	2027	1374	1093	929	819	739	677	628	587	553	524	498	476	456	439
	14	6382	4351	2963	2013	1604	1364	1203	1086	995	923	863	813	771	734	701	672	646
	21	10290	7038	4816	3283	2620	2232	1970	1779	1632	1514	1417	1335	1266	1205	1152	1105	1063
35	7	7841	5345	3638	2470	1968	1674	1476	1332	1221	1132	1059	998	945	900	860	825	793
	14	11380	7775	5307	3612	2881	2453	2165	1954	1792	1662	1556	1466	1389	1323	1264	1212	1166
	21	18271	12528	8599	5876	4696	4003	3536	3194	2931	2720	2547	2401	2276	2168	2073	1988	1912
42	7	13128	8964	6113	4159	3316	2823	2490	2248	2061	1912	1789	1686	1598	1521	1454	1394	1341
	14	19010	13012	8901	6070	4847	4129	3646	3293	3021	2803	2624	2473	2344	2232	2134	2047	1969
	21	30392	20892	14381	9849	7881	6723	5942	5371	4930	4577	4286	4042	3833	3651	3491	3349	3223

Medical gas supply system testing

The concern regulatory authorities inspect and test the following major components and segments of a medical gas supply system for mainly ensuring the purity of gas and pressure of the supply system:

1 Piping
2 Master and area alarms
3 Outlets
4 Flexible connectors
5 Automatic pressure switches
6 Shut-off valves

The following compulsory tests are carried out:

i Pressure drop test to ensure a leak-proof system
ii Valve tightness and correct valve zoning
iii Satisfactory mechanical operation and non-interchangeability of each terminal outlet by means of probes
iv Anti-confusion test for checking cross connection
v Flow rate and pressure at terminal outlets
vi Satisfactory changeover mechanism of gas-control panels
vii Satisfactory functioning of compressors, vacuum pumps, and waste anesthetic gas disposal (WAGD) system
viii Satisfactory performance of alarm system

When the whole system is in place, all piping should be subjected to a test pressure of about 1.5 times the working pressure, or 1035 kPa, for 24 hours to check if it can withstand the pressure as well as leakages.

References

[1] www.wikipedia.org "Medical gas supply" Retrieved from https://en.wikipedia.org/wiki/Medical_gas_supply on 11 January, 2025.
[2] Miguel Gómez-Chaparro, Justo García-Sanz-Calcedo, and Luis Armenta Márquez. (2018) "Analytical determination of medical gases consumption and their impact on hospital sustainability" *Sustainability* 10, no. 8: 2948. https://doi.org/10.3390/su10082948
[3] Hamzah Meraj (2017) "Central medical gas distribution system" Retrieved from https://www.slideshare.net/hamzaaaaaah/central-medical-gas-distribution-system on 12 January, 2025.
[4] Pattons Medical (2025) "Overview of medical gas manifolds" Retrieved from https://www.pattonsmedical.com/overview-of-medical-gas-manifolds-3/ on 12 January, 2025.
[5] The Lincoln Electric Company (2025) "Two-stage regulators" Retrieved from https://www.harrisproductsgroup.com/en/Resources/Knowledge-Center/Tech-Tips/Tech-Tips-OLD/Two-Stage-Regulators on 12 January 2025.
[6] Uniform Plumbing Code "Sizing procedures" Hawaii Plumbing Code 2021, Retrieved from https://up.codes/s/sizing-procedures on 12 January, 2025.
[7] www.oregon.gov "Health care facilities and medical gas and vacuum systems" Chapter 13, Part 1, Retrieved from https://www.oregon.gov/bcd/codes-stand/Documents/pl-14opsc-chapter13.pdf on 19 July 2025.
[8] Engineering Calculators (2019) "Building calculators" Retrieved from https://www.buildingcalculators.com/medical-gas-pressure-drop-calculator on 13 January 2025.

11 Healthcare plumbing

Introduction

Nowadays, in many cases, life of majority humans starts and ends in hospitals or healthcare centers, where there are some healthcare facilities. In those facilities, the use of water is a must, and as a result, there is a generation of wastewater that needs a drainage system for proper disposal. So, plumbing is there, which is expected to be more health-friendly than those normally developed in buildings of other occupancy, not only for the patients' safety but for the staff and visitors as well. There are a number of plumbing issues that are unique to hospitals, healthcare facilities, medical laboratories, etc. Plumbing in healthcare centers is therefore special and considered to be more complex than those of other occupancy-buildings.

In the ever-developing world of healthcare services, reliable, flexible, and easy-to-maintain plumbing systems are the primary demand of the time, for which acquiring additional knowledge in this regard is the key to supporting sustainable healthcare service. With a view to achieving this objective, the specialties and the complexities of healthcare plumbing will be discussed exclusively, in this chapter.

Healthcare plumbing

Healthcare plumbing is more or less unique in terms of safety with respect to health and environment and quality assurance, both in water supply and drainage systems developed in dedicated healthcare service areas, in addition to very common areas where plumbing systems are installed in almost similar ways as practiced in buildings of all other occupancies. Health care service needs special fixtures, appliances, and machines, which require plumbing of a special nature.

Furthermore, in healthcare buildings, plumbing installation for medical gas supply for the patient, in addition to fuel gas supply for its various uses, is required; those are discussed in previous chapters, separately.

Special fixtures and appliances

In the buildings having healthcare facilities, there are some special fixtures and appliances used, other than very common plumbing fixtures and appliances used in buildings of other occupancies. The following is a brief discussion on some of those special fixtures and appliances used in a building providing healthcare services.

DOI: 10.1201/9781003724056-11

Hand-washing clinical sink

Tempered water is the best preferred for hand-washing. The tempered water used is at 40°C, which is obtained by mixing 60°C hot water maintained in the system and normal water developed along with the hot water system. For a very remotely located sink, a tankless, also known as a point-of-use electric water heater producing warm water of desired temperature, can be used.

Elbow- or foot-operated or non-touch sensor-activated mixer faucets are recommended for all the clinical hand-wash sinks, producing laminar flow outlets, as a measure of infection control; soft flow outlets are not preferred in the sinks. Sensor-activated mixer faucets need frequent battery maintenance. If a mixture faucet with a wrist-blade handle, as shown in Figure 11.1, is used, the handle shall be at least 150 mm long.

Such a wash area should be marked as a hand-washing station where a "sink" instead of a "lavatory" is installed. A sink requires a 40 mm drain pipe and trap.

Laboratory glass washer

In healthcare services, generally in laboratories, reusable glass items are needed to be decontaminated; so, they are required to be washed and dried properly. For such washing and then drying a wide range of different glassware and other materials, glass washers are used.

Laboratory glass washers may be an automatic type, as shown in Figure 11.2, which requires hot water, usually 60°C boosted to 82°C distilled or deionized water, and compressed air. These washers possess a rapid and efficient drying system. There is also a manual type glass washer for washing bottles and tubes using cold water only.

Most washers require a flow rate of at least 7.6 lpm to operate properly. Water supply pressure should be 2 to 3.5 bar, having the 19 mm NPT pipe and befitting fittings and accessories on the water supply line.

Air supply flow where required, should be maintained about 0.8 cum/min at pressure ≥0.8 mPa [2].

All residual moisture, chemical reagents, and cleaning agents are thoroughly discharged by the machine. The machine must be installed reliably ensuring the water inlet and drainage hose fixed and installed properly. This machine drains by gravity, so the drain pipe of 30 mm standard size must be lower than the machine. Ensure the drain hose has no bends or constrictions in it.

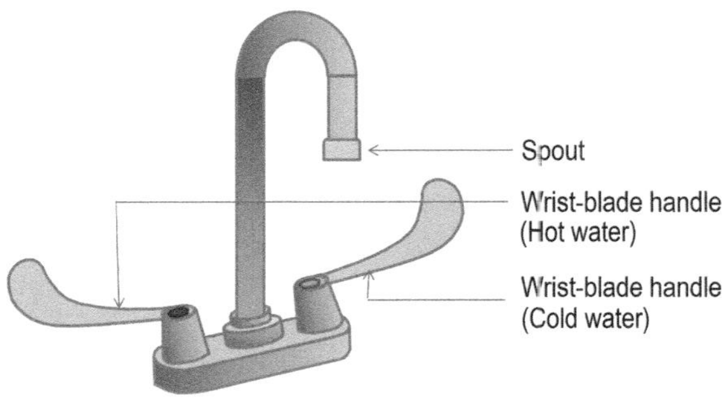

Figure 11.1 Mixture faucet with wrist-blade handle.

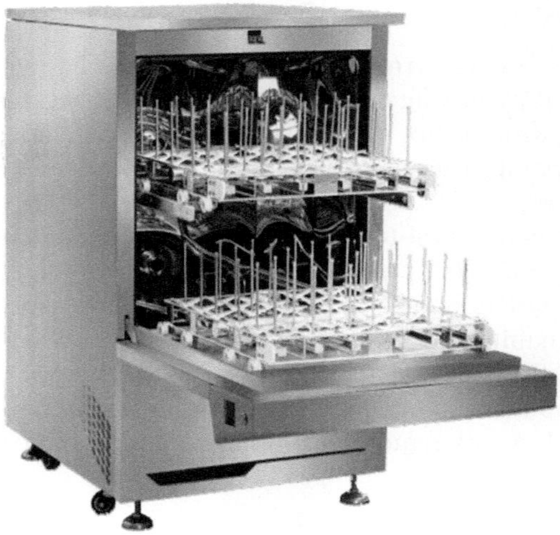

Figure 11.2 A laboratory glass washer [1].

Emergency shower

Provision of a safety shower is needed wherever there is a possibility that either highly corrosive or highly toxic substances may splash over substantial areas of the body of a person while working. Emergency showers cause the hazardous material splashed over the body to flush away, which will cause injury for delayed removal.

Emergency showers are designed to flush the user's head and body but not the eyes, as its huge rate of flow under high pressure could damage the eyes in some instances. Separate eyewash stations are designed to flush the eye and face area only. There are combination units' providing both the facilities: emergency shower and eyewash, as shown in Figure 11.3.

Emergency or safety showers should be located in the adjacent corridors or at the exit door. The showers must be accessible, require no more than 10 seconds to reach, and be within a travel distance of no greater than 30.5 m from a working station in a hazard room. The shower head shall be a deluge-type shower on a pipe of 25.4 mm nominal diameter supplying cold water, a stay-open design, and a supply valve operated by a hanging pull chain.

For drainage, a floor drain may be provided nearby, if required. If floor drains are provided, trap primers should be incorporated.

Eye and face-wash fountain

Eye and face-wash fountains are commonly integrated with an emergency shower but are also found either wall- or counter-mounted with a water-supply arrangement to facilitate quick and abundant water flow to the eyes and face through double side-mounted, full face-wash outlets. This unit is operated either with a wrist-blade handle or a foot pedal. It may also be a deck-mounted unit with a hand-held face and body spray fitted with a hose. Sometimes a tempered water supply may be considered.

Emergency showers must flow at a minimum rate of 76 lpm of potable water at 207 kPa, for 15 minutes. Likewise, emergency eyewashes must deliver at least 11.4 lpm at 207 kPa, for

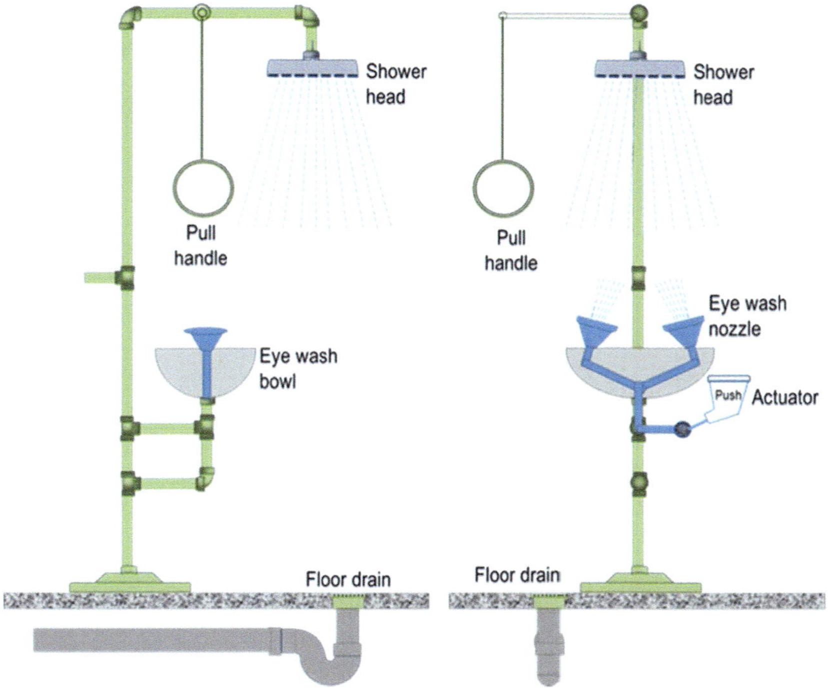

Figure 11.3 Combined emergency shower and eye wash.

15 minutes. The combination eye and face wash required flow as "a minimum of 11.4 lpm at 207 kPa. Emergency shower and eyewash stations are intended to provide tepid water in the range of 16 to 38°C.

Mop-service basin

A mop-service basin is generally used to facilitate dumping dirty mops and cleaning so that wastewater is contained in the basin and drained out ultimately.

Mop-service basins are generally preferred to be floor-mounted, as shown in Figure 11.4, to facilitate cleansing of floor mopping devices. These basins can be obtained in precast or (terrazzo) molded-stone units of various sizes. Rim guards are generally provided to protect the rims from damage, and wall guards are provided to protect the surrounding walls from splashing and developing chemical stains. Service faucet providing water, usually a two-handle mixing faucet mounted on the wall with a wall brace, vacuum breaker, and hose adapter.

Water flow for the faucet should be 5 to 8.0 lpm at 414 kPa. The drain pipe size shall be 76 mm.

Bedpan washer

Generally, a bedridden patient needs a bedpan to defecate, or a urine bottle to urinate, and other similar items, which need cleaning sanitarily for reuse. A full bedpan can contaminate

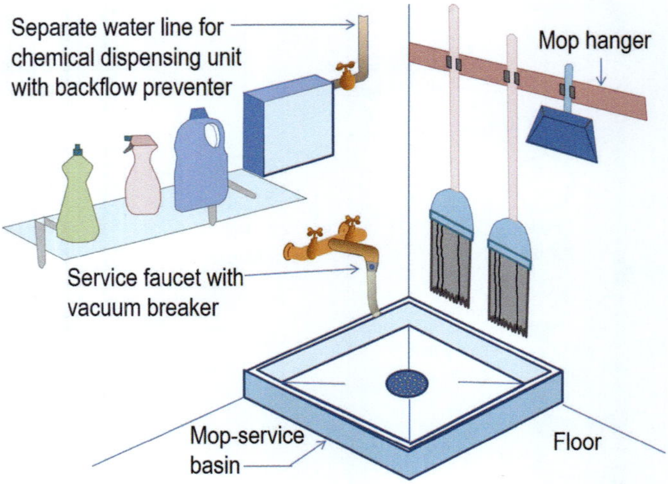

Figure 11.4 Floor-mounted mop service basin.

hands and the environment. Bedpans and infected hands can transmit microorganisms, whereas an unclean bedpan can be a reservoir for the germs.

Used bedpans, urine bottles, and other similar reusable items must be cleaned meticulously before they are reused. Cleaning bedpans thoroughly manually is, to some extent, difficult for maintaining sanitation; so, automatic washers and disinfectors are preferred in the healthcare services. The plumbing installations for manually bedpan washing have been shown in Figure 11.5.

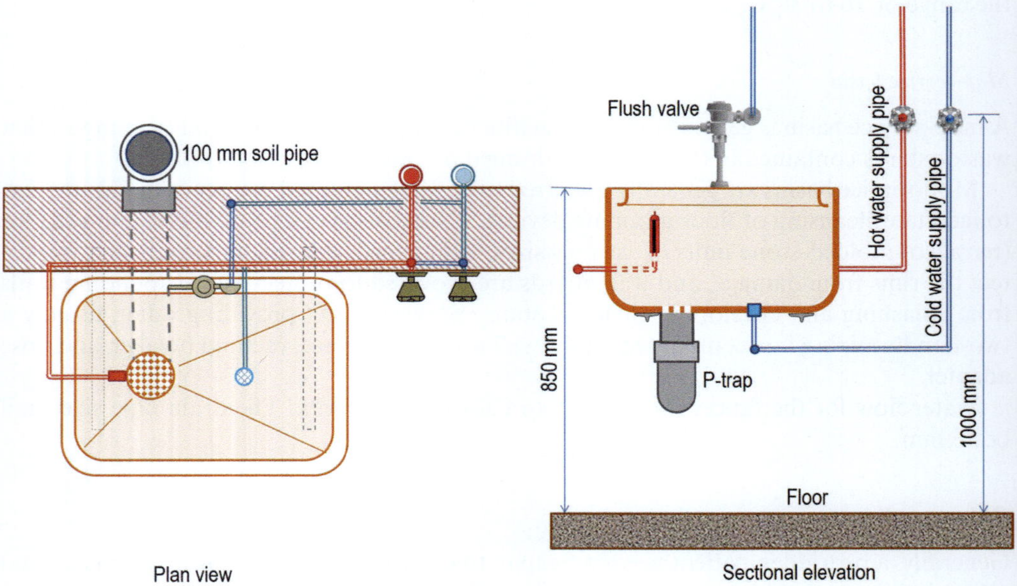

Figure 11.5 Plumbing for manual bedpan washer.

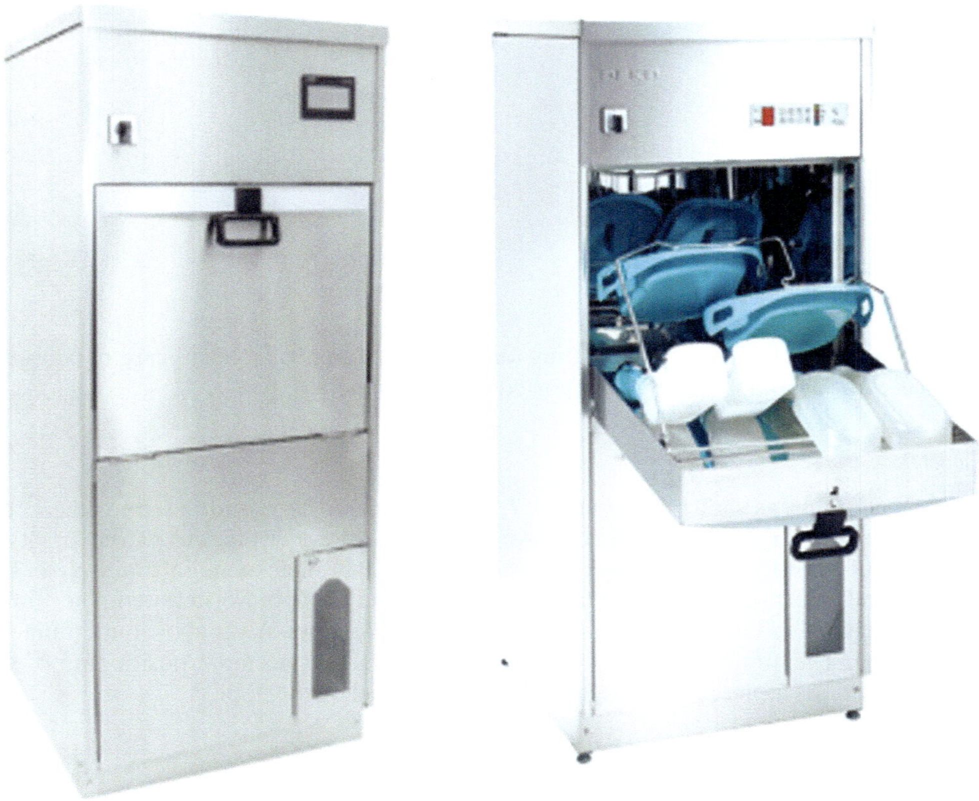

Figure 11.6 Automatic bedpan washer. Courtesy: Dekomed ltd. UK [3].

Automatic bedpan washers and disinfectors, as shown in Figure 11.6, empty, flush, clean, and thermally at about 80°C disinfect reusable human waste containers such as bedpans, commode pots, and urine bottles, etc., intended for reuse. These are designed to hold multiple items at the same time depending upon their size and capacity.

Bedpan washers require a standard electrical and hot water connection nearby. Minimum flow pressure is 100 kPa (1 bar), and the maximum permissible water pressure is 800 kPa (8 bar). A 20 mm water connection is required. In some washers, a steam supply pipe connection might be needed if steam is used for sterilizing purpose.

There is a steady flow of wastewater from the washer to the drain, to remove human waste. Bedpan washers shall connect to the sanitary drainage with venting. The minimum size of a sterilizer vent serving a bedpan steamer shall be 38 mm in diameter.

Floor drain

Floor drains are fixtures installed to facilitate draining water accumulated on a floor. In toilet rooms floor drains are mostly provided; however, there are many other areas where provision of a floor drain is considered according to the applicable codes. The plumbing engineer should give consideration to providing a floor drain with a trap seal in. Floor

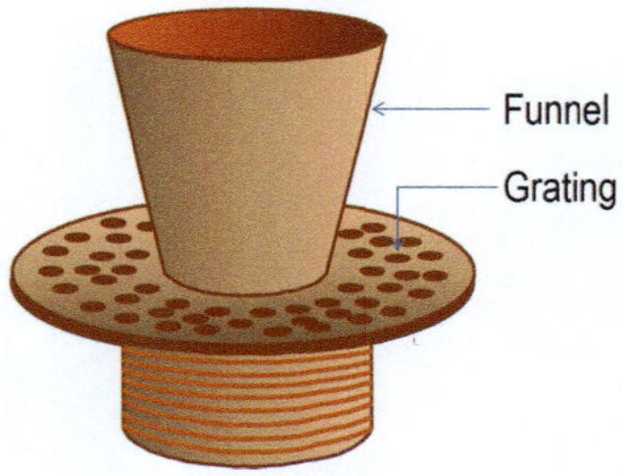

Figure 11.7 Funnel floor drain.

drains shall not be installed in operating and delivery rooms of a healthcare building. In some cases, floor drains are to be supported by trap-primers.

Funnel floor drain

The funnel drains, as shown in Figure 11.7, are one type of floor drain used to receive the condensate, drip, or wastewater directly from indirect wastewater pipes to prevent any type of splash of wastewater and direct the waste into the branch or main drain pipe. The exposed grate portion facilitates drainage for any type of floor wastewater. The funnel remains attached to the grate with the help of concealed screws. An oval or round funnel is mostly used. The adjustable round top or funnel top allows accurate positioning of the grate on the floor as needed for meeting the finished level of the floor.

Splash-free floor drain

A splash-free floor drain is a type of floor drain designed to minimize water splashing when wastewater enters into it. These floor drains are purposefully used in healthcare facilities, anywhere hygiene and safety are priorities. Anti-splash design includes sloped grates, or baffles to reduce water splash. These drains possess a deep drain basin that helps control water flow to reduce turbulence and splashing, the bottom of which is sloped to guides water flow smoothly into the drainage system.

Medical equipment of plumbing concern

There are some medical equipment that needs water to operate and function and so generates wastewater to be drained out. Qualities and characteristics of water required and the wastewater generated vary with the type of equipment for particular service. So, plumbing development, especially the drainage system, would vary subject to the special requirements for any medical equipment or machine. The following medical equipment needs plumbing installations for their operation.

Dialysis machine

Dialysis is a process to remove waste products, toxins, and excess fluid from the blood when the kidney fails to effectively remove those. For this purpose, a dialysis machine is used for diverting blood from the patient's body to the machine to be cleaned. A dialysis machine has a dialyzer, basically a blood filtering unit that performs hemodialysis by removing waste, excess fluids, and toxins from the blood.

In a hemodialysis station dialyzer requires one service box called dialysis-box that provides essential connections for water, dialysis concentrate, drainage, and power. It includes backflow preventer, trap primer, thermostatic mixing valve, hose bib and drain connection.

There is dialysis-wall-boxes those are recessed into the wall, which contain connections for the dialysis machine to receive acid and base concentrates and treated water, and dispose of waste products. Dialysis box for dialysis machines utilizes reverse osmosis (RO) or deionized (DI) water, mixed with dialysate saline concentrate. RO water is preferred for dialysis over DI water because of its superior ability to remove contaminants and ensure patient safety. Dialysis machines also require a water softener, a charcoal filter, and some type of chlorine destructor to reduce the amount of chlorine content in the water. While widely used, the RO system requires greater capital costs, monitoring, and maintenance.

Dialysis wall boxes can become easily contaminated with microorganisms, which can subsequently be transferred to the patients. So, the wall boxes need to be cleaned, disinfected periodically, and properly maintained to decrease the risk of patient infections.

During an average week of hemodialysis, a patient can be exposed to 300 to 600 liters of water, Water used for producing dialysate, substitution fluid, or to reprocess hemodialyzers for renal replacement therapy must be free from chemicals and microorganisms. Water must be free from chlorine or chloramine.

There is also an all-in-one dialysis box system that eliminates the need for excess dedicated cold water distribution piping by concealing both a backflow prevention device and a trap primer inside of the access box. Thereby, the cost and space requirements for additional wall space intended for remote backflow preventers, trap primers, and supplementary access doors are eliminated. It increases patients' safety by protecting the water source at the point of connection.

Water supply for dialysis machine

A storage tank holding purified water and a pump is used to supply water to the dialyzer. Non-pigmented polypropylene, stainless steel, and glass pipes are chosen for dialysis water distribution systems. The piping shall be a single-loop system of 6.4 mm inside dimension, maintaining a minimum velocity of 0.46 mps.

According to the Association for the Advancement of Medical Instrumentation (AAMI), the microbiological standards for the water to be used for the process of dialysis have been furnished in Table 11.1.

Table 11.1 AAMI water standards for preparing dialysate for dialysis [4]

Reference document	*Allowable water TVC*	*Action level water TVC*	*Allowable level water EU*	*Action level water EU*
AAMI RD52:2004 (Minimum regulatory requirement)	<200	≥50	<2	≥1
ANSI/AAMI/ISO 13959:2014 (Preferred recommendation)	<100	≥50	<0.25	≥0.125

EU, Endotoxin unit; TVC, Total viable count.

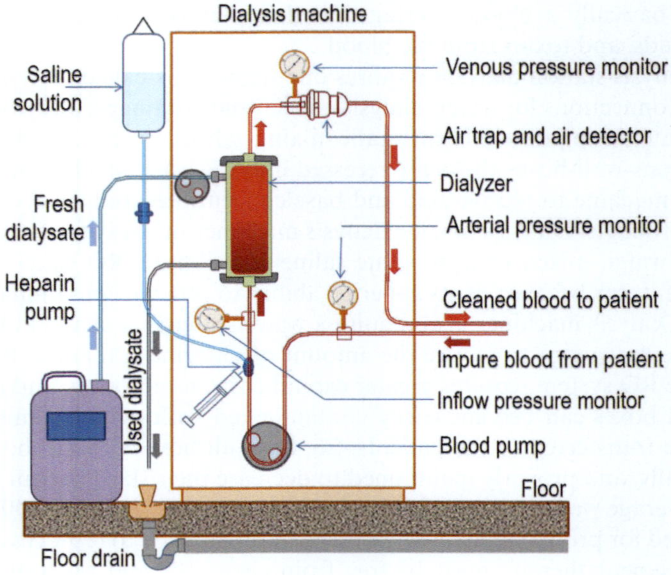

Figure 11.8 A dialysis machine including its major components.

Drainage system for dialysis machine

Services that facilitate the drainage of waste fluids from the hemodialysis (HD) machines must be well ventilated to prevent condensation and the subsequent growth of mold. The dialysate and the reprocessing effluent should drain into a separate horizontal drainage system with adequate capacity to handle the volume of waste. Covers or screens for the drainage systems should be constructed of a chemically resistant material. If plastic is used, then high-density polyethylene (HDPE) pipe should be chosen.

The number of bends in horizontal pipelines shall be kept to a minimum and be connected directly to the nearest drainage stacks without bends or offsets. Dialysis machines require a funnel drain or floor sink to drain out their wastewaters, as shown in Figure 11.8.

Heart-lung machine

A heart-lung machine is an apparatus that functions as both the heart to pump blood and the lungs to oxygenate the blood, used during the open-heart surgery. The important components of this machine include pumps, an oxygenator, a temperature regulator, and a filter. The oxygenator removes accumulated carbon dioxide and supplies oxygen to the blood that is pumped into the arterial system. Oxygenator requires water flow for their integral heat exchanger. Wall water systems, as shown in Figure 11.9, are an option for oxygenator heat exchangers. For this purpose, Oxygenator specifies heat-exchanger water to be provided at 8 to 15 lpm and a pressure of 193 kPa [5]. In present days, low-pressure, standalone heater-cooler unit (HCU) technology effectively used instead of more expensive wall water systems that required significant infrastructure and high-pressure water connections in the operating room and suffered from fewer temperature regulation safeguards. Independent temperature control of the patient and cardioplegia using separated water circuits is also used in several critical cardiac surgery or prolonged cardiac surgeries

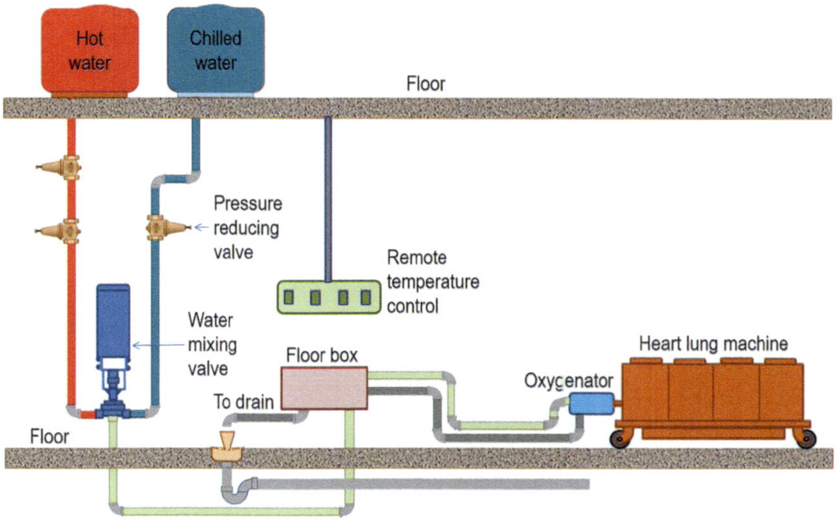

Figure 11.9 A heart-lung machine and its water supply and drainage system.

Water after use in the heat exchanger is drained out. Heart-lung machines also require a funnel-type floor drain for its drainage. If the apparatus is located in the operating room, indirect wastewater removal is required.

Electron microscope

Electron microscopes that produce higher-resolution images than a light microscope require filtered, backflow-protected, cold water or circulated chilled water. Chilled water is required to cool the diffusion pumps and to keep the electronics' and column temperature stable. Generally, recirculating laboratory chillers are used to cool such laboratory equipment. For these purposes, there are air-cooled or water-cooled packaged chillers in various configurations. Around 1.75 kW horsepower air-cooled, stand-alone chiller is generally designed for indoor operation purpose.

Generally, there is a drain port on the back of the chiller through which wastewater is drained out, as shown in Figure 11.10.

A closed-circuit water cooling system is essential if the water is acidic, has comparatively high chloride concentration, floating particles, and temperature that fluctuates and is uncontrollable; then there might be specimen drift problems. If water temperature is very cold, it leads to condensation problems or diffusion pumps not functioning properly in the microscope.

Still

Water distillation unit is a machine which is used for the purification of water by a process called distillation for removing contaminants such as heavy metals, salts etc. present in water. The impure water is boiled to steam, after which the condensed liquid is accumulated in a separate chamber from where distilled water is collected.

The typical large quantity distillation apparatus consists of an evaporator section, an internal baffle system, a water-cooled condenser, and a storage tank, as shown in Figure 11.11.

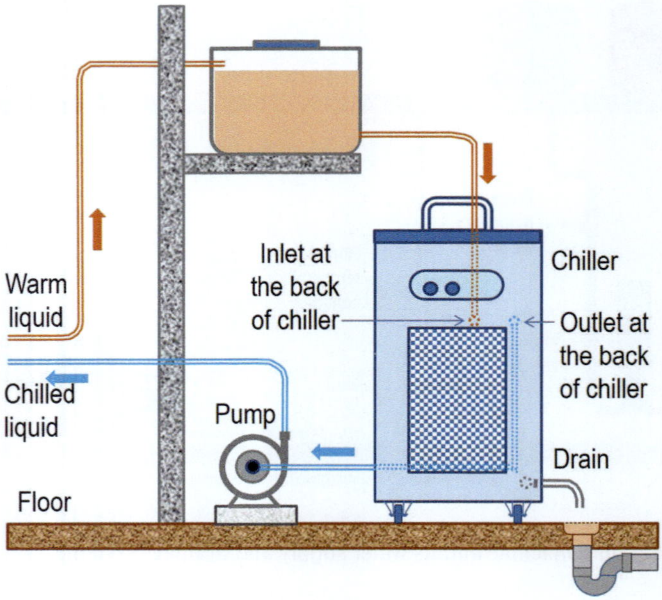

Figure 11.10 Chilling water recirculating system and a laboratory chiller.

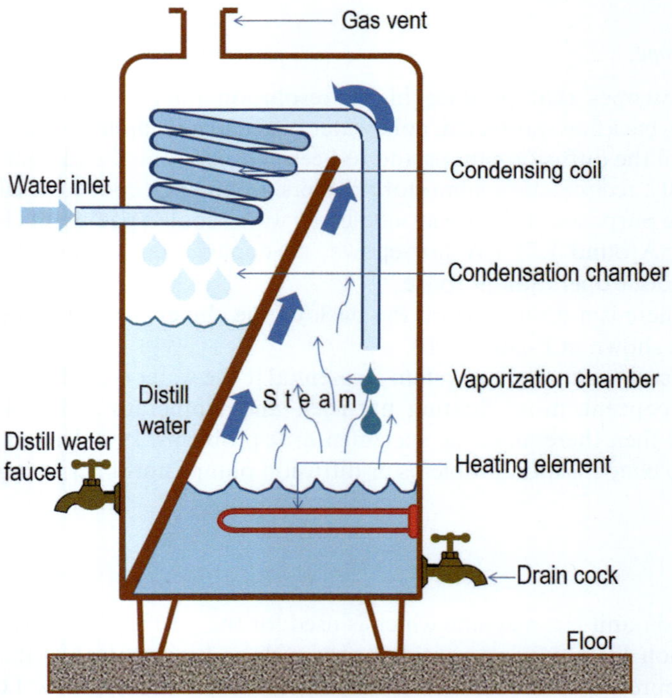

Figure 11.11 The components and process of a water distillation unit.

The best material for its construction is a pure block-tin coating for both the still and the tank. The heat sources, in order of preference based on economy and maintenance, may be steam, gas, and electricity. The still may be operated manually or automatically. The distilled water may be distributed from the storage tank by gravity or by a pump. A drain is required for system drainage and flushing. On stills larger than 189.3 lph, a cooling tower should be considered for the condenser water.

So, water distills need supply of normal water through a built-in solenoid valve with connection hose 12 mm supported by a vacuum breaker. The minimum flow rate maintained is about 1 lpm at a minimum pressure of 20.7 kPa, from where an output of 4 lph distilled water is obtained.

Water stills require regular maintenance to keep the unit operating properly. Un-evaporated pollutants remaining in the boiling chamber need to be regularly flushed to the sanitary drainage system. The residual water may contain calcium and magnesium scale accumulated at the bottom of the boiling chamber. Eventually, this scale is to be removed, either by scrubbing by hand or by applying acid. Wastewater should be flushed into floor sinks or funnel floor drains.

Sterilizer

Most medical and surgical devices used in healthcare facilities are mostly made of heat-stable materials. So, these are sterilized by heat, primarily by steam. Through sterilization, microorganisms on the metal surface or in a fluid are killed. For the operation of sterilization, a machine termed "autoclave" has already been discussed in Chapter 8.

The sterilization cabinet generate hot condensate water. Hot effluents damage plastic drainage piping. So, the drain pipe is supported by a drain tempering valve (DTV) as shown in Figure 11.12. Operating the valve, condensate wastewater can be drained out from the reservoir. Residue builds up in the drain pipe, which turns into sediment that may cause

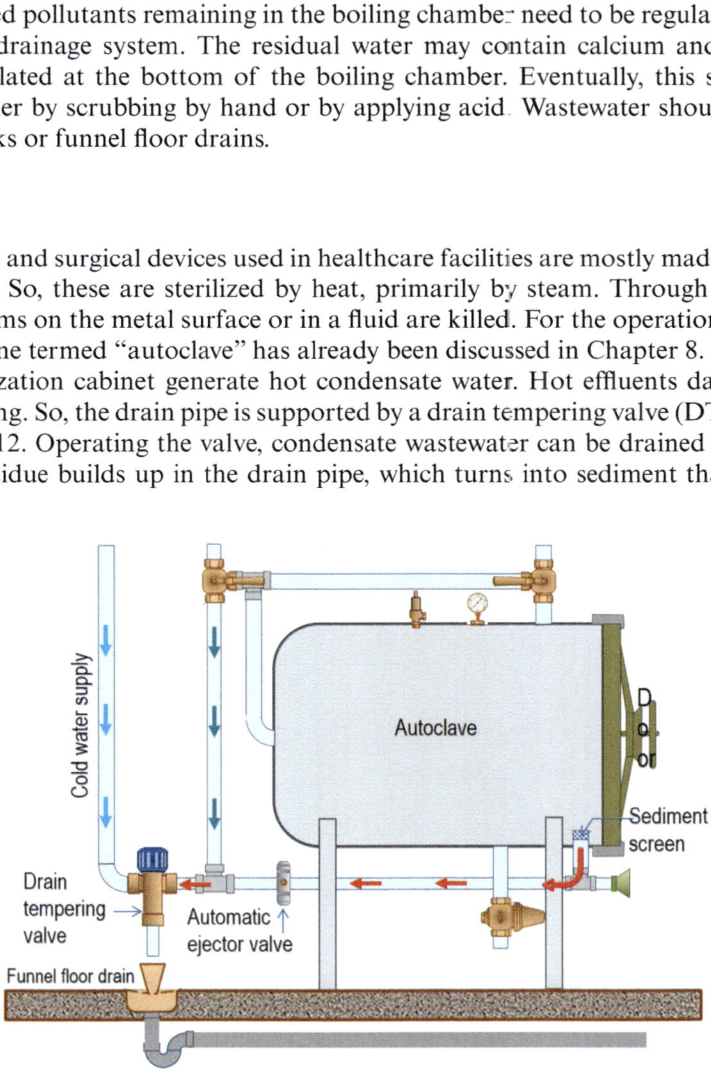

Figure 11.12 Autoclave drainage incorporating a drain-tempering valve.

clogging of the valve for which regular maintenance is required. Sterilizers require an acid-resistant floor sink or funnel drains to drain their wastewater.

Dental lavatory

Dental care areas need water supply and wastewater drainage as well; in addition, it requires air, medical gas, and nitrous oxide supply. Where there is provision of oral surgery, a separate surgical vacuum system should be arranged.

Water supply

A dental lavatory is kept installed by the side of a dental chair, as shown in Figure 11.13. So, each chair requires its own individual water supply to its lavatory. In the room housing a dental chair, there is need of an additional basin for hygienists and dentists to wash their hands. A pressure-boosting pump may be required to direct water with adequate pressure to the lavatories. If there is a centralized reverse osmosis (RO) filtered water supply installed in the center, then a storage tank for RO water supported by a pressure booster pump needed to be installed to supply water to the lavatory faucet. In the dental chair, RO-filtered water is needed to use as rinsing water and scalar water.

Compressed air

Dental hospitals, clinics, and surgeries require compressed air to power dental instruments. A dental air compressor is a machine that creates high-pressure, good-quality air, which is then delivered through specialized flexible pipes to the dental operatory, where it runs the

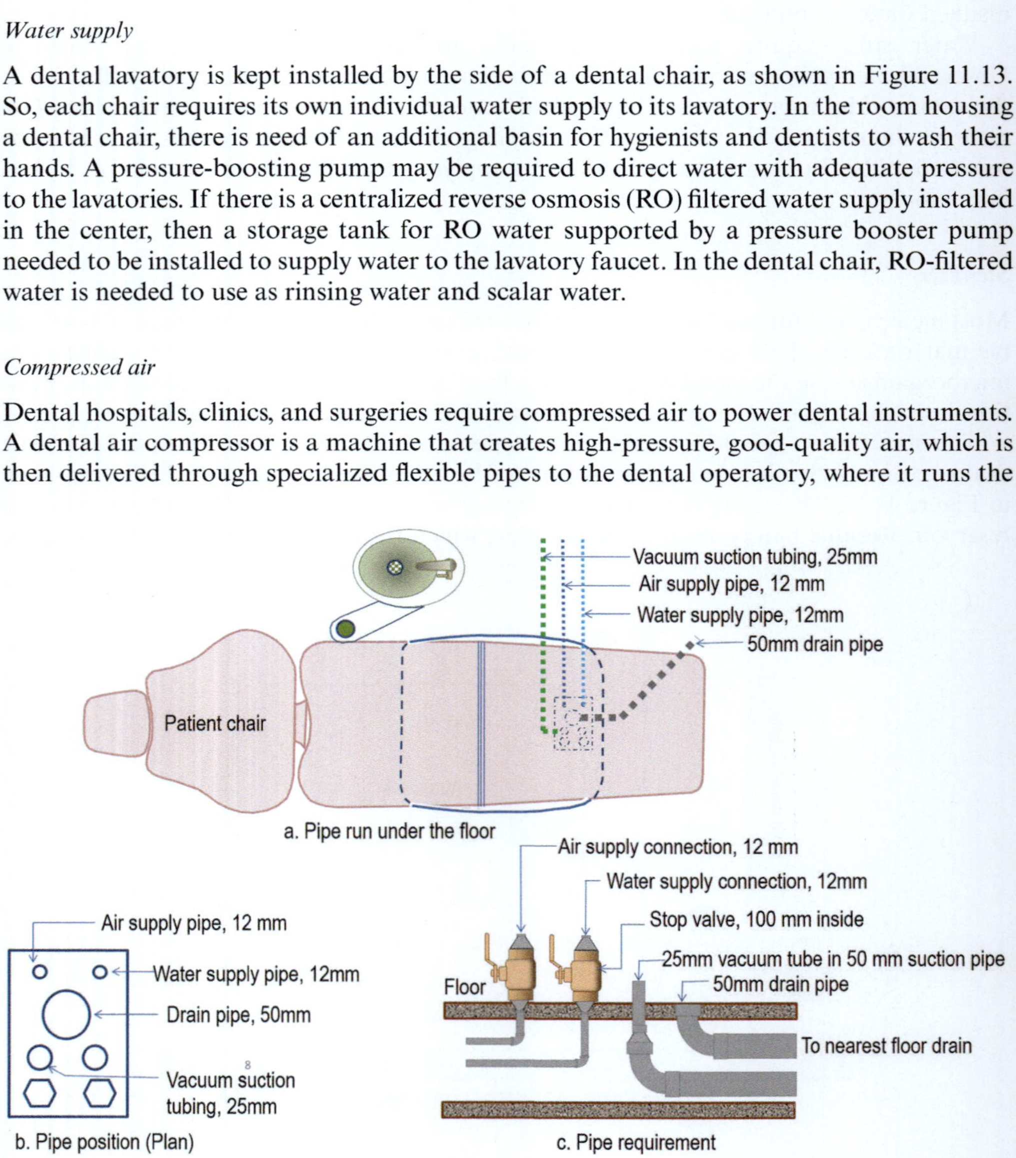

Figure 11.13 A dental care area with a supporting plumbing system.

air-driven hand-pieces that a dentist or hygienist uses to clean and treat teeth. Electric-driven hand-pieces do not require compressed air.

Separate pipes should be installed from the air compressor to each chair to supply compressed air. This pipe should have a minimum number of joints to avoid leakage. Instead of regular CPVC pipes, composite plastic-aluminum pipe may be used to make minor bending avoiding fittings.

If the compressor is placed inside the clinic, enough ventilation is a must for compressor storage. If placed outside, the compressor should be protected from direct rain and should be placed on a raised platform to avoid being submerged in water.

Vacuum system

A vacuum system is required to remove detritus from the operation site. In dental care treatment, a suction device is used to draw a good amount of air and saliva into the mouth over a short period of time during the process of dental cleanings, oral surgeries, and cosmetic treatments to keep teeth and mouth clean and dry.

Dental vacuum systems typically use a vacuum pump to cause variation in the suction pressure depending on the number of suction tools activated on the operatory head at the same time. For improved vacuum control, a precision vacuum regulator (PVR) is used.

In the vacuum-suction system, a separator is used for separating the mixture of solids, liquids, and suction air used in the suction process to draw air and saliva from the mouth. It has a housing that incorporates an inlet for the mixture to get in from the suction nozzles, an air separator chamber having a deflector to deflect clean air toward its outlet leading from this to the suction pump, a settling tank arranged beneath the air separator chamber to settle the solids up to a predetermined level, and there is an outlet for the liquid remaining above the sediment solids accumulated. The outlets are kept secure against the ingress of unwanted air. A secondary tank is associated with the settling tank, opens into the air separator chamber, and has the liquid outlet that is secured against the ingress of unwanted air. A system for transferring liquid is configured between the settling tank and the secondary tank, the entrance level for this system being above the maximum level to which solids are settled down.

Drain pipe

Drainage is required from dental lavatory and amalgam separator in vacuum system used in oral or dental surgery and treatment. Due to the use of cements and plasters in various procedures of dental treatment, the waste of these materials might create a blockage in the drain pipe. So, a minimum size of 50 mm diameter drain pipe should be laid on the maximum allowable slope.

Service pipes installation

To make a pipe-free floor, plumbing pipes are generally run under the floor, and a small chamber is made just below the chair as shown in Figure 11.13. All the pipes will directly get connected to the chair from this chamber. Plumbing inlet, outlet pipes, and floor trap are joined inside the chamber. This chamber will be later covered by the chair box in a way that none of the pipes will be seen outside. The exact position and size of the dental chair should be coordinated with the installer of the chair. In a few cases, the chair may have to be customized a bit to connect the services from the bottom. From the chamber box drain pipe is laid under the floor, so at least 150 mm clear space would be needed to run the drain pipe while maintaining the slope.

Special valves and accessories

Various special valves and accessories, other than the commonly used similar items, are used in healthcare plumbing. Among those, some special as mentioned below, will be discussed herein.

1 Trap primer
2 Atmospheric vacuum breaker
3 Drain tempering valve

Trap primer

A trap primer is a device that causes a small amount of water to be supplied to the trap of a drain automatically, as shown in Figure 11.14, to prevent the trap from losing its water seal, which generally happens due to evaporation. There is another type of trap primer that causes the flow of water from the water supply system to multiple floor drains. This type of trap-primer is supported by an automatic timer, electric solenoid valve, vacuum breaker, and a piping manifold having valve installed on each outlets supplying to a floor drain.

A water supply shall be provided for supplying water to the trap primer to cause resealing of the trap.

Atmospheric vacuum breaker

Vacuum breakers eliminate the possibility of contaminating waters with any other impure liquids that are in use, particularly in a laboratory. Virtually, an atmospheric vacuum

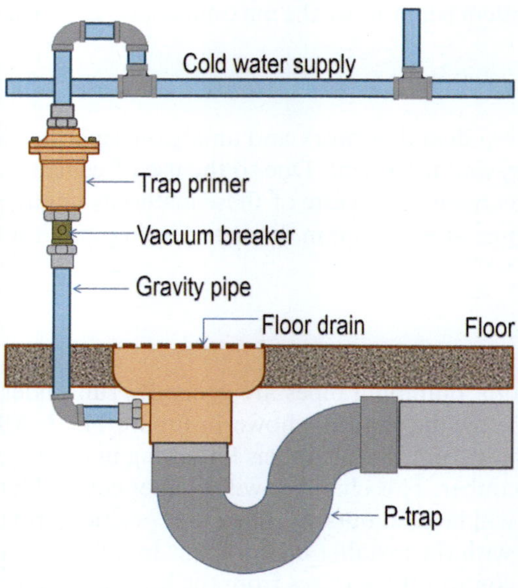

Figure 11.14 A trap-primer connected to a floor-drain.

breaker prevents back-siphonage that may occur when there is possibility of developing negative pressure in pipes, allowing water to flow opposite to the usual flow direction, thus eliminate the chances of contaminated water running through a faucet.

When the vacuum breaker is operating as it should, it would be found that water flows out of the faucet after it has been turned off. This small amount of wastewater drainage is a usual phenomenon ensuring that the vacuum breaker is functioning well. This is because, when the faucet is turned off, gravity causes the water remaining inside the spout portion to flow downstream out of the faucet. It is to be kept in mind that vacuum breakers do require occasional maintenance. Figure 11.15 illustrates the atmospheric vacuum breakers and their installation.

Vacuum breakers shall be installed not less than 150 mm above the flood level rim of the water retaining fixture or device. The flood level rim of the hose connections shall be the maximum height at which any hose is operated.

Drain tempering valve

The drain tempering valve (DTV), as shown in Figure 11.16, is an automatic cooler of draining wastewater, used in applications where a high-temperature wastewater discharge is required to be cooled down by mixing with cold water before discharging into a drain or sewer. The thermostatic drain tempering valve operates automatically to keep drain water within appropriate temperature limits and prevent damage of drain piping, plastic pipe in particular, from over-temperature conditions. Codes suggest wastewater temperature entering the drainage system does not exceed 60°C. The drain tempering valves are available with 12, 20, and 25 mm NPT connections.

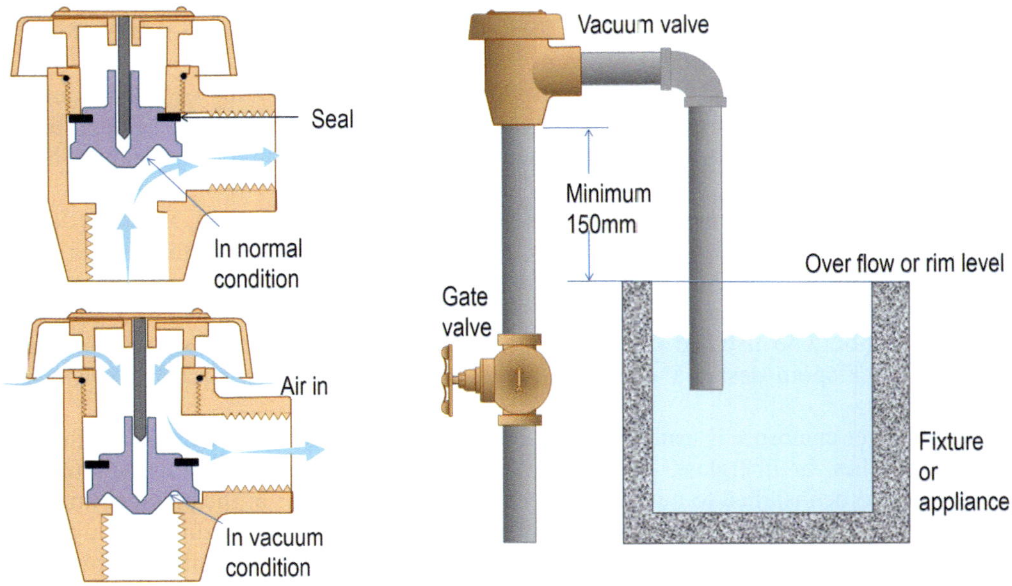

Figure 11.15 Atmospheric vacuum breakers and their installation.

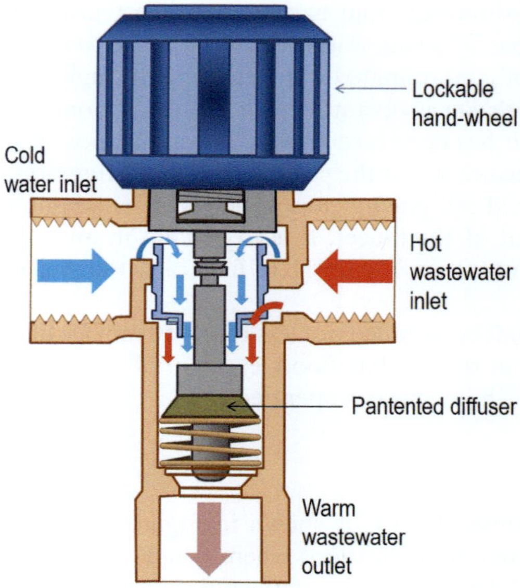

Figure 11.16 A drain tempering valve (DTV).

Areas where plumbing is installed

In addition to the common toilets and kitchen areas, plumbing installations are desired in the following areas of a building providing healthcare services.

Patient room

Patient rooms, single or group of patient rooms are provided with a toilet room containing at least a water closet, a lavatory, and a shower. The plumbing fixtures should conform to the following recommendations:

The water closet should be of vitreous china, preferably wall-hung or floor-mounted, having an elongated shape bowl. All water closets or urinals are preferred to be operated by a flush valve. Water closets should have open-front seats on the rim for maintaining better sanitary conditions.

Bedpan lugs and bedpan washers are often required by the local codes. Flushing devices shall be provided with bedpans in each inpatient toilet room; however, installation of flushing devices is optional in the toilets for ambulatory psychiatric and alcohol-abused patients.

The lavatory should be a minimum of 508 by 457.2 mm in size and 190.5 mm deep. Lavatories should be installed at least 864 mm above the floor. Mixing faucets should be of the gooseneck-spout design and provided with wrist-blade handles, electronic, or hands-free controls.

The shower enclosure is usually constructed of masonry, making inner-surface finished by tile, acrylics, or fiberglass. The shower bases should be nonslip surfaces. Automatic valves should be installed to automatically manage the variations in the pressure and temperature to deliver water at the desired temperature that will prevent scalding. Local plumbing codes require the installation of grab bars within the shower enclosure. The plumbing engineer should check the designs conforming to the local regulatory authority guidelines regarding approval.

Bathtubs, if provided, can be constructed of cast iron, fiberglass, acrylics, or steel. Bathtub faucets should be as alike as one for showers in operation. Shower heads may be of the stationary type, but handheld showers are also required in some cases.

According to the local plumbing code, installation of lavatories is sometimes required for use by the doctors, nurses, and other hospital staff. Lavatory for this purpose is usually located on the wall near the door and fitted with a gooseneck spout faucet, either elbow or leg operated.

Ward room

Ward rooms are infrequently found in healthcare facilities, particularly in the private hospital and healthcare facilities. These rooms require at least one lavatory. This lavatory should be a minimum of 508 by 457.2 mm and made of vitreous china or stainless steel. The lavatory faucet should be of the gooseneck-spout type, maneuvered by wrist-blade handles or leg-operated.

Nurse station

The hospital's nurse station is usually provided with a lavatory of minimum size 508 by 457.2 mm having a high gooseneck spout faucet with hands-free maneuvering options. An infant's bathtub, wall- or counter-mounted with an integral large drain-board and rinsing basin, is provided. Water-supply faucet spout is installed over the basins, which is operated by a separate hand-valve controls. The faucet and the flow are usually supplied and controlled through a thermostatic mixing valve. With a view to maintaining a safe water temperature, a separate hot water supply tank should be used.

Intensive-care room

These rooms usually have utility sinks supported by hands-free controls and faucets with high gooseneck spouts. There should be a water-supply fitting including a gooseneck spout, provision for washing bedpan, either at an immediately adjacent water closet or at a separate bedpan washing station within an enclosure in the room. It is desirable to include combination lavatory-water closets for patient use, especially in cardiac-care units.

Emergency (triage) room

In an emergency room, the plumbing fixtures to be installed include a utility sink with an integral tray, on which there is by a water-supply fitting supported by a gooseneck spout operated by wrist-blade handles. In addition, there should be a clinic sink preferably of vitreous china or a flushing-rim sink, for the disposal of solids, supported by a water-supply fitting comprising of a flush valve and a separate combination faucet with vacuum breaker mounted on the wall above the plumbing fixture.

Examination and treatment room

These rooms are generally served by the lavatories, either of vitreous china or stainless steel. The supporting water-supply faucet should be a hands-free valve type, equipped with a high, rigid, gooseneck spout. For a particular examination room or a group of patient rooms, an adjacent toilet room is provided containing a specimen-type water closet, as

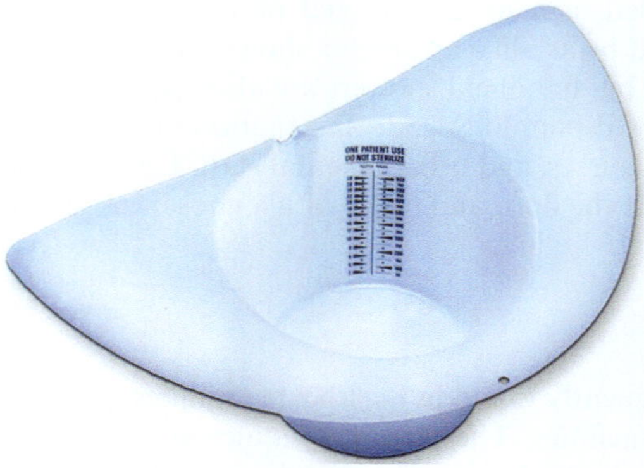

Figure 11.17 A specimen-collector. Courtesy: Online Medical Supply [6].

shown in Figure 11.17, for inserting a specimen-collector. The toilet room should include a lavatory supported by a water supply faucet having wrist-blade handles or hands-free controls along with a gooseneck spout.

Physical therapy treatment room

The plumbing fixtures and related equipment required in a physical therapy treatment room generally include hydrotherapy immersion baths and leg, hip, and arm baths. These units are generally furnished with electric-motor-driven whirlpool equipment, as shown in Figure 11.18. The water is poured into the stainless-steel tank enclosure through a faucet, including a thermostatic control valve to prevent scalding, which is usually installed wall-mounted, adjacent to the bath for operation by a hospital attendant. The water supply system is sized for minimum tub-fill time. A hydrotherapy shower is sometimes required. These showers generally comprise of multiple shower heads, includes as many as 12–16, mounted vertically in order to direct the water-streams toward a standing patient by means of a sophisticated control console operated by a hospital attendant.

The immersion baths are also provided with overhead hoists and canvas slings to facilitate the lifting in and out of the bath of a completely disabled patient.

Cystoscopic room

In cystoscopic rooms, cystoscopy is done to investigate problems in the urinary tract and bladder, such as a blockage in the urethra or tumors in the bladder of the patient. A wall-mounted clinic sink equipped with flush valves and bedpan washers and combination faucets; lavatories provided with water-supply fittings and gooseneck spouts; and, in a separate adjacent room, a specimen water closet and a lavatory are to be installed. If a floor drain is installed in cystoscopy, it shall contain a non-splash, horizontal-flow flushing bowl beneath the drain plate.

Nourishment station

These stations are generally developed on each floor, having patient rooms and being near the nurse station, with a view to serving nourishment between regularly scheduled meals.

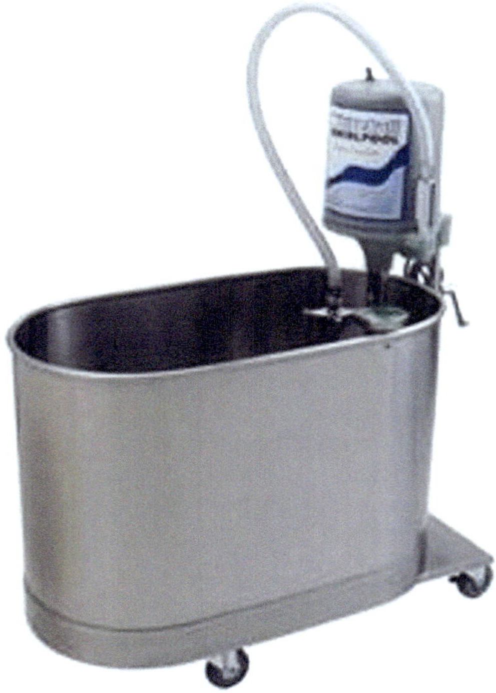

Figure 11.18 Hydrotherapy whirlpool immersion baths. Courtesy: prohealthcareproducts.com [7].

These stations are usually provided with a sink equipped with hands-free controls for hand washing, an ice maker, and an optional hot-water dispenser to give service and treatment for the patients.

Operating room

In the operating rooms of a hospital, no plumbing fixtures or floor drains are installed. However, the scrubbing station located adjacent to the operating room should have at least two scrub sinks, usually made of vitreous china or stainless steel, furnished with hands-free water-supply fittings, and equipped with gooseneck spouts. These sinks should be large and deep enough to allow the scrubbing of hands and arms to the elbow. A soiled workroom is generally made for the exclusive use of the surgical staff, which is located near the operating room area. This workroom should have a flushing-rim clinical sink of vitreous china—for the disposal of solids, with the help of water-supply fittings supported by a flush-valve bedpan washer, a faucet mounted on the wall above the fixture, along with hand-washing facilities in a vitreous china or stainless-steel lavatory having a gooseneck faucet operated by wrist-blade handles. Sub-sterile rooms should be equipped with an instrument sterilizer and a general-purpose sink. Concerned plumbing engineers should consult with the manufacturer of the desired instrument sterilizer for any special plumbing-related requirements. The general-purpose sink shall be supported by a hands-free water-supply fitting with a gooseneck spout and can be installed countertop-mounted.

Recovery room

The post-anesthesia recovery rooms for the surgical patients should consist of a hand-washing facility, like a lavatory equipped with a gooseneck spout operated by wrist-blade handles and made of vitreous china or stainless steel, and a flushing-rim clinical sink of vitreous china for the disposal of solids, equipped with the water-supply fitting supported by a flush valve and a faucet mounted on the wall above the fixture with a vacuum breaker. A bed-pan-washer should also be installed next to the clinic sink. The type of bedpan washer depends upon the method of washing adopted by the hospital and sterilizing bed-pans.

Birthing room

A birthing room should include a lavatory preferably of vitreous china supported by a gooseneck spout operated by wrist-blade handles or hands-free maneuvering. All labor rooms should have independent toilets, including one water closet and one lavatory. A shower enclosure, including showering facilities, should be provided for the labor room patients. The shower controls, including the pressure or thermostatic mixing valve; should be located outside the wet area for use by the hospital's nursing staff. A water closet should be accessible from the shower facility.

Anesthesia workroom

These rooms are generally provided for the cleaning, testing, and storing of the anesthesia equipment. These rooms should have a work counter-mounted sink and necessary water supply facilities. The sink is usually made of stainless steel. The faucet should be of the gooseneck spout design with hands-free controls.

Fracture room

In this room, a sufficiently large-sized plaster work sink, preferably of vitreous china, equipped with a combination water-supply fitting like a gooseneck spout operated by wrist-blade handles, is to be provided. There shall be a plaster trap on the wastewater drain line, which should be located at a convenient location allowing easy access.

Pathology laboratory

Pathology laboratories differ from general laboratories in many respects, including plumbing aspects. The pathological laboratory operations in hospitals and other medical facilities have significant potential sources of pollutant discharges to the sanitary sewer. As commonly understood, these laboratories not only generate chemical contaminants but also metal contaminants. Rinsate from hazardous chemical containers generates hazardous wastewater. Many testing equipment may generate obnoxious wastewater that should not be discharged directly into the sanitary sewer. Discharge of laboratory waste into sewer may be prohibited due to pH, temperature, explosivity, radioactivity, etc., as specified in the concern regulations. There are chemicals used in laboratories that can foam up in the drain pipe for which extra measures are to be taken.

The pathology laboratories require different forms of water, for which various conditioning machinery would be required that would again require exclusive drainage systems as well.

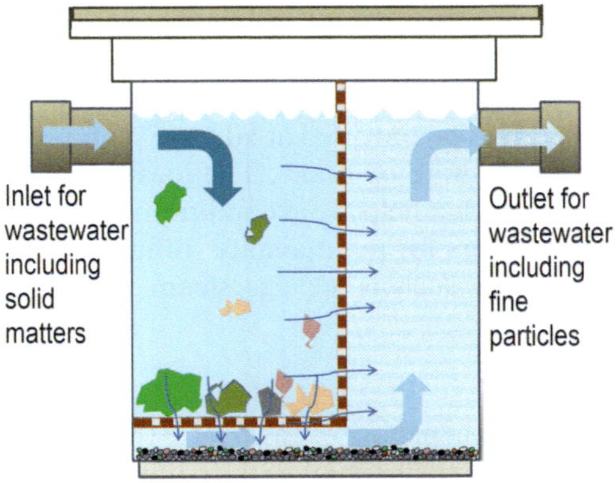

Figure 11.19 A solids interceptor.

Natural gas, reverse osmosis or deionized water, compressed air, and vacuum are the most common supply systems. Side-by-side acidic waste drainage systems should be developed along with the normal plumbing system, in medical pathology laboratories.

Pharmacy and drug room

The plumbing fixtures for pharmacy and drug rooms include medicine and solution sinks. These sinks can be counter-type, supported by a mixing faucet and a swing spout, and preferably made of stainless steel or vitreous china. A solids interceptor, as shown in Figure 11.19, should be considered for areas where necessary to prevent harmful solid materials from entering the drainage system on a continuing basis. A solid interceptor is to be used instead of a "P" trap on drainage installation.

Autopsy room

In the autopsy or postmortem room, where the medical examination of a dead body and the internal organs is done to find the cause of death, a table is usually provided with cold- and hot-water supplies, a vacuum breaker or backflow preventer, and a waste line. It is obligatory that the concerned plumbing engineer consult with the table manufacturer and the controlling authority regarding the requirements of the table to carry out autopsy activities easily. Drain systems for the autopsy table shall be designed to positively avoid splatter or overflow onto the floor or back siphonage and for easy cleaning and trap flushing. The autopsy room is also usually equipped with a stainless steel or vitreous china sink with hands-free fittings, a clinic sink, and an anti-flash floor drain. A water closet and a shower room are usually provided adjacent to the autopsy room. Many autopsy rooms are equipped with waste-disposal unit integral to the sink.

Kitchen

Normally, kitchens require toilet fixtures for kitchen staff, food preparation sinks, hand-wash sinks, pot and pan-wash sinks, dishwashers, glassware washers, floor drains, hose bibs, mixing stations, and grease interceptors. Kitchen grease traps shall be located and arranged to permit easy access without the necessity of entering food preparation or storage areas.

Grease traps shall be of the capacity required and shall be accessible from outside the building without the necessity of interrupting any services. In dietary areas, floor drains and or floor sinks shall be of a type that can be easily cleaned by the removal of a cover. Floor drains or floor sinks shall have to be provided at all "wet" equipment, such as ice machines, and as required for the wet cleaning of floors. The location of floor sinks and the floor drains shall be in such positions that help avoid conditions where the location of equipment makes the operation of the covers for maintenance difficult. Also, the kitchen equipment may require other utility services, such as fuel gas, steam, and condensate.

Laundry

Laundry facilities require large-capacity washers or extractors and dryers, presses, and folding machines. Wastewater drainage may require lint interceptors. These facilities are prime candidates for heat and water-recovery systems. Also, the laundry equipment may require other utility services, such as fuel gas, steam, and condensate. The hot-water temperatures required for laundries are 38, 60, and 82°C.

Water supply systems

Water supply is the most essential component of providing basic health services, which again need special care than the water supply system developed for other common occupancies. The special requirements for this service to be developed in healthcare facilities are stated below.

Various qualities of waters supplied

A healthcare water supply system is developed for adequate flow of desired quality of water needed at the points of use. These water supply systems typically planned for the following quality of waters:

1 Potable water

    a Normal (Cold) water
    b Hot water (at various temperatures).
    c Chilled water. 7.2 to 10°C for drinking fountains

2 Non-potable water
3 Pure-water

    a Distilled water
    b Deionized (or demineralized) water
    c Reverse osmosis filtered water

Water service connection

A building of health care facilities having medical, surgical, psychiatric, nursing, or custodial care on a 24-hour basis for more than five persons who are not capable of self-preservation shall have two water service pipe connections of equal size, so that with the loss of one service pipe, the remaining service pipe can meet the water demand for the entire facility; i.e. to ensure the provision of an uninterrupted supply of water.

Each water service pipe shall have a shutoff valve inside the building premise and a shutoff valve at the utility-provided point of connection to the water main or other source of potable water.

Water source

Normal cold water should be supplied at all locations where water is needed. For the development of a water supply system, it is needed to collect water from the potential sources of water. Though there are various sources of water for its collection, for healthcare facilities it is recommended to collect water from the following two common sources of water.

1 Water supply main
2 Underground water

Water storage

From the water sources, water is collected and stored in reservoirs built mostly underground. It is generally advised to store water for one-day consumption, but some codes suggest 12-hour storage to prevent any contamination of water. The daily water consumption for hospitals based on the number of beds has been furnished in Table 11.2.

The underground storage reservoir will also act as a buffer during some unusual demands, pressure fluctuations, and interruptions in supply in the main. In the planning of water supply system provisions for water conservation should be considered with a view to finding the lowest volume of water to be stored, thereby determining the minimum size of the water reservoir for storage. It is to be kept in mind that unnecessary big water reservoirs play a role in contaminating stored water.

Water distribution systems

Pipes are the major components of a water distribution system. Proper pipe material and size are important considerations in healthcare water supply system. Use of diversity

Table 11.2 The daily water consumption for various types of healthcare facility with respect to number of beds [8]

Sl. no.	*Bed number*	*Average water demand l/d/b (liter per day per bed) for facilities of*						
		Acute health-care	*Specialist acute health-care*	*Long term care*	*Recovery and convale-scent*	*Geriatric and chronic illness*	*Psychiatric*	*Teaching*
1	0–25	299	319	180	216	246	180	680
2	26–50	299	347	180	206	246	180	680
3	51–100	398	362	269	185	203	180	680
4	101–200	490	479	247	181	164	269	866
5	201–300	590	530	217	181	127	247	830
6	301–400	590	530	306	181	127	247	904
7	401–500	599	530	306	181	127	350	904
8	501–600	599	530	306	181	127	350	1228
9	Over 600	978	530	306	181	127	350	1228

factors for sizing the pipes for water supply systems must be carefully analyzed, particularly for medical laboratory classrooms, having comparatively higher rates of simultaneous use than general laboratories. In addition, emergency rooms, outpatient treatment rooms, and operating room wash-up areas also have high rates of simultaneous use.

Extreme care must be taken in order to protect the potable-water supply from contamination due to cross connection with any non-potable water. For this purpose, an appropriate vacuum breaker would be necessary for each fixture to be protected, a hose-end outlet, and an aspirator or other serrated-tip laboratory outlet, subject to the requirement by the local plumbing regulations. The vacuum breakers provided for fume-hood outlets should be located outside the hood. Built-in or integral vacuum breakers are preferred to the hose-end type units.

Excessive pressure fluctuations in water distribution are highly undesirable, particularly in medical-research laboratories. With a view to ensuring constant pressure in the system, pressure-sustaining valves on the supply pipe should be provided, most importantly in gravity-fed systems.

Hot water supply system

Hot water supply is developed by producing hot water adopting the most energy saving methodologies and then it is supplied to the locations requiring hot water. In the buildings of healthcare facilities, special medical fixtures and equipment need hot water supply for their use and operation.

When sudden and large-volume hot water demand i.e. "dump" loads are anticipated from kitchens and laundries, storage of hot water is recommended. Hot-water usage in patient-care areas requires consideration of water temperature and bacterial growth. The most common water-borne bacterium of concern in hot water supply system is the growth of *Legionella pneumophila*.

Recommended water temperatures for specific applications are as follows:

1 Patient-care and hospital general usage require water temperatures between 40 and 49°C, except where the local plumbing codes or other regulations require other maximum temperatures. Circulation-type hot-water distribution systems serving patient-care areas shall be developed to ensure constant recirculation of hot water to provide hot water instantly when desired at each hot-water outlet. The temperature of hot water for bathing fixtures and hand-wash lavatories shall be appropriate for comfortable use but shall not exceed 49°C.
2 Kitchen generally requires water at 60°C for fixtures, except the dishwasher's sanitizing cycle. The sanitizing cycle requires 82 to 88°C for the dishwasher, with 82°C minimum required at the dish rack. Also, some health departments set a maximum temperature of 40 to 49°C for hand-washing lavatories.
3 Laundry facilities should be supplied with two water temperatures, 60°C for general usage and 71°C minimum to washers or extractors for laundry sterilizations.
4 A circulation-type hot water supply system is recommended to supply hot water to the faucet with a pipe length exceeding 15 m, for shorter pipe lengths, a non-circulation system can be allowed. In the system, a temperature of 60°C is to be maintained, and in the return pipe, it is preferred to have a minimum temperature of 50°C, where the cold water temperature should be greater than 20°C.

Providing a point-of-use booster heater for high-temperature applications instead of a central water-heater system is often more economical. A thermometer should also be provided on the outlets of water heaters and thermostatically controlled valves. A pressure regulator, gauge, and flow meter may also be desired on the inlet side of pressure-sensitive equipment.

Chilled water supply system

In a healthcare center, in many areas, chilled water is used generally to cool machines and maintain the required operating temperatures; but it is mandatory in some areas like in the radiology and imaging room. Advanced medical imaging equipment, such as MRI and CT scanners, generate a significant amount of heat during operation. Chilled water may be required for the cooling of electron microscopes and x-ray tubes. Film processors operate at a normal range of 4.4 to −1.1°C. Some models do require controlled water temperature for film processing.

In healthcare facilities, chilled water supplied to instruments or equipment is typically part of a circulating type supply system. In addition, where cooling of space or for drinking purposes, chilled water supply is required, the chilled water supply system should be of a recirculating type design.

Pure water supply system

Pure water, also called purified water, is that water free from all sorts of particulate matters, minerals, soluble ions, bacteria, pyrogens, organic matters, and dissolved gases, which frequently exist in water. In the pharmacy of hospitals, central-supply room, laboratories, and laboratory-glassware washing facilities, pure-water supply systems are generally required.

There are two basic types of pure water required in hospital facilities: bio-pure water that contains no microorganisms and high-purity water that is free from minerals, dissolved gases, and particulate matters, etc.

Water purity is most easily measured as specific resistance in ohm-centimeter (Ω-cm) units, or expressed as parts per million (ppm) of an ionized salt (NaCl). The theoretical maximum specific resistance of the pure water is given as 18.3 MΩ-cm at 25°C. This water of such purity is troublesome to produce, store, and even distribute. This water gets "starved" for impurities and constantly tries to absorb contaminants from surrounding. It is to be noted that the specific resistance of the pure water is indicative only due to its mineral contents, but in no way does it show the bacterial, pyrogenic, or organic contamination level. So, an independent laboratory analysis should be made whenever required.

Pure water is generated from an effective purifying system that removes all impurities. There are four basic methods of producing pure water, those are as follows:

1 Distillation
2 Reverse osmosis
3 Demineralization
4 Filtration,

Depending upon the level of pure water required in a particular use, one or more of these methods will be effective. Under certain conditions, a combination of several methods may be necessary. In this system, utmost care is taken in maintaining makeup water connections and the water purification system in a way that prevents backflow. There must not be human contact with the water after it leaves the purifier.

High-purity water supply systems should be provided with a circulation loop. Dead-end legs should be avoided where possible; in unavoidable cases it must be limited to. 1.52 m. System design velocity should be between 1.22 and 2.13 mps; so as to discourage bacteria accumulation and provide transport back to an ultraviolet sterilizer and filtration for removal.

Bio-pure water is needed in the hospital's pharmacy, central supply room, and other areas where there may be patient contact. Bio-pure water may also be desired in certain specific laboratories at the owner's request and as a final rinse in the laboratory's glassware washer.

Distillation

Distilled water produced by the method called distillation, is considered to be the most available form of pure water. Distillation produces bio-pure water, which is completely free from particulate matters, minerals, organics, bacteria, pyrogens, and most of the dissolved gases, and has a minimum specific resistance of 300,000 Ω. In this case, an important consideration is that the water is free from bacteria and pyrogen contamination for being dangerous to the patients, most importantly, where intravenous solutions are concerned.

Small distilled water is generated through boiling water into steam. The contaminants and impurities settle at the base of the distillation system while the steam collects in a condenser and cools down as temperature drops and produces distilled water droplets. Finally, the pure water droplets pass through the outlet pipe to generate distilled water.

Distillation cannot eliminate volatile chemicals, which can be removed by reverse osmosis (RO) filtration process.

Reverse osmosis

Reverse osmosis, in short RO, is an ultra-nano filtration system that produces high-purity water, not having high resistivity of demineralized water, but is not bio-pure. Filtering membranes remove both suspended and dissolved matter in water. Under certain conditions an RO process can offer economic advantages over demineralized water. In areas that have high mineral contents, an RO process can be used as a pretreatment for a demineralizer or still. Filtered water through RO filter membrane is purified water. In many cases of healthcare treatment, particularly for the dialysis machine, RO-filtered water is required.

There are several types of reverse osmosis units currently available. Units consist of a multiple layer of fine semipermeable membrane, in either a roll form as shown in Figure 11.20, or a tube containing numerous hollow fibers. The water is then forced through the semipermeable membrane under high pressure.

The RO system requires a drainage system to drain out the filtrates generated after filtering water.

Water to be filtered through RO must be free from the chlorine that can destroy the RO membrane.

Demineralization

Demineralization, also called "deionization", is the process that produces high-purity water that is completely free from positively and negatively charged mineral ions, using ion exchange; such as cations from sodium, iron, calcium, and copper, as well as anions of

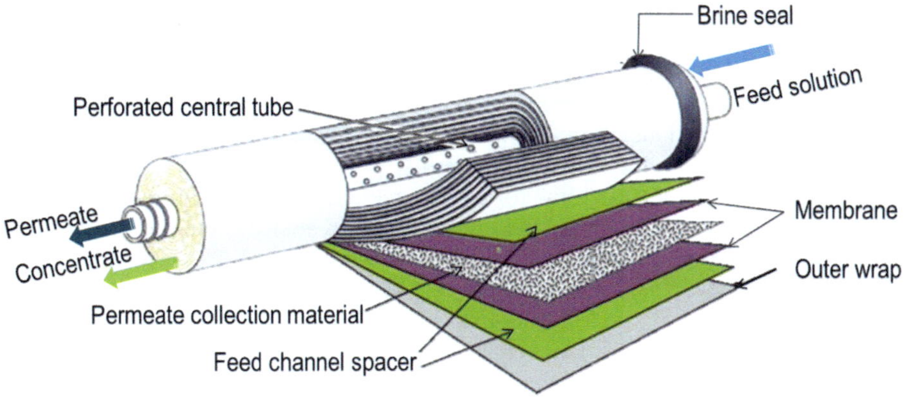

Figure 11.20 RO membrane details [9].

chloride and sulfate, particulate matters, and dissolved gases. Demineralized water can be employed in most laboratories, the laboratory's glass-washing facilities, and as the pre-treatment for still feed water.

Ion exchange resins are organic polymers with long chains and cross linked structure having functional groups through which various ions are exchanged. The resins are porous and insoluble in water.

In the physical process of demineralization water is flown through a bed of ion exchange resins in a Deionizer machine. Cation resins in a column exchanges hydrogen ions ($H+$) for positive ions, while Anion resin in another column exchanges hydroxide ions ($OH-$) for negative ions. The cation and anion columns are placed in series, and water passes through the synthetic resin in cation column first, followed by the anion column, as shown in Figure 11.21. Thus the minerals from the water are removed, and finally deionized water comes out through an outlet pipe. Deionized water thus produced is therefore comparatively

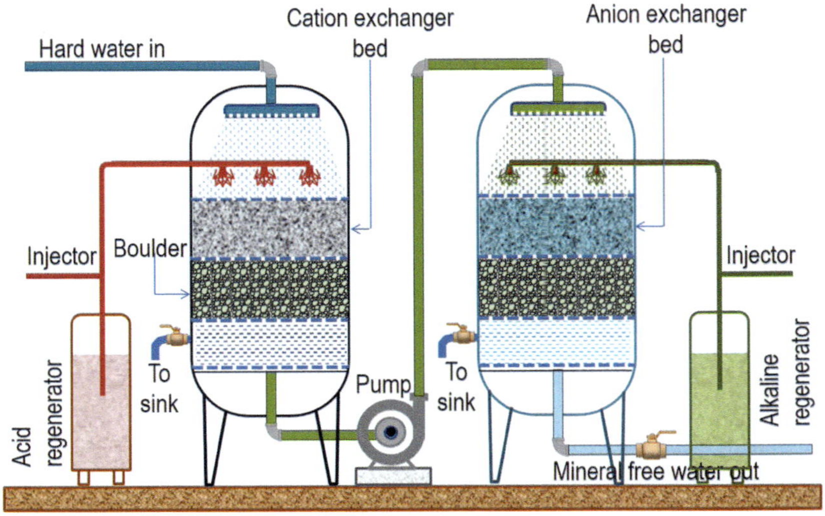

Figure 11.21 Ion-exchange processes.

costly. An outlet is provided to both the cation and anion columns to ensure better monitoring and flow control during operation and regeneration cycles, allowing efficient and complete treatment of the water.

There is also mixed bed demineralizer that combines both cation and anion exchange resins in a single vessel. This allows removing all dissolved ions both positively and negatively charged, from water, making it highly pure.

There are separate outlets to allow for easier regeneration and maintenance. The cation and anion resins require different regeneration processes, such as acidic solutions for cation regeneration and alkaline solutions for anion regeneration.

In a deionization (DI) machine, resistivity is a key indicator of the quality of the treated water and the efficiency of the ion-exchange process. Resistivity measures how much the water resists the flow of electrical current, and it is directly related to the concentration of dissolved ions in the water. The higher the resistivity, the fewer ions are present, and the purer the water is. The typical demineralizer apparatus of a two-bed deionizing unit may possess a resistivity range of 50,000Ω to 1 MΩ, or a mixed-bed deionizing unit with a resistivity range of 1 to 18 MΩ.

In a demineralizer, pressurized gravity down-flow is typically preferred over up-flow. A demineralizer must be chemically regenerated periodically, and during that regeneration time, no pure water is produced. If a continuous supply of pure water is needed, a backup unit should be considered, as the regeneration process takes several hours. The regeneration process can be done manually or automatically. An atmospheric, chemical-resistant drain is required. High-flow water is required for backwash during the regeneration.

Demineralized water could be contaminated with bacteria, pyrogens, and organics which may be produced inside the demineralizer itself.

Filtration

Various types of filters are currently available to remove the suspended particulate matter from the water. Mostly, depending upon the rate of filtration, type of filters is chosen. Pressure filters are generally used for high rate of filtration. A draining system is to be employed for these filters. To remove or kill pathogens from filtered water, ultraviolet sterilization is employed. Activated carbon filtration removes chemicals like chlorine, volatile organic compounds (VOCs), and foul odors. In Chapter 12 pressure filtration has been discussed.

Pure-water supply piping materials

Water treatment and conditioning systems employ various components to remove various impurities from the water. For this purpose, majority components are connected by installing interconnecting pipes. These interconnecting pipes should not contribute to adding any impurity back into the water in the process of treatment or conditioning. Selection of piping material is therefore very important in this regard, as the piping material or its coating may play a role in contaminating water. Pure water supply in the healthcare industry can be very sensitive to the piping material selected. General pure-water supply piping requirements include:

1 Inert materials must not leach contamination into water
2 In the joining method, avoid solvents, lubricants, and crevices
3 Pipe surface erosion must not flake off particles
4 Inner surface should not allow microorganism growth

5 Inner surface should be smooth, crack- and crevice-free, and nonporous
6 Avoid dead-leg systems; there should be continuous flow through piping
7 Provide chemical cleaning connections
8 Install pipe maintaining slope for ease of cleaning and disinfection

A wide variety of piping materials are available, but the common pure-water piping materials are as follows:

1 Stainless steel of various grades (304L and 316L)
2 Aluminum
3 Tin-lined copper
4 Glass or glass-lined pipe
5 Polypropylene (PP)
6 Polyvinylidene fluoride (PVDF)

Metal pipe

Aluminum and stainless-steel pipes have all been used in pure water treatment systems and supply systems as well.

Aluminum pipe has also been used in pure-water systems. Pure water creates an oxide layer inside the pipe that continually erodes, producing particles and aluminum in the water.

Stainless steel has been used extensively in high-purity water systems. It can be joined with threads; butt welded, flanged, or manufactured with sanitary-type connection ends are also allowed. Steel pipe can use sanitary pipe joints using mostly tri-clamp fittings, with a view to achieving a secure, easily dismountable connection while maintaining the necessary level of purity required for clean steam applications. The sanitary pipe jointing method has been used popularly in many pharmaceutical applications.

However, experience has shown that even the best grades of stainless steel, with the best joints, still leach material from the metal that can cause problems in critical water systems.

Glass or glass-lined pipe

Glass piping has been used in some special laboratory applications, though it is fragile. It leaches material to a very little extent, into the pure water. Under extreme conditions like high temperature, strong acids or bases, or demineralized water, a little leaching of silica or trace alkali ions can occur. Under normal conditions, glass does not significantly leach.

Plastic pipe

Plastic pipe is generally not recommended for pure water supply in healthcare services, especially where ultrapure or high-purity water is required, such as in laboratories, dialysis units, or for pharmaceutical-grade water.

Plastic pipes contain color pigments, plasticizers, stabilizers, and antioxidants that can all leach out of the plastic and get into pure water.

When pure water supply system pressures exceed 482.6 kPa or temperatures exceed 24°C, (Polypropylene) PP- or (Polyvinylidene fluoride) PVDF-lined metal piping may be used to meet high pressures up to 1034.2 kPa [10].

Non-potable water supply system

Non-potable water supply systems are usually employed in areas having multiple natural water requirements in very large quantities. In this system, high precautionary measures are taken so that non-potable water could not contaminate the potable-water supply. Areas where non-potable water systems are usually employed include flushing-rim floor drains in animal rooms, all outlets in autopsy rooms, outlets in isolation rooms, and all outlets in infectious-disease and tissue-culture rooms. These systems normally use reduced-pressure-type backflow preventers as the means to protect the potable-water system. Hot water, when required, may be provided by a separate generator supplied from the non-potable water system.

Disinfecting water supply system

The water supply piping system, after its installation and repair works involving changes in pipes in particular, needs complete disinfecting, to ensure safe water for use. There are various methodologies of disinfecting water supply piping systems. In health care establishments the accepted methods of disinfecting are followed as prescribed in the IPC 202, those are mentioned as below.

1 The pipe system, starting from start to the end including all appurtenances, shall be flushed with clean, potable water until dirty water does not appear at the points of outlet.
2 The system or part thereof shall be filled with a chlorine-water solution containing not less than 50 ppm (mg/l) of chlorine, and the system or part thereof shall be valved off and allowed to stand for 24 hours; or the system or part thereof shall be filled with a chlorine-water solution containing not less than 200 ppm (mg/l) of chlorine and allowed to stand for 3 hours.
3 After the required standing time, the piping shall be flushed with clean potable water until the chlorine is almost purged from the entire piping.
4 The procedure shall be repeated where shown by a bacteriological examination to ensure that no pathogenic contamination is present in the system.

Drainage system

Where there is use of water, installed water conveyers, containers, appliances, equipment etc., befitting drainage system is to be provided there for draining out wastewaters generated after water-use, and flushing, cleansing and dripping or leaking water generated from the plumbing elements.

Special drainage system

In addition to the conventional sanitary drainage systems that are found in most of the buildings, special sanitary drainage systems may be required in health-care facilities.

Insofar as possible, drainage piping shall not be installed within the ceiling or exposed in operating or delivery rooms, nurseries, food-preparation, food-serving and food-storage areas, central services, electronic data-processing areas, electric closets, and other sensitive areas. Where exposed overhead drain piping is unavoidable in these areas, special measures are to be taken to protect the underlying area from leakage, condensation, or dust particles.

Acid-waste drainage

Acid-waste drainage systems require special design criteria because the corrosive solutions demand special handling from the actual work area to an approved point at which such acid waste can be safely neutralized and then discharged. Concerns must exercise extreme care in protecting themselves from inhaling acid fumes.

Wastewater possessing the lower pH number below 7, the higher the concentration of acid in the wastewater. Discharging high concentrations of acid into a public sewer may cause considerable corrosion to piping systems and eventual failure. Most local authorities do not allow acid wastes to be discharged to a public sewer without some form of treatment. Acid-resistant waste and vent systems are necessary where acids with a pH lower than 6.5 or alkalis with a pH greater than 8.5 are present. These special conditions are commonly encountered in hospitals, research facilities, and laboratories. Since acid fumes are often more corrosive than the liquid acids themselves, proper drainage and venting is imperative.

Strong acids and caustics may enter the sanitary-waste system in large quantities and at elevated temperatures. These substances can mix to form highly corrosive and even dangerous compounds. Common laboratory procedures encourage neutralization or flushing with copious amounts of water in order to dilute and cool these chemicals to more acceptable levels.

Nationally recognized standards for sanitary systems that handle acid wastes and other reagents are set forth in model plumbing codes; such systems are often further regulated by local building and safety or health department requirements.

Corrosive-waste handling pipe materials

Borosilicate glass pipe

These glass pipes are highly corrosion- and flame-resistant. Sizes are available from 40 to 150 mm. A mechanical jointing system is adopted for joining pipes. Clear pipes allow easy visual inspection.

High-silicon cast iron

These types of cast-iron pipes are highly corrosion- and flame-resistant. Sizes are available from 40 to 100 mm. A mechanical joint system is adopted for joining pipes. These pipes are more fragile and heavier than standard-weight cast iron pipe. These pipes are susceptible to breakage while handling and transporting. So, these are preferred for moderate- to high-budget projects.

Polypropylene

This plastic pipe can be used in draining wastewater of moderate acids at low temperature. This pipe is inexpensive compared to borosilicate glass or high-silicone cast iron, easy to install for being lightweight.

Sizes range from 40 to 150 mm in diameter. A mechanical or heat-fusion jointing system can be adopted. Mechanical joints are not recommended for straight runs or sizes over 50 mm; they should be used to access P-traps or other maintenance areas. Polypropylene pipes are flame resistant and acceptable within most jurisdictions.

Like acid-resistant drain pipe, approved corrosion-resistant piping should be used for vent piping as well.

Precautionary measures

Double-containment waste piping

In a double-containment system, the drain pipe is put within another pipe for more protection. Usually a polypropylene drain pipe is put inside a PVC or fiberglass outside. The drain pipe should be sloped toward a containment vault for the collection of leaking fluid.

Alarm system

An alarm system can be employed to detect leaks at the collection basin or, if the nature of the liquid allows, sensors can be installed between the pipe walls that can locate the original leak location. In underground installation, the latter can reduce the amount of excavation or exploration required to locate the leak.

Discharge to sewer

The building's sanitary-sewer discharge should be at an acceptable pH level before it can be admitted into an outside sanitary sewer system. In such cases, it is recommended that a clarifying or neutralizing tank be added to the sanitary system. A clarifier with limestone is located under a sewer of low flow rates; however, sufficient space must be allowed above the unit for servicing. Unless properly maintained and monitored, this type of system can be rendered ineffective.

Acidic waste neutralization

The neutralization of acidic wastes is generally and most economically done through an acid-neutralization tank. An acid-neutralization tank may be constructed of polyethylene, molded stone, stainless steel, or another acid-resistant material. Tanks are sized to provide a detention time of 2 to 3 hr. Limestone or marble chips fill the interior of the tank, which causes it to neutralize acid wastes. Chips may be 25 to 75 mm in diameter and should have calcium carbonate content in excess of 90 percent. A discharge pH sensor and routine maintenance schedule must be provided to ensure that the system operates properly. An example of a neutralization tank is depicted in Figure 11.22. Table 11.3 can be considered in determining the proper size of the neutralizing tank to be used.

If building wastewater is not well controlled, many unwanted materials, such as glass fragments, plastic items, and needles, will find their way into the neutralization tank, thereby clogging the limestone or marble chips. When this happens, replacement of the chips is required. One way to prolong chip life is to install an acid-waste solids interceptor immediately upstream of the neutralization tank, as shown in Figure 11.23, although maintenance of the interceptor may have to be done quite frequently.

Acid-waste metering detail

Many local authorities require some means of sampling effluent from industrial, institutional, and laboratory buildings. An example of a device used for this purpose is a sampling

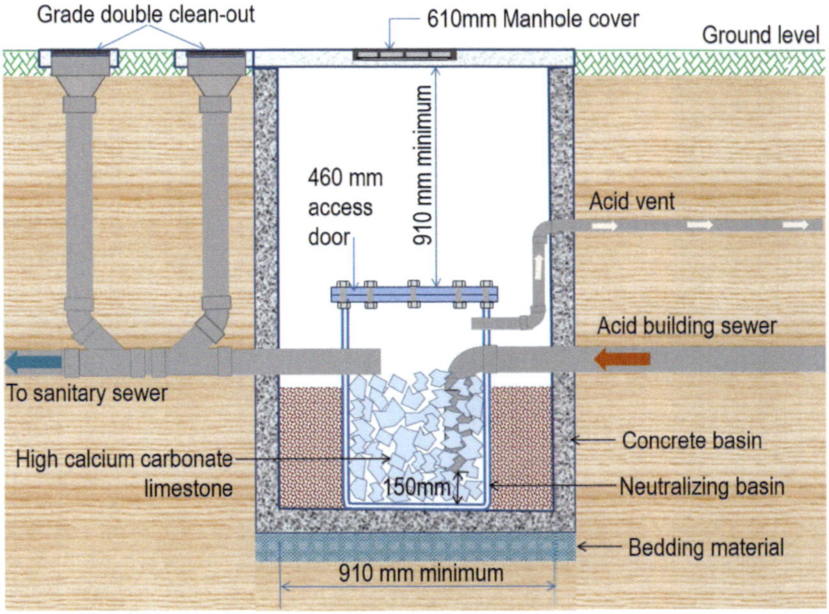

Figure 11.22 Acid-neutralizing tank installation detail [10].

Table 11.3 Acidic-waste neutralization tank sizing [10].

Sl. no.	*Number of sinks*	*Tank size (l)*
1	2	18.9
2	4	56.8
3	8	113.6
4	16	208.2
5	22	283.9
6	27	340.7
7	30	408.8
8	4	567.8
9	5	662.4
10	60	757.0
11	75	1040.9
12	110	1362.6
13	15	1898.5
14	175	2081.8
15	200	2460.3
16	300	4542
17	500	7570
18	600	11355

Note: For commercial and industrial laboratories, the number of lab-sinks should be multiplied by a 0.5 use factor.

manhole, depicted in Figure 11.24. This unit is the last component, installed before neutralized acidic wastes or treated industrial wastes are discharged to a public sewer. There are as many types of sampling point requirements as there are municipal sewer authorities.

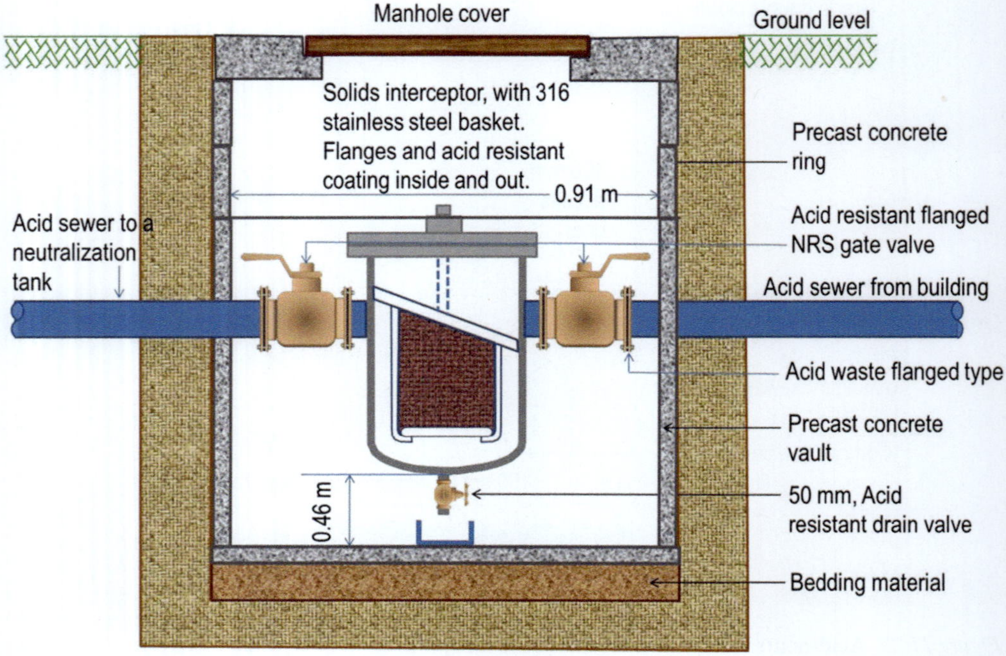

Figure 11.23 Acid-waste solids interceptor detail.

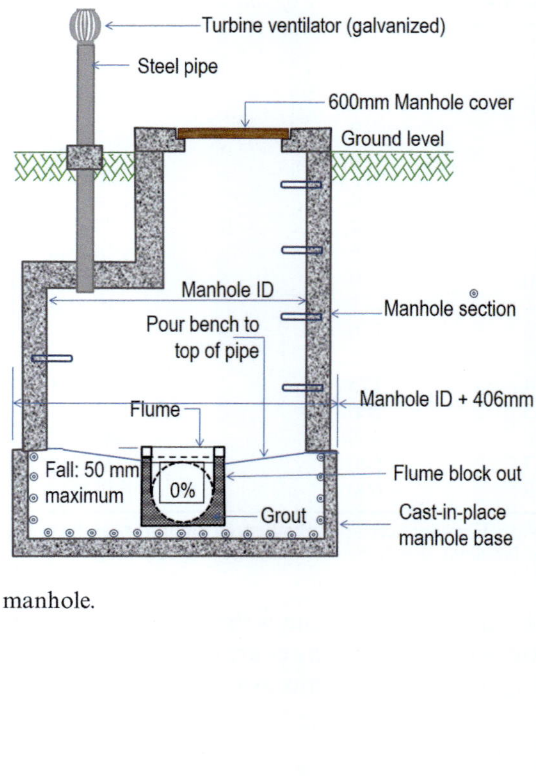

Figure 11.24 Sampling manhole.

Indirect wastewater drainage

An indirect waste drain is a drainage piping that runs from a plumbing fixture or appliance and discharges into a separate waste pipe, fixture, receptor, or receptacle; that waste piping or receptacle is connected to the building drainage system for final disposal.

An indirect waste pipe cannot be connected directly to the drainage system; at first, wastewater is to be discharged into a fixture, receptor, or receptacle, which in turn will discharge to the drainage system. To discharge wastewaters from various types of fixtures, appliances, and equipment, the indirect wastewater discharging method is to be employed for the proper drainage.

Sterilizers, steamers, and condensers shall be discharged to the drainage through an indirect waste drainage pipe and then to the building drainage system, maintaining a separation by means of an air gap. Where, a battery of not more than three sterilizers discharges to an individual receptor, the distance between the nearest receptor and the sterilizer shall not exceed 2.44 m. The indirect wastewater drain pipe on a bedpan steamer shall be supported by a trap.

There are two types of indirect waste piping:

1 Indirect wastewater drainage that discharges into the receptor through an air break
2 Indirect wastewater drainage that discharges into the receptor through an air gap

There are two types of air gaps:

a Water distribution air gap
b Drainage air gap

Air gap and air break

Indirect waste piping shall discharge into the building drainage system through an air gap or air break as set forth in plumbing code. The minimum drainage air gap that shall be maintained is the vertical distance measured from the lowest point of the indirect waste pipe or the fixture outlet to the flood-level rim of the receptor, which shall be not less than 25.4 mm, as shown in Figure 11.25.

An air break basically provides a physical separation or gap between the fixture's indirect waste pipe and the receptor. If codes allow, the fixture's indirect waste pipe may be terminated below the flood level rim of the receptor, thereby creating a low inlet into the receptor. An air break simply requires that there be no direct physical connection between the fixture's indirect waste pipe and the receptor, as shown in Figure 11.25b.

The air gap for the drainage is not the same as the air gap provided in the water distribution system. The minimum air gap in the water distribution system to be maintained is also 25 mm or a minimum of two times the diameter of the indirect wastewater pipe if the size is over 12 mm, whichever is greater.

The drainage air gap is defined as the vertical distance as measured from the lowest point of the indirect waste pipe or fixture outlet to the overflow rim of the receptor. This distance is a minimum of 25 mm. There are no other requirements, unlike those that are followed for the air gap in water distribution.

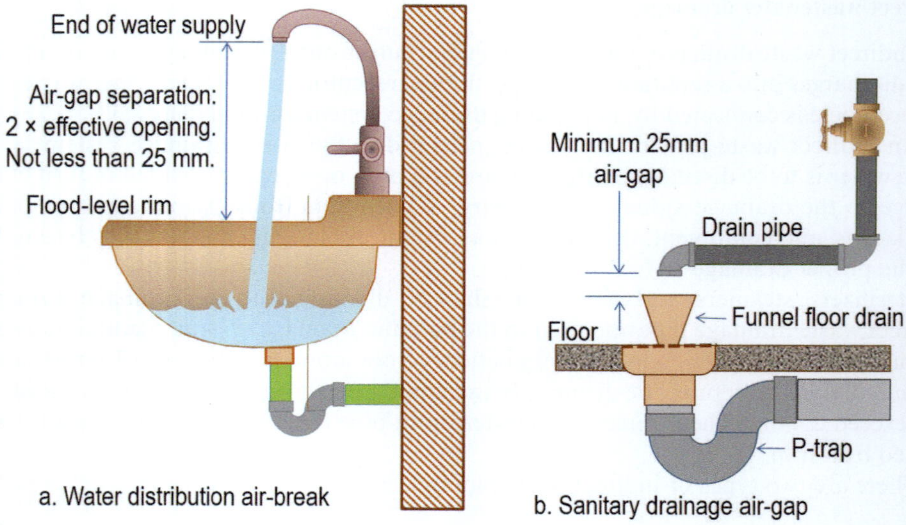

Figure 11.25 Air break and air-gap.

Efficient healthcare plumbing system planning

Because of the sensitivity of the healthcare environment and the importance of providing quality water to a hospital's patient care mission, there are many unique and special requirements that professionals must consider when planning a healthcare plumbing system. In today's ever-changing world of health care, flexible, reliable, and easy-to-maintain plumbing systems are the antidotes to many hospital-shutdowns being unable to provide due patient care. The planning concerns of plumbing include the following:

1 Accessibility
2 Maintainability
3 Redundancy
4 Serviceability

Accessibility

All the plumbing elements, including its equipment and controlling accessories, shall be located in an easily accessible area, not critical to maintaining it over time. Locating plumbing elements in easily accessible spaces will minimize disruption to patients and hospital staff and will help to ensure regular maintenance as required.

Maintainability

Maintenance must be considered prior to the selection and installation of the plumbing elements. Sufficient funds must be allocated and manpower employed for the purpose.

Redundancy

At least one redundancy for some important equipment like pumps, boilers, etc. should be kept in the planning of the plumbing system of healthcare facilities to ensure uninterrupted service. Such redundancy may also help make future expansion and program changes.

Serviceability

All the plumbing elements shall not be placed or located in such areas where it's all operations, including monitoring, maintenance, replacement, etc., and the movement of persons or accessories for such operations hinders, contaminates the sensitive areas, or paths in a hospital or healthcare facility. Movement of persons for any plumbing operation should not be through the sterile area of healthcare establishments.

References

[1] Infitek Co., Ltd. (2022) "Automatic glassware washer" Retrieved from https://infitek.com/products/automatic-glassware-washer-lwd-200i/ on 13 July 2024.

[2] EASTGLAZ, Glass Processing Machines (2025) "CNC flat glass washing machine (vertical/horizontal)" Retrieved from https://www.eastglassmachine.com/glass-production-line/glass-washing-machine/ on 24 February, 2025.

[3] www.dekomed.co.uk "Why choose a bedpan washer?" Retrieved from https://www.dekomed.co.uk/washer-disinfector/why-choose-a-bedpan-washer/ on 24 February, 2025.

[4] Center for disease control and Prevention (2024) "Dialysis safety" Retrieved from https://www.cdc.gov/dialysis-safety/hcp/recommendations-resources/water-use-in-dialysis.html?CDC_AAref_Val=https://www.cdc.gov/dialysis/guidelines/water-use.html on 25 February, 2025.

[5] Gregory S. Matte, Thomas J. Sandora, Robert J. Howe, William L. Regan, Gail Potter-Bynoe, James R. Neal, Pedro J. del Nido (2018) "A novel wall water system for cardiopulmonary bypass may reduce the risk of aerosolized infection" *The Journal of Thoracic and Cardiovascular Surgery*, Vol. 156, No. 1, 318–324. Retrieved from https://www.sciencedirect.com/science/article/pii/S0022522318306184 on 22 July 2025.

[6] Online Medical Supply (2025) "CURITY™ SPECIPAN™ measuring system" Retrieved from https://onlinemedicalsupply.com/products/curity%E2%84%A2-specipan%E2%84%A2-measuring-system?pr_prod_strat=e5_desc&pr_rec_id=0565ef1fc&pr_rec_pid=6915332079664&pr_ref_pid=6915000664112&pr_seq=uniform on 1 March 2025.

[7] www.prohealthcareproducts.com "Extremity Whirlpool tubs" Retrieved from https://www.prohealthcareproducts.com/extremity-whirlpool-tubs/ on 1 March, 2025.

[8] Dubai Health Authority (DHA) "Public health – water systems design (WS)" Version 1.0, pp. 170–172, Retrieved from https://services.dha.gov.ae/tahpi-app/HealthFacilityGuidelines/Guidelines/FileContent/Preview/DHAHFG/Part%20E_5%20-%20Public%20Health%20-%20Water%20Systems%20Design%20(WS) on 25 July 2025.

[9] LUMINOR Environmental Inc. (2025), "Reverse osmosis overview" Retrieved from https://www.luminoruv.com/education/reverse-osmosis-systems/ on 3 March 2025.

[10] American Society of Plumbing Engineers (ASPE) "Plumbing design for health-care facilities" Chapter 2, ASPE Data Book- Volume 3.

12 Swimming pool plumbing

Introduction

In the primitive age, humans used to love bathing in surface water bodies like rivers, lakes, ponds, etc. Such love for bathing in natural surface water bodies still remains in present days, but access to these water bodies has now become very limited for various obvious consequences.

As people now remain engaged most of their time in buildings, they want to fulfill their desire of having a bath, swimming, or playing in a large pond-like water body. So, to fulfill such desire, a hydro-mechanized structural element is built inside or outside of a building, termed a swimming pool. These swimming pools are a luxurious amenity of a building for its occupants.

In developing a swimming pool, three major aspects of technology are involved—architectural, structural, and electro-mechanical. As these pools contain water, there is a need for various plumbing elements for the management of this water and its wastewater as well. So, this chapter is dedicated to that part of the swimming pool that falls under the purview of plumbing, as specialized plumbing.

Swimming pool

A swimming pool is a structure designed to hold water to enable swimming or other leisure activities in water. Though swimming pools are mostly built into the ground, nowadays, swimming pools are built inside a building, maybe on the top roof, on the ground floor, or even on any intermediate floors. In buildings, pools are most commonly constructed from materials such as concrete and natural stone, but small pools might be made of metal, plastic, or fiberglass. In a building, a swimming pool might be built to a standardized size or in a customized size and shape.

The primary objective of maintaining a swimming pool is to keep its water clean, comfortable and healthy. To keep pool water healthy, stagnation of water at any location of the pool is to be avoided. So, pool water is to be kept flowing, for which the pool plumbing system must create a continuous loop of water flow that conveys pool water through the filter and sanitizing appliances and returns the filtered and sanitized water to the pool again. For this circulation of water, a pumping system is employed. Once, generally, about every five to seven years [1], when the pool water becomes ineffective with respect to cost for its filtration and sanitization, the whole water is drained out and refilled with fresh water.

DOI: 10.1201/9781003724056-12

There are two types of development of a swimming pool as named below. Depending upon the level and surroundings of pool water, swimming pools are termed differently as mentioned below:

1 Infinity edge pool
2 Zero-edge pool
3 Below-edge pool

The circulation system of infinity edge or zero edge and below-edge pools differs, which is discussed later on.

Types and sizes

Swimming pools are designed in various shapes and sizes for use in various purposes, ranging from small to vast ones, varying in area and depth. The right size pool for a particular purpose depends on the plan to use it, personal aesthetics, budget, and the area available for it.

Popular pool sizes on the shape and type of pool based on the location of construction, type of use, aesthetic features, etc. The most common sizes of various types of pools have been furnished in Table 12.1, which will be helpful for planning.

With respect to the method of collecting overflown pool water for recirculation, swimming pools may be grouped into two types as below:

1 Skimmer type
2 Balance tank type

Components of swimming pool

A typical swimming pool is developed incorporating a number of major and minor components. Of them, the following are the major components:

1 Basin
2 Skimmer
3 Balancing tank
4 Pool pump
5 Filter

Table 12.1 Sizes of various types of pools [2]

Sl. No.	*Pool type*	*Common depths (m)*	*Popular sizes: width (m) × length (m)*
1	In-ground	0.90–3.00	4.88 × 9.75, 5.49 × 10.97, 6.10 × 12.19
2	Above-ground	0.90–1.50	3, 4.88 and 7.32 m, in diameter
3	Olympic	2.13–3.00	25 × 47
4	Lap	1.50–3.00	3 × 12.19, 2.44 × 15.24, 1.83 ×12.19
5	Plunge	1.50–2.13	1.83 × 3, 2.44 × 3.66, 2.44 × 6.10
6	Endless	1.22–2.44	2.74 × 4.57, 3.66 × 5.49, 4.57 × 6.10
7	Freeform	2.44–3.00	3.66 × 9.144, 4.57 × 10.67, 5.49 × 12.19
8	Small	2.44–1.83	1.83 × 3, 2.44 × 3.66, 3.66 × 6.10

6 Heater
7 Disinfector
8 Water inlets
9 Drain suction inlet
10 Piping
11 Valves

A schematic diagram of a skimmer type swimming pool, including its major components, has been shown in Figure 12.1.

Basin of swimming pool

The basin is the main structural element that holds the whole water for swimming or other use. Most public and private swimming pool basins are generally designed rectangular but can be formed in various shapes. They can vary in size, shape, and depth as furnished.

The standard depth of a swimming pool varies between 1 and 3 m deep. A swimming pool might have multiple depths. For diving in the pool, a depth of 3 m would be required. The length may be as long as 50 m in the case of long race course swimming in Olympic games. Width may be varied, which depends upon the comfortable space between the simultaneous swimmers keeping, which is 2.5 m width for each. So, for 6 swimmers swimming at a time, it should have a width of 15 m. A swimming pool of 5 m long 2.5m width and 1m deep is considered to be standard for soaking and floating, might be the comfortable smallest size of a swimming pool.

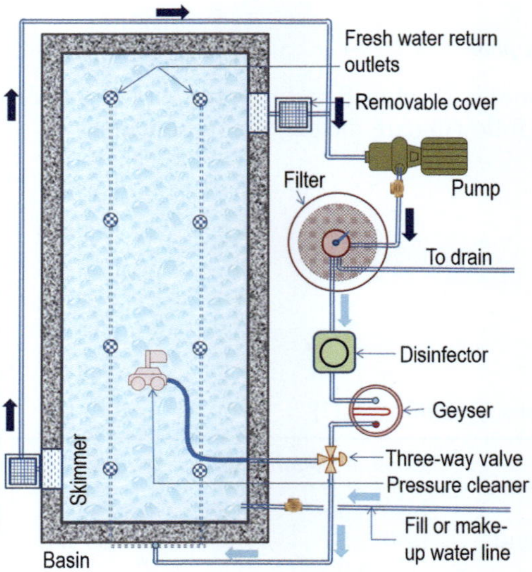

Figure 12.1 A schematic diagram of a skimmer type swimming pool incorporating a skimmer and other major components.

Materials of construction

The type of pool basin may be made of the following materials, depending mostly upon its size and shape:

1 Concrete and Gunite
2 Vinyl
3 Fiberglass

Concrete pool

Larger and deeper swimming pools are generally constructed by reinforced cement concrete (RCC) materials. At first, reinforcing bars are laid and bounded according to the shape and thickness of the elements of the basin of the pool. Concrete is then poured on the floor of the swimming pool up to its thickness height and in between the timber or steel-sheet formworks for the side walls. For small-depth swimming pools, side walls can be made of masonry bricks or blocks. The pool surface is then finished with cement-sand plaster, paint, tile, exposed decorative aggregate, or fiberglass, etc. The surface finish material shall have maximum 0.5 percent water absorbing capacity.

Gunite pool

Gunite is basically a mixture of cement and sand, which is generally sprayed to coat around the rebar, assembling a framework grid according to the shape of the pool. Before that, a hole has been excavated to accommodate the basin of the pool, and the plumbing is in place. The framework grid of rebar is assembled by binding rebar at certain space intervals. The sprayer unit mixes dry gunite mix with water just before spraying to make it suitable for spraying. After completing spraying, the inner surface is troweled smooth and left to cure. Gunite pools are found to be highly durable, which can be constructed in any desired shape and size. The surface can be finished in the same way as done in concrete pools.

Vinyl-lined pool

A hole is excavated in the ground according to the pool desired, and a metal, plastic, or timber frame wall is assembled around the perimeter. Sand filler is laid at the bottom of the hole, and then a vinyl lining is secured to the perimeter wall. These materials tend to be cheaper than other in-ground designs but are less durable, as the liner requires replacement typically every ten years.

Fiberglass pool

Fiberglass-reinforced plastic is molded into a basin shape to the required specifications on the ground. A hole is excavated in shape as the pool is molded, the necessary plumbing pipes are installed, and the pool is lowered into the hole; sand filler is put in the gap between the earth and the pool wall to fit the exact shape. Usually, a deck structure is constructed around the perimeter of the pool.

Basin construction

Pool basins must be constructed and installed water-tight. To meet the criteria set out in British Standard 8007:1987, the construction is such that during the 7-day test period, the total permissible drop in water level after allowing for evaporation should not exceed 1/500th of the average water depth of the full tank, 10 mm, or another specified amount [2].

There are two common types of sealant used in pools as a means of sealing leaks. These are as follows:

1 Epoxy-based sealants
2 Unplasticized polysulfide sealants

Skimmer

The primary function of a skimmer is to prevent floating debris on the water surface from entering into the pool-water conditioning system. The skimmer's front is of rectangular shape, as shown in Figure 12.2, located inside the pool, at the top water surface level on the side walls; those would likely receive more debris being pushed into the skimmer by the blowing wind. For effective operation, it requires being fitted to the middle of the skimmer front with pool water.

A surface skimmer is typically made of PVC plastic and has a tank or box with a projecting throat on its upper side. There is a swinging door or flap known as a weir that acts as a low head dam, as shown in Figure 12.3. The weir is made on the water entranceway into the skimmer box. The flap opens and closes according to the water level and the movement of the water. While the water is moving and the equipment is on, the weir will flap open. As the water passes through it, the floating debris gets in and accumulates in the screen-like basket. Once the equipment turns off, the weir closes so that all trapped debris cannot go back into the pool. To remove the debris collected into the skimmer basket, its lid, located in the swimming pool deck, is to be opened, the skimmer basket is to be lifted out of the skimmer box and emptied off the basket, and afterward the basket is replaced back into the skimmer box.

The pool skimmer shall be placed at such a height so that the water level is at about the halfway point on the skimmer. This condition helps the pool skimmer and filter pump system to work effectively.

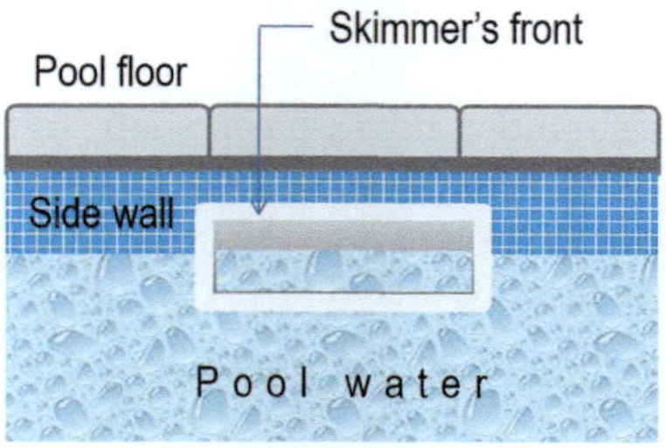

Figure 12.2 A skimmer on the side wall of a swimming pool.

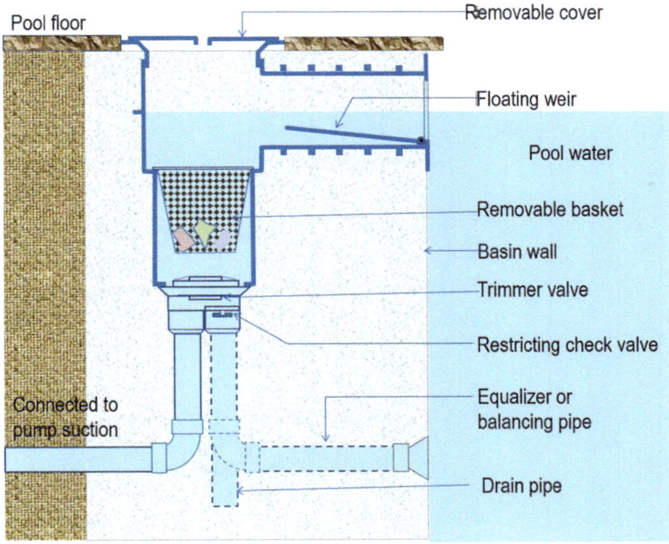

Figure 12.3 Detail of a pool skimmer.

Table 12.2 Skimmer sizing, based on surface area of pool [3]

Pool or spa	*Area per skimmer (sqm)*
Public pool	46.45
Residential pool	74.32
Spas (all types)	13.94

For Imperial: 1 square foot = 0.0929 sqm.
Source: UP Code 2021.

Skimmer number

According to UP Code 2021, where automatic surface skimmers are used as the sole overflow system, not less than one surface skimmer shall be provided for the square meter areas, or fractions thereof, indicated in Table 12.2.

In a small pool, at least two skimmers should be provided. Skimmers shall have a flow-through rate of 1.9 lps, or 93 mlps per lineal cm of weir, whichever is greater [4].

Balance tank

A balance tank is used in a swimming pool that does not use skimmers. Its primary use is for the holding of excess water overflow due to the displacement of water caused by the immersion of the body of the users. That excess overflowing water from the pool falls into a gutter covered by gratings, built on the deck near and around the pool wall, which then flows by gravity into the balancing tank. The balance tank helps maintain a constant depth in the pool regardless of the number of people in the pool. The balancing tank also plays a role in compensating for variations in the water level in the swimming pool.

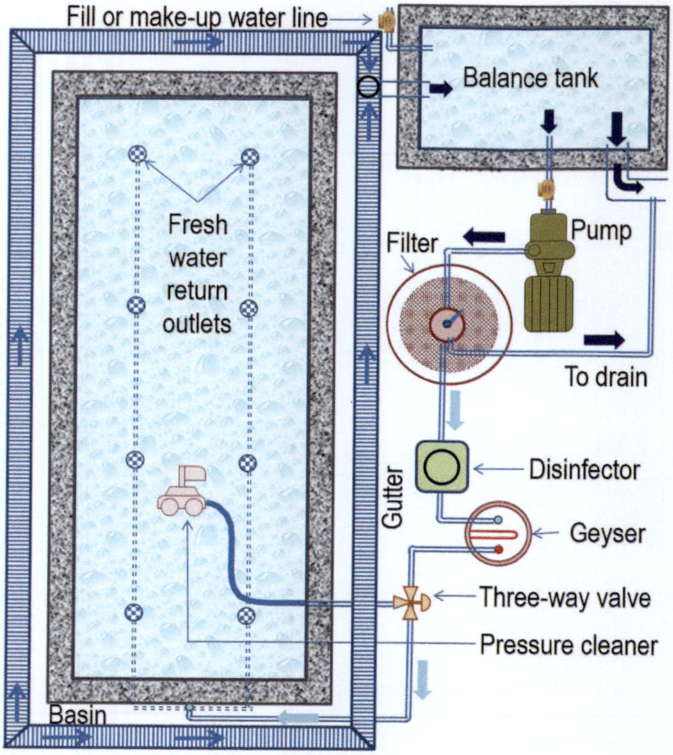

Figure 12.4 A swimming pool incorporating a balancing tank.

The balance tank is placed right before the filtration system. The balancing tank has a drain at its bottom to allow the pump to draw out the water and send it into the filtration system and then to the pool again. Thus the recirculation of water is made in such overflow swimming pools with balancing tanks, as shown in Figure 12.4.

Piping and accessories of balance tank

The balancing tank is supported by the following pipes and accessories for its intended performance:

1 It has a water inlet equipped with a solenoid valve controlled by the regulation system. The top water level is regained automatically, as there occur water losses due to evaporation, splashes from swimmers, etc.
2 It has a suction hose between the balancing tank and the pump.
3 It has an overflow pipe to prevent the tank from overflowing, which is usually connected to the sewer system.
4 It has a non-return valve on the suction line, before the pump but after the balancing tank, to prevent the pool from draining by this balancing tank overflow when the pump is stopped.
5 It has a water level control system to manage the reduction of water or excess of water in the balancing tank.

Balance tank sizing

In rational sizing of a balance tank for a swimming pool, several factors come into play, as mentioned below:

1 The volume of the swimming pool
2 The nature of the perimeter overflow and fall into the collection drain
3 The length of the overflow run
4 The maximum expected number of swimmers
5 The flow rate of the circulation pump
6 The presence of an auxiliary pump, which improves the overflow effect by increasing the thickness of the overflowing water

But to simplify the way of determining the size, a general rule of thumb is to take 10 percent of the total volume of the swimming pool as the volume of the balance tank [5].

Pool pump

The circulation of swimming pool water is done mainly by a pump. The pump is basically a centrifugal pump driven by an electric motor. The specialty or uniqueness of a swimming pool pump is that it is a type of pump that is supported by an additional element, which is a strainer, as shown in Figure 12.5. The strainer is essentially a screen-like basket having a lid. The pool water is sucked through the basket where debris is retained, and then the debris-free water is pushed through a filter bed by the pump. The strainer is normally made of plastic or stainless steel to prohibit rust.

Pool water might become salty, so the pump material shall be good against corrosion, which requires pumps made of stainless steel, brass, or aluminum for durability. A general rule of thumb is to run the pool pump 8–10 hours a day.

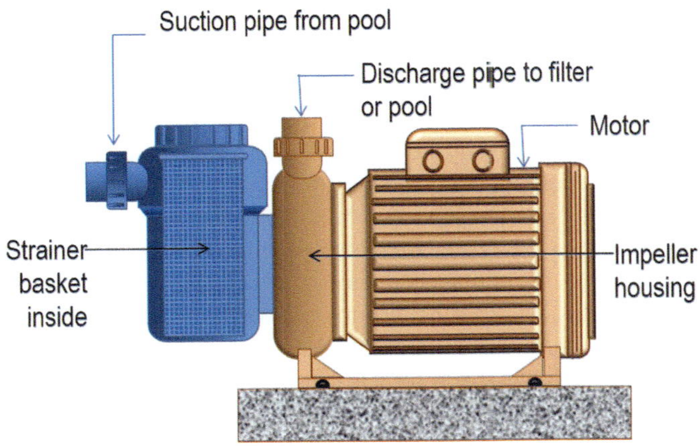

Figure 12.5 A swimming pool pump with strainer.

Types of pool pumps

A type of swimming pool pump based on speed can be one of the following three types:

1 Single-speed pump
2 Dual-speed pump
3 Variable-speed pump

Single-speed pump

When all the water supply-related activities are planned to be performed at a particular constant flow rate, then the supply system can be done by a constant-speed or single-speed pump. In a constant single-speed pump, the impellers rotate at a constant speed, and so the pump delivers water at a constant rate. Single-speed pumps are much cheaper but will cause more operating costs when compared to other types of pumps. These single-speed pool pumps are typically found to be generating more sound than variable-speed types.

Dual-speed pump

For a swimming pool to run in a comparatively energy-efficient way, it is recommended to run the pump at a low-speed setting to meet the general, day-to-day swimming pool water circulation and chlorination and a high-speed setting while heating the water, cleaning the pool water features, or during a large bather load. So, to meet these varying requirements for operation of a swimming pool water flow, dual pumps, also called two-speed pumps, are preferred.

In a dual or two-speed pump, to run at two fixed speeds—high and low—it requires an automatic control device to adjust between the two speeds. Dual-speed pool pumps are more energy-efficient than a single-speed pump. The energy usage will drop when the pool pump speed slows down to the lower speed. In a dual-speed pump, reducing the RPM by 50 percent, an energy saving of 75 percent can be achieved, and by reducing the RPM by 66 percent, an 89 percent reduction in energy consumption can be achieved [6].

Variable-speed pump

In the operation and maintenance of a swimming pool, various functions like filtering, heating, cleaning, and spa or water features are to be performed; each function requires water flow at a particular rate for efficient performance. So, to maximize the efficiency of all the elements and not to overwork the pump by supplying water at an over rate for some time, varying the flow rate can only be done as and when needed by employing variable speed pumps instead of single or dual speed pumps.

The more energy-saving variable speed motors virtually sense the pressure in the system and automatically adjust to the most efficient speeds to overcome the pressure. A variable-speed pump is driven by a variable-speed motor, which uses a permanent magnet motor. The magnet motor creates less friction compared to the induction motor that uses constant-speed motors. The reduced friction increases the efficiency of the variable speed motor. In addition, the variable-speed motor is quieter in operation. A pump with a variable-speed motor is comparably more expensive than those having a single or dual-speed motor. However, it is more cost-efficient in its lifetime operation period, as a variable-speed pump can reduce energy consumption by 90 percent compared to a single-speed pump [6].

Pool water flow

The whole water of a swimming pool requires continuous maintenance, by regular cleaning and disinfecting, for making it safe to swim in. To do so, the water requires circulation. Circulation of pool water enables cleaning by filtering, conditioning by heating, and disinfecting by chemical dosing in flowing pool water.

Circulation of pool water is made by pulling in water through the skimmers located along the top of a pool or from a balancing tank and the main drains located at the deepest bottom. Pulled water is then pumped through a filter, heater, and disinfector and finally back into the pool through the return jets. Pool water is circulated at a planned turnover rate.

The circulation flow rate of a pool is measured in liters per minute (lpm) and is the rate of water flow within the circulation system (Figure 12.6). There shall be sufficient flow rate to achieve required turnover rates, as determined using the following formula:

$$\text{Flow rate} = \text{Pool volume} / \text{turnover rate} / 60\,\text{min}/\,\text{hour} \tag{12.1}$$

Turnover

The turnover rate is the time that a pool takes to circulate all the water in the pool. The length of time it takes for the whole water of the swimming pool to be circulated through the filtering, conditioning, and sanitary system is the turnover period. If the whole water is circulated twice, i.e. 2 turnovers in a day, then the turnover rate is 12 hours. Similarly, if the water is circulated 4 times, i.e. 4 turnovers in a day, then the turnover rate will be 6 hours. The turnover rate to be adopted depends upon the quality of pool water, but it must be adequate to meet the following water quality parameters: pH between 7.2 and 8.0.

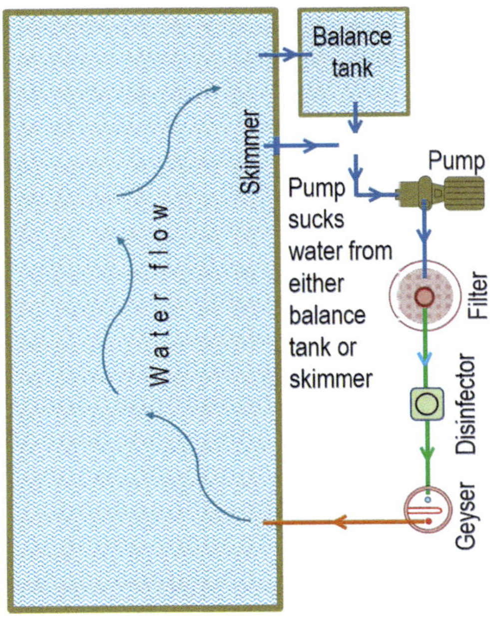

Figure 12.6 Recirculation flow of a swimming pool.

Table 12.3 Turnover rate (hours) for different types of pools [7, 8]

Sl. no	*Pool type*		*Turnover period (hours)*
1	Swimming pool: Domestic, diving		4–8, Average 6
2	Competition pools 50m long		3–4
3	Public pools up to 25m long, 1m shallow end		2.5–3
4	Leisure pools, wave pool, activity pool	Less than 0.5m depth	0.17–0.75
		0.5–1 m deep	0.5–1.25
		1–1.5 m deep	1–2
		over 1.5 m deep	2–2.5
5	Spas	Whirl pool	0.5
		Residential	0.25
6	Teacher, learning and training pools		0.5–1.5
7	Waterside splash pools:		0.5–1

According to ANSI, the recirculation system shall be designed to provide a minimum of four turnovers of the pool volume per day. Pools that are less than 93 sqm at health clubs shall be required to provide eight turnovers per day. On an average, the turnover rate of a pool's water can be practiced 6 hours or less, i.e., the swimming pool recirculation system shall be capable of processing one pool volume of water in 6 hours or less.

The recommended flow rate of different types of pools has been shown in Table 12.3. It is to be understood that a drop in flow rate may be for the filters not performing as desired, for which maintenance is needed by backwashing the filter unit, skimmers need cleaning, or the pump is running inefficiently.

Flow velocity

More generally, the very effective way of reducing pressure drops is by reducing water flow velocity in the pipe. For this purpose, pipes of larger diameter can be chosen. In the water circulation of a swimming pool, the following water velocity should be maintained, and accordingly, concerned pipes are to be sized.

1 In the suction: 0.5 and 2 mps; ideally 1.5 mps
2 At the discharge: lower than 3 mps; ideally 2 mps

In general, a flow velocity of 1.5 mps can be maintained in all respects of flow.

Pump flow

The circulation flow rate by a pool pump is the pool water volume multiplied by 3, or 4 for a 98 percent filtration, divided by the pump run time in hours for flow rate in hours, or minutes for flow rate in minutes.

Head of pump

The head of the pump is its total dynamic head (TDH) which refers to the total equivalent height at which a fluid will be pumped by taking into consideration all friction losses or head losses in the pipe lengths, including fittings, valves, equipment, etc., and the rise in elevation, i.e., the static head. The head loss in the pipe for different flow rates can be obtained from the nomograph shown as Figure 12.7.

Head loss is the reduction in pressure that occurs when fluid flows through a pipe or any hydraulic elements due to friction, turbulence, or other factors. It also represents the energy loss in a fluid system, which is typically expressed in units of length in meters.

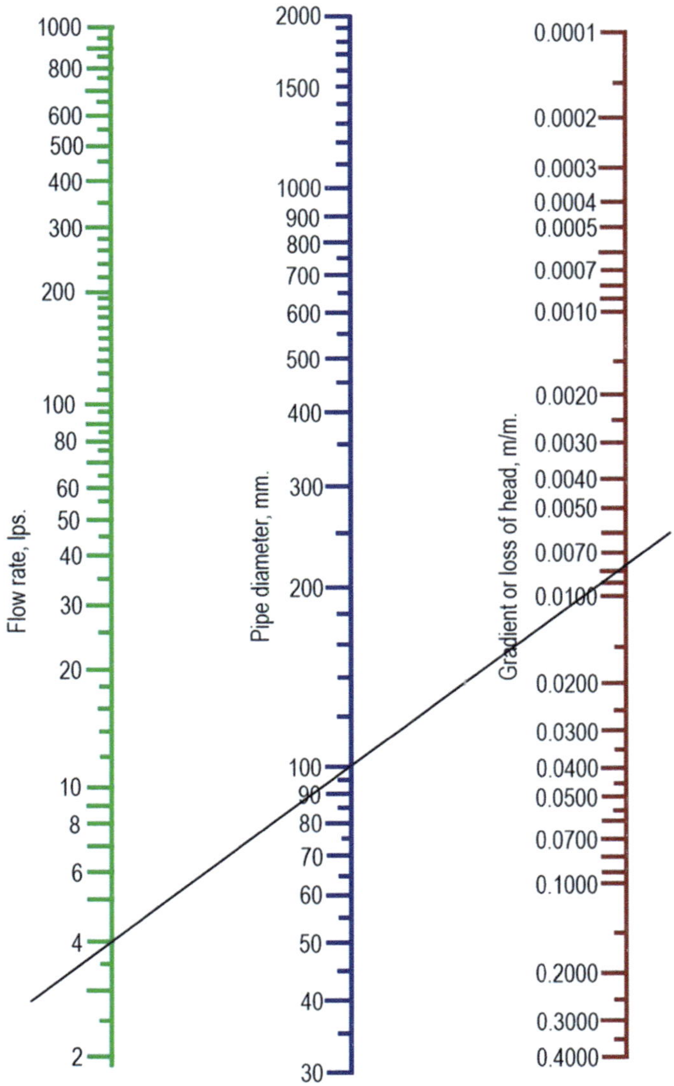

Figure 12.7 Nomograph for Hazen William's formula, based on C = 100.

Table 12.4 Head loss for flowing water through various elements of swimming pool plumbing [9, 10]

Sl. no	*Elements*	*Head loss (m)*
1	Pool filter	
	a DE filter	2 at 265 lpm flow [10]
	b Sand filter	7 at 265 lpm flow [10]
	c Cartridge filter	5 at 265 lpm flow [10]
2	Pool heater	3
3	Pool cleaner	3
4	Pool piping	0.1 m per meter
5	90 degree fittings	0.3 m each
6	Inline chlorinator or purifier etc.	0.6
7	Directional valves	0.3
8	Skimmers	1
9	Main drain	0.6
10	Wall or floor returns,	0.6

The resistance of flow through various equipment and appliances used in swimming pool plumbing has been shown in Table 12.4 for understanding and estimation only.

Power of pump

A swimming pool pump mostly uses electricity. So, the use of energy-efficient pumps helps save energy and reduces power costs for operation. Pool pumps should, therefore, be around 80 percent efficient. The output power of a swimming pool pump is as given below:

$$\text{Pump power requirement}\left(\text{kW}\right) = \frac{Q \times H \times S}{360 \times E} \tag{12.2}$$

where, Q = Flow rate in cum/h

H = Head in m
S = Specific gravity of pool water
E = Efficiency.

Pool water quality and condition

Water in a swimming pool, an outdoor swimming pool in particular, gets contaminated by mostly air-blown dust and debris, carried in pollutants by the bathers, filled water from unsanitary sources, rain containing microscopic algae spores and droppings from birds, etc. So, to keep pool water healthy for the swimmers, pool water is to be filtered and disinfected as well.

During low atmospheric temperatures, pool water temperature drops to atmospheric temperature, turning pool water discomforting for swimming. To raise pool water temperature, heating would be needed to raise water temperature to comfort level.

Table 12.5 Recommended chemical concentrations in swimming pool water [8, 11]

Sl. no.	*Parameters*	*Recommended range*
1	Temperature (°C)	24–30
2	pH	7–7.8 (Optimum: 7.2–7.6)
3	Total alkalinity (ppm)	60–200
4	Isocyanuric acid (ppm)	30–50
5	Calcium hardness (ppm)	0–500
6	Total dissolved solids (ppm)	1000–2000
7	Turbidity (NTU)	0.5; The bottom drain must be clearly visible [8]
8	Free chlorine (ppm)	a. Pool temperature < 26°C >2 b. Pool temperature ≥ 26°C >3
9	Residual combined chlorine (ppm)	≤ 0.6
10	Total bromine (ppm)	2–5 [8]
11	REDOX potential (mV)	250–900 [8]
12	Legionella spp (UFC/liter)	<100 [8]

So, for conditioning swimming pool water for the healthiness and comfort of the bathers by maintaining the chemical concentration of pool water at a level furnished in Table 12.5, the following measures are to be employed in the swimming pool plumbing system:

1 Filtration
2 Heating
3 Disinfection

Filtering pool water

Filters are one kind of water-treating appliance that removes suspended impurities from water. These filtering appliances consist of a tank, usually made of fiberglass, steel, or concrete, in which there is a thick bed of filtering media or cartridge. Filter media is composed of trapping particles ranging from 1 to 40 microns in size

When water is pushed through this filtering media, suspended matters that are not retained by the skimmer or pump strainer are mostly removed, and almost turbid-free, clear water is delivered out. Through a good filter media, more than 99 percent of the feed water turbidity can be removed to produce clear water with turbidity as low as 0.1 NTU or less [12].

Various types of filters can be used for filtering pool water. Filters are classified based on the following filtering conditions:

1 Mode of operation
2 Type of filtering media and
3 Type of construction material

Depending on the mode of operation, swimming pool water filters may be of the following types:

1 Gravity filters
2 Pressure filter

Gravity filters operate under gravity to filter water by flowing through the filtering media under gravity. These are of simple design, consuming lower energy, as no pumps are required, so they are economical for large-scale water purification.

On the other hand, pressure filters use pumps to force water to flow through the filtering media under pressure. So, higher flow rates can be achieved in a more compact design. Pressure filters are preferred for producing more filtered water in less time and where space is limited.

As the filtration continues, the filter media gets clogged with the retained particles. This results in a continuous increase of pressure drop across the filter. When a predetermined pressure drop level is achieved, the filter is shut down for cleaning the clogged filter media.

Types of filters used

Depending on the filtering media used, swimming pool water filters may be of the following types:

1 Sand filters
2 Diatom filter
3 Dual media filter
4 Multi-grade media filters

Gravity sand filter

The water is filtered by passing it through a layer of sand bed placed over a gravel bed. At the bottom of the filter bed, there are bottom-half perforated pipes through which filtered water is collected. There are two types of sand filters: slow and rapid sand filters. Rapid sand filters are mostly preferred for filtering pool water.

The considered filtration rate of rapid sand filters shall not exceed 2.0 lps/sqm of filter area. High-rate sand filters shall not exceed a filtration rate of 600 lpm/sqm, though some filters are designed for 800 lpm/sqm.

Pressure sand filter

Traditionally, pressure sand filters as shown in Figure 12.8 work under pressure, use multi-grade silica sand as a filter media for filtering water for trapping suspended and coagulated matters up to 30–50 microns, more rapidly than conventional gravity sand filters. Multi-grade silica media is made by mixing coarse and fine silica together in a fixed proportion. The advantage of these types of filters is that they can be repeatedly used after backwashing.

Generally, a pressure of around 3.5 bar (kg/sqcm) is applied for filtering in a pressure filter, though the maximum operating pressure can be escalated around 10 bar. The water filtration velocity is around 30–40 mph.

Pressure filters of high filtration rate are available; those operate between 25 and 50 cum/sqm/h, though pressure filters of higher filtering rate found to be as high as 75 cum/sqm/h [13].

When pressure increases by 0.5 kg/sqcm from the initial operating filter pressure, it is recommended to backwash the filter bed. The normal life of these media is 5–7 years [14]. The pressure filter shall be installed with adequate clearance and facilities for ready and safe inspection, maintenance, disassembly, and repair.

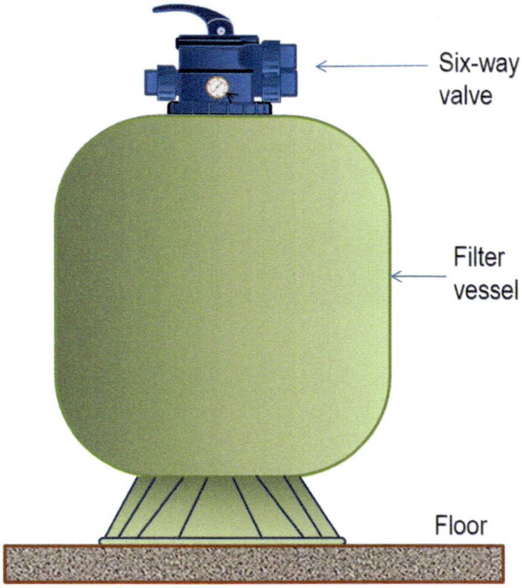

Figure 12.8 A pressure sand filter for swimming pools.

Dual media filters

To enhance filtration efficiency, dual media filters, a combination of different filter media, are employed. These filters are also termed as multimedia filters. Typically, a layer of anthracite coal is placed on top of the sand layer with a view to achieving better filtration and reducing pressure drop as well. These filters are commonly used in applications when the finest particulates down to about 5 microns are desired to be filtered.

Diatom filter

Pool water is filtered using extremely fine, porous silica made from fossilized single-celled algae skeletons, known as diatoms or diatomaceous earth (DE). A thin layer of diatoms is spread onto a nylon or polyester membrane, which is stretched over a supporting filtering bed called a septum. So, DE filtration is referred to as precoat-filtration. The flow rate may be 0.35–1.4 lps/sqm. Working pressure is about 0.2 mPa to a max of 3.5 bar.

Diatomaceous earth pool filters can filter particles of 1 micron. Backwashing is suggested when pressure increases by 0.5 bar.

There are two types of DE filters that exist: (1) pressure filters, in which a pump pushes water at high pressure through the inlet side and the filter media and water are enclosed within pressure vessels; and (2) vacuum filters, which have a pump on the outlet side that sucks water through the outlet side while water and filter media are open to the atmosphere, as the container vessel is open at the top.

Diatom possesses the best filtration ability, allows longer intervals for backwashing, and offers longer life. On the opposite, it is costlier in maintenance and operating; most importantly, DE is carcinogenic, and handlers should refrain from inhaling it.

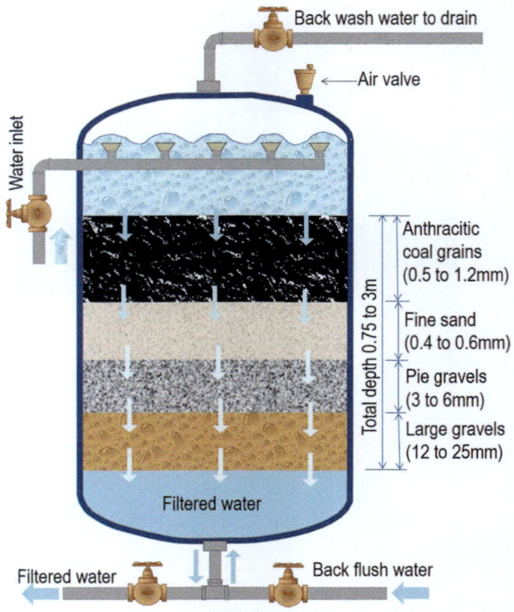

Figure 12.9 A multi-grade media gravity water filter.

Multi-grade media filter

This type of filter consists of multiple layers of filter media, including a fine layer of anthracite coal or activated carbon, a graded fine sand layer, and pebble layers as shown in Figure 12.9. The multi-grade media filter retains primarily the suspended particles and secondarily adsorbs free chlorine, organic compounds, color, and odor onto the surfaces of activated carbon.

Multi-grade media filters work on comparatively higher flow rates, not only filtering suspended solids up to 10–20 microns but also reducing turbidity.

Various materials of construction are available for pressure sand filters, as listed below:

1 Fiber reinforced plastics (FRP)
2 Mild or carbon steel (MS)
3 Stainless steel (SS)

Six-way valve

A six-way valve, also known as a multiport valve, as shown in Figure 12.10, is mostly used on swimming pool pressure sand filters to control and divert the flow of water through the filter to run filtration and cleaning processes as well. The typical positions on a six-way valve are as follows:

1 Filter: Incoming water through the inlet pipe flows through the filter bed and gets out through the outlet pipe connected to the pool, filtering out suspended matters.
2 Backwash: Reverses the flow of water through the filter, cleaning the sand and removing trapped matters by sending it through the waste outlet pipe to drain.

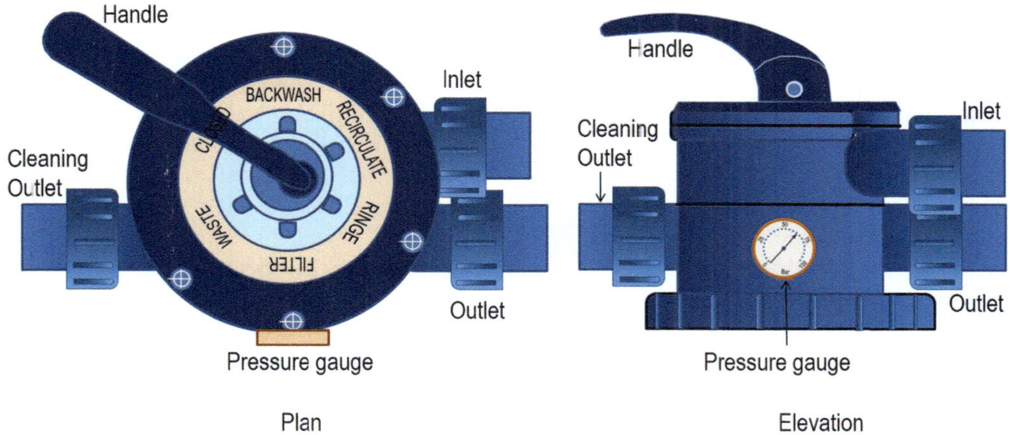

Figure 12.10 A six-way valve.

3 Rinse: After backwashing, this setting flushes out the filter's internal pipes and helps resettle the sand before returning to the filter mode.
4 Waste: Directs the water directly out of the pool without going through the filter, useful for lowering the pool water level or vacuuming out large debris.
5 Recirculate: Bypasses the filter, circulating the water back to the pool without filtering, often used when the filter is malfunctioning or during chemical treatments.
6 Closed: Stops all water flow through the valve. Generally used when performing maintenance works.

Sizing a pressure filter

The size of the swimming pool pressure filters is directly related to the flow of the pump to be maintained and the filtration rate of the filtering media to be used. The filtration rate is related to the flow of water in liters per minute filtered through each square meter of filter media area, as formulated below. The filtration rate is generally found in the range of 80–400 lpm/sqm.

From the above relation, the filter surface area, i.e., the size of the filter unit with respect to its diameter, is determined. For quick reference in determining the size of a filter, Table 12.6 can be used.

The size of the filter is directly related to the size of the pool pump. If a filter is too small for the pump, additional unwanted strain due to improper size or installation of piping is added to the pump, which shortens its lifespan.

Table 12.6 Pressure sand filter dimension [15]

Type of filter	*Filtration surface (sqm)*	*Flow rate (cum/h)*	*diameter (m)*	*height (m)*
Pressure sand filter	0.19–7.00	2.00–170.00	0.50–3.00	2.00–3.70
Continuous sand filter	0.28–7.05	4.00–97.00	0.60–3.00	2.38–5.55

Backwashing

To start backwashing, first the pool pump is turned off. The multi (six) port valve is set to the "Backwash" position, and the handle is locked. Then the pump is turned on and the sand filter is allowed to backwash for 2–10 minutes or until the water in the sight glass, typically located on the top of the filter, is found clear. When clear water is observed, the pool pump is turned off. Then the valve handle is set to the "Rinse" position and locked in place. Then the pump is turned off and the rinse process is run for 1–5 minutes or again until the water in the sight glass is visibly clearer.

When it is found that the water is very clear after rinsing, the pool pump is turned off, and then the valve handle is returned to the "Filter" position and locked into place. Finally, the pump is turned on and continued to run under normal operating conditions.

Backwash rate is occasionally determined as the velocity of the water flow during backwashing, expressed as meters per hour. Generally, it is maintained 12–15 mph, i.e. 200–250 lpm per sqm, in sand filters and 25–30 mph, i.e. 415–500 lpm per sqm, in multimedia layered filters. To determine the volume of water required for backwashing, both the desired backwash flow rate (lpm) and the duration of backwash (minute) must be known. The desired backwash pumping rate (lpm) for a filter depends on the desired backwash rate (lpm/sqm) and area of the filter (sqm), which is the recommended backwash velocity multiplied by the surface area of the filter bed.

Heating pool water

In natural conditions, the normal room temperature in a building is no different from the water temperature in a pool inside the building; thus, outside atmospheric temperature governs the temperature of the pool water built outside or on the roof of the building.

Without heating, the pool water temperature typically cannot be maintained above 22 °C, or at most 25 °C during very hot periods. Sports swimmers like to swim in pool water that is at least 26°C, while pools are even heated to 30°C.

Swimming pool water can be heated using several different types of water heaters, depending on various factors like the pool size, location, budget, etc. Heating swimming pool water consumes huge energy that adds up to high costs. It is found that for each degree rise in pool water temperature, it causes an approximate 15–18 percent escalation in operation cost [16]. So, choosing the right size and type of pool water heating system is one of the major important jobs in pool plumbing planning and designing.

Types of pool water heaters

According to the Department of Energy, there are three energy-efficient ways of heating swimming pool water. These are as follows:

1 Gas pool heaters
2 Solar pool heaters
3 Heat-pump pool heaters

In addition, there are electric pool heaters. Its high running cost makes them uncommon for large pools. So, this type may be used for small pools or spas.

Gas pool heater

Gas-fired pool heaters are generally found to be the most popular, being an economic system for heating the pool water. Fuel gas is used to heat water in a vessel. So a fuel gas connection system is to be developed.

Natural gas and propane heaters are supposed to be the most powerful, followed by electric heaters and heat pumps, which are less responsive, less powerful, and, to operate, they usually require temperatures in excess of about 10°C.

Solar pool heater

Solar pool heaters use heat energy from solar panels. The solar heater's panels absorb heat and energy from the sun. Pool water is circulated through the heater to get heated, which is then conveyed to the pool.

Solar pool heaters are environmentally friendly, have a lower cost to run, and have a long life of approximately 15–20 years [17]. But a large space is needed for installing solar panels, which may be about the surface area of the pool.

Heat pump pool heater

A heat pump is a reliable and efficient option compared to other types of pool heating systems like gas-powered and electric-resistance heaters. The coefficient of performance of heat pumps generally measures between 5 and 6. A coefficient of performance is an efficiency rating that measures the ratio of energy output to energy input. Due to this high coefficient of performance, heat pumps have very low operating costs. It is found that the operation cost per month of a heat pump is about six times less than that of a gas heater and electric resistance heater. Furthermore, if properly installed and maintained, a heat pump's lifespan can exceed twice that of a gas heater that typically lasts around 5 years, but improper water flow can severely damage internal components, and a much longer time would be required than a gas heater to heat water.

A heat pump draws in the outside air to heat water, so once the air temperature reaches below 10°C, they become very inefficient [18].

Heat pump (electric) pool heaters are more expensive than the traditional gas-powered heaters, but they consume less energy than the propane gas-powered type. In the long run, electric pool heaters will cost a lot less than traditional gas heaters. The positive aspect is that an electric heater does not emit carbon dioxide like the gas heater, so it helps reduce the amount of carbon in the air.

Sizing a pool heater

There are two general methods used in sizing a pool heater:

1 Surface area and ambient air temperature
2 Pool capacity and time rise

Surface area and ambient air temperature

Though there are various factors to be considered for sizing a pool heater, basically, a heater for the purpose is sized based on the surface area of the pool and the difference between the pool water temperature desired and the average air temperatures. Other factors that affect the heating load, mostly for outdoor pools, are wind speed, humidity levels, and cool night temperatures. Therefore, pools subject to higher wind speeds, lower humidity, and cool nights require a larger heater. To get the correct size of a pool heater, do the following:

1 Determine the mean average atmospheric temperature for the coldest month the pool will be used.
2 Determine the desired pool water temperature during use.
3 Find the temperature rise of water needed which is the temperature difference between the above-mentioned two temperatures.
4 Determine the pool surface area.

This sizing is based on heat loss from the surface with an assumed wind velocity of 5.6 km/h. This is normally suitable for sheltered pools surrounded by fences, buildings, or shrubs protecting pool water from strong winds. To maintain the desired temperature in more exposed areas, the heater capacity requirement is increased by wind factors 1 for 5.6 kmh, 1.25 for an 8 km/h wind velocity, and by 2.0 for a 16 km/h velocity.

The heater capacity can be found from the formula below:

$$\text{Heater capacity}\left(\text{kW}\right) = 0.06 \times A_s \times \Delta T \times W_f \tag{12.3}$$

where,

A_s = Pool surface area (sqm)
ΔT = Temperature differential between desired water temperature and ambient air temperature (°C)
W_f = Wind factor.

Pool capacity and time rise

Swimming pool capacity refers to the total volume of water a pool can hold, which is usually measured in cubic meter (cum). Simply determine the pool capacity by measuring its surface area in square meter, height in meters, and multiply them. The heat-up load is the heat load required to increase the pool temperature. The sizing on this method is based on raising the pool temperature approximately 10°C per hour. If it is required to raise the temperature faster, the heater capacity requirement is to be increased accordingly.

The heat-up load can be calculated using the following formula:

$$\text{h}_{\text{heat-up}} = \rho \text{C}_\text{p} \text{V}_\text{s} \frac{\text{dT}}{\text{dt}} \tag{12.4}$$

where,

Heat-up = Heat flow rate required (kW, kJ/s)
ρ = Density of water (1000 kg/cum)
C_p = Specific heat of water (4.2 kJ/kg°C)

V_s = Volume of water in swimming pool (cum)
dT = Temperature to be raised = Difference between initial and final temperature (°C)
dt = heat-up time (sec).

Example: Determining heat-up load for a swimming pool with dimensions of 10 m by 3 m and a depth of 1.25 m, heated from 10°C to 20°C in 10 hours.

Volume of water to be heated = 10 × 3 × 1.25 = 37.5 cum
Temperature to be raised = 20°C−10°C = 10°C
Heat-up time = 10 hr = 36000 sec

So, Heat-up = 1000 kg/cum × 4.2 kJ/kg°C × 37.5 cum × 10°C/36000 sec = 43.75 kW.

Disinfecting pool water

Swimming pool water is greatly contaminated by the microorganisms mostly transmitted by the swimmers. Every swimmer adds around 10^6–10^9 microorganisms to the pool water [19]. The water itself may contain microorganisms coming from the source of water. Among the microorganisms, bacteria like *Salmonella* and *E. coli*, viruses like Norovirus, parasites like *Giardia*, *Cryptosporidium*, and *Shigella*, etc., and insects like head lice are the major concerns for health, which need to be disinfected for ensuring good health of the users.

Among the number of disinfection techniques available for disinfecting swimming pool water, chlorination techniques greatly enhance the pool water quality and destruction of the majority type of organisms and oxidation for swimming pool water conditioning, though it cannot kill head lice. Chlorine is usually added to the pool water as hypochlorous acid (HOCl) or hypochlorite (OCl^-).

Application of chlorine is made limited. If too much chlorine is applied, then excess chlorine can cause eye and mucous membrane irritation of the bathers due to the formation of chloramine. Therefore, the available chlorine concentration in swimming pools is set to 0.5 mg/l (ppm), while the maximum level is set to 1.5 mg/l.

The proper pH level in the pool should be 7–7.4. When the water has a higher pH, water is alkaline, or with a lower pH, then it is acidic. In both conditions, chlorine does not work properly; it might not work at all.

Calcium hypochlorite is chlorine in granule, pellet, or tablet form. It is typically 66 percent available chlorine with a pH of approximately 10. When added to the pool, the pH levels will rise, and it is necessary for it to be corrected by the addition of an acid.

Cyanuric acid is sometimes used as a stabilizer for chlorine-based water disinfectants. When using cyanuric acid (stabilizer), minimum and maximum values are set to 2.0 and 5.0 mg/l, respectively. For outdoor swimming pools and indoor pools smaller than 20 sqm, the maximum level is set to 5.0 mg/l.

Inline chlorination

An inline automatic chlorine dispenser chemical feeder, shown in Figure 12.11, is used on a pipe conveying water from the filter or heater, if used, to the pool, as shown in Figures 12.14 and 12.15. The dosing rate is to be determined on the basis of the flow rate maintained. The dosing rate is calculated using the following formula:

$$\text{Chlorine dose rate}(\text{ml/h}) = \frac{\text{Chlorine dosage}(\text{mg/l}) \times \text{Flow rate}(\text{cum/h})}{\text{Chlorine fluid concentration}(\%)/100} \quad (12.5)$$

Example: Required chlorine dose 3 mg/l, flow rate in the supply pipe 100 cum/h, chlorine liquid concentration is 2 percent; then the chlorine rate to be ensured is as below:

$$\frac{3 \times 100}{2(\%)/100} = 15000\,\text{ml/h} = 250\,\text{ml/min}$$

Water supply and circulation

A reliable and well-designed water supply system is essential for the proper operation and use of a swimming pool. It ensures that the pool is filled with the desired quality of water and maintains the required water level during use. The supply system must meet the pool's demand for makeup water, which is needed due to evaporation loss, as well as for periodic cleaning or backwashing. The system should also facilitate efficient and smooth flow in integration with filtration, heating, and disinfecting units incorporated in the system. Backup or emergency water provisions may be necessary in large or public pools.

Water supply

After construction of a swimming pool or draining it fully, it needs to be refilled with potable water. For the first time filling and refilling a swimming pool, a water supply system is incorporated with the pool.

For the first time filling, a hose for conveying water from a temporary source of water for the pool will be required. A permanent inlet pipe in the basin can be installed for this purpose to draw water from a permanent source of water.

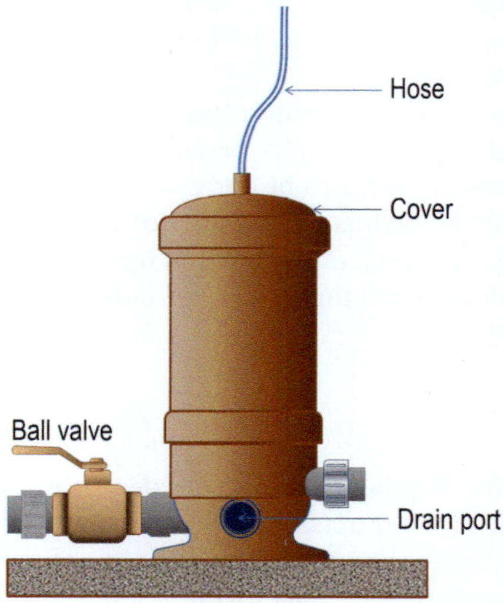

Figure 12.11 Automatic chlorine dispenser (chemical feeder) for a swimming pool.

In addition, pools shall be equipped with a supply pipe for adding make-up water as and when needed.

A dedicated fill or make-up water line can be connected from the potable water supply to the pool system. This pipe is usually connected to the balance tank in overflow-type pools, or directly to the pool basin in skimmer-type pools.

A float valve or automatic level controller to maintain constant pool water level. Backflow prevention techniques like installing check valve or vacuum breaker and or providing air gap are adopted to prevent pool water contaminating the potable supply. The invert of make-up or fill line should be above the overflow level with a true air gap, which is at least two times the fill-pipe diameter.

Water recirculation

Pool water is recycled, drawing it from the pool and then conditioning it properly, finally returned to the pool through the water inlets, efficiently by positioning those in some strategically ideal locations. A piping system is therefore needed to be well planned for the circulation of water. For comfortable use and functioning of a swimming pool, the following maximum flow rates are recommended:

a Suction over outlet grilles: 0.5 mps
b Suction pipework: 1.2 mps
c Return water pipework: 2 mps

Water inlets

Through water inlets, recirculating water is introduced in the pool in the form of spraying. These inlets shall be located in such strategic locations so as to cause uniform circulation of water throughout the pool without the existence of dead spots. To achieve the objectives, the guidelines below are to be followed:

1 The inlets shall discharge at a minimum depth of 250 mm below the pool overflow level.
2 According to UP Code, one return inlet shall be provided for every 27.9 sqm of pool surface area, or fraction thereof [20].
3 It is important that the inlets are placed directly opposite the skimmers. Otherwise, there is a risk that some of the water will not be filtered.

Multiport valves

A swimming pool plumbing system is commonly comprised of pumps, filters, and various other appliances for pool water operation, control, maintenance, and sanitation. To facilitate operation, control, and economize the system, all of these services are interconnected by a multiport valve, also known as a diverter valve.

A multiport valve consists of a valve body generally machined from a solid block material with a multiple number of tube ends. Multiport valves offer multidirectional control along with shut-off capacity. A six-way multiport valve has already been discussed; the others used in swimming pool are as given below and shown in Figure 12.12:

1 Two-way valve
2 Three-way valve

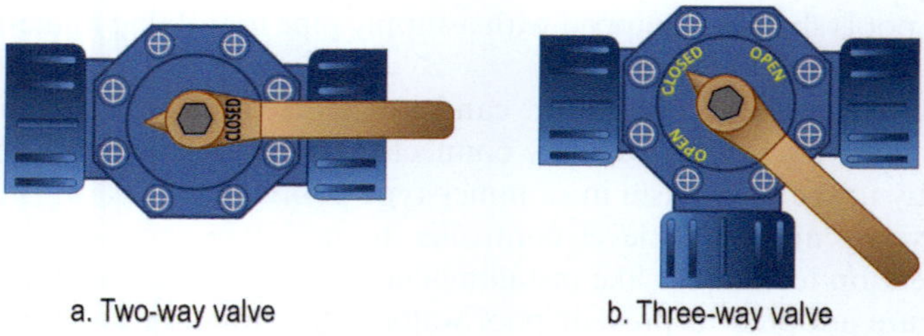

Figure 12.12 A two-way and a three-way valve.

Two-way valve

A two-way valve, also known as the diverter valve, is installed onto a piping system in the swimming pool on the suction pipe of the pump to divert or control the flow of water through one pipe to suck water from the pool and to another to suck water through the pipe from the skimmers.

Three-way valve

In addition to commonly used gate valves and check valves, one special type of valve termed a three-way valve is used. In a swimming pool, a three-way valve shut off the water supply to a skimmer and a pool drain while keeping the flow to the pool pump open, a control system needed when filling or topping up the water in the pool.

A three-way valve has three ports, one at the center and two at the opposite ends. These have been designed so that it's possible to close off the water supply from either end while keeping the center port open. The center port cannot be shut off anyway. The handle rotates 180° on the side opposite the center port. Flow to the pump cannot be restricted, so the flow to the pool pump remains open.

Swimming pool drainage

Swimming pool drainage is basically designed to divert excess or unwanted water to an approved location of the disposal system, with a view to preventing the pooling of unhygienic water into the pool basin.

Generally, highly contaminated or polluted water in in-ground large pools' needs to be drained out completely and refilled. Generally, these pools need to be drained and refilled every 5 years on average or when the total dissolved solids (TDS) reach over 2500 ppm [21], or if there is a major necessary repair, cleaning, etc., and when filtering is unwise due to greening of pool water for algae growth. Sometimes partial draining and then refilling is also done to avoid costs and troubles encountered in complete draining and refilling.

Drain suction inlet

Swimming pools are provided with a drain system, the inlet of which is located in the lowest level of the floor of the pool. The drain inlets are designed to be either round or square-shaped. The major function of the drain of a swimming pool is to pull water from the pool basin, convey it mostly along to the pump being sucked and pushed through a filtering and

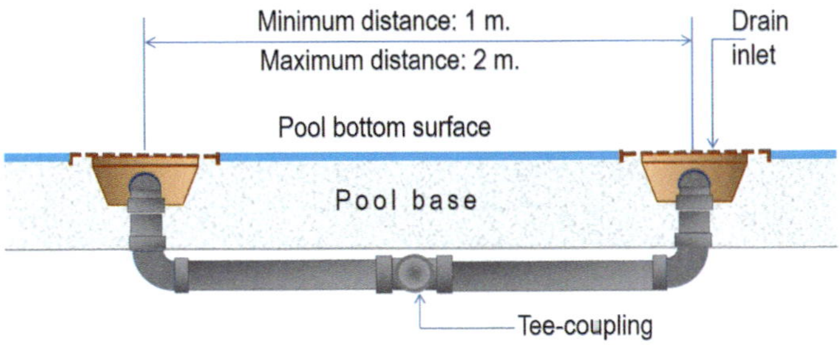

Figure 12.13 Drain inlets.

other conditioning system, or drain it out into the disposal system. This drain may be optional for above-ground pools, but it makes cleaning the pool easier.

As a safety measure to reduce the risk of entrapment accidents of the swimmers, there must be at least two inlets for the draining; generally, for a 10 m width one inlet is provided [22]. The distance between the draining inlets and from a wall shall be 1 m minimum and maximum 2 m, and the same criteria of distance is followed between the two drainage inlets, as mentioned in Figure 12.13. In addition, the inlets should have grates or an anti-vortex cover to prevent suction entrapment. The anti-vortex cover diverts the flow of water to prevent a vortex from forming that causes a risk of suction, holding bathers against the drain inlets. One outlet shall be provided for each 74 sqm of pool surface area [22].

Pools water shall be drained properly and the discharge of the pool water to the sewer shall be at a rate not exceeding 45.5 cum/hr [22].

Swimming pool plumbing

Swimming pool plumbing is its lifeline. The functionality of a swimming pool is fully dependent on the development and maintenance of the plumbing system that addresses the management of water quality and quantity in the swimming pool. So, the plumbing system shall be well planned and developed following some principles as mentioned below. The schematic diagrams of plumbing system for both the swimming pools incorporating balancing tank and skimmers have been shown in Figures 12.14 and 12.15, respectively.

Swimming pool plumbing principles

Hidden plumbing work is to be completed prior to concrete finishing. Before and during casting of concrete all openings of pipes are to be kept covered to ensure that concrete does not enter into the pipe to block the pipes. The basic plan is to push water in one general direction.

Recycling and filter water returning to the pool through the water inlets is to be performed efficiently through placing them in some strategically effective locations. Pool water is sucked out and recycled through the filter. The more inlets or outlets will be provided, the less pressure from each individual outlet will exist. The number is to be determined based on the size and shape of the pool and the size of the pool pump. Before covering the pipes, it is important to have a pressure test to ensure that there are no leaks in the system where water can be lost.

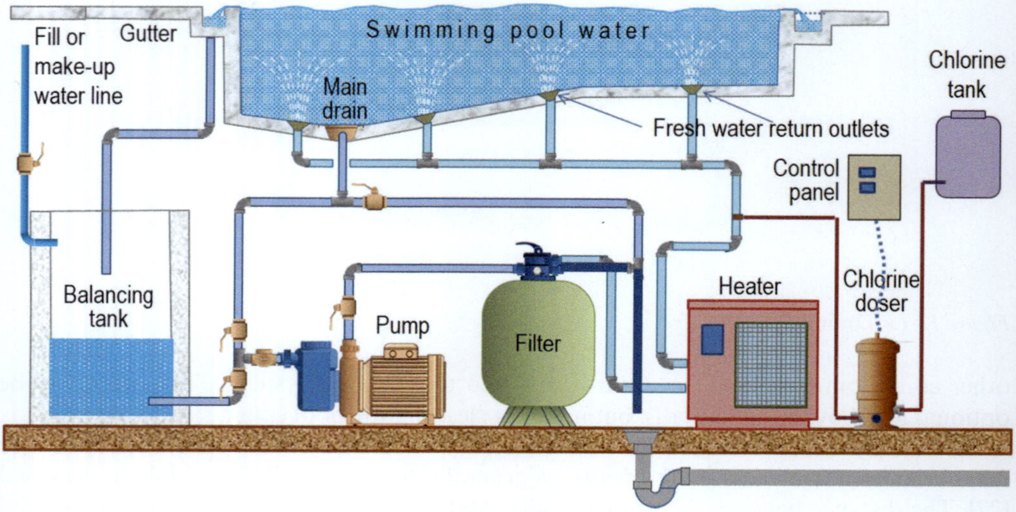

Figure 12.14 Schematic diagram of plumbing of a swimming pool incorporating balancing tank.

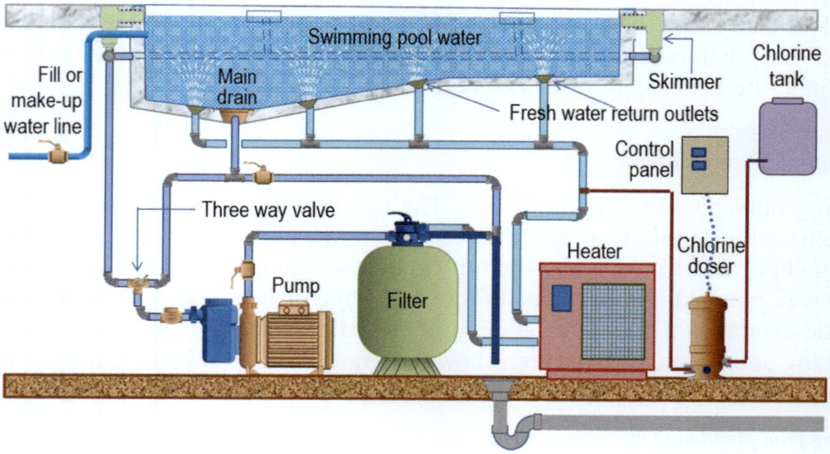

Figure 12.15 Schematic diagram of plumbing of a swimming pool incorporating skimmer.

Piping of swimming pool

In old days, piping used in swimming pool construction mostly had copper pipes. Nowadays plastic is a very popular choice in most cases, and so in the case of piping for swimming pools. As there are various types of plastic pipes, among those, unplasticized Polyvinyl Chloride (uPVC) pipes can be used for water supply, and chlorinated Polyvinyl Chloride (CPVC) pipes would be the best choice when warm water is to be supplied to the pool. For draining pool water, acrylonitrile butadiene styrene (ABS) plastic pipes are recommended.

Swimming pool maintenance

To ensure a long lifespan, regular maintenance of a swimming pool is essential. Maintaining a swimming pool includes regular checking, cleaning and water testing, chemical treatments, and machinery servicing.

1 Cleaning:

   i Cleaning the area around the pool terrace
   ii Cleaning the filter beds by backwashing
   iii Cleaning out the skimmer baskets

2 Vacuuming the floor of the pool
3 Brushing the wall of the pool at least once a week, particularly the dead spots:

   i Behind the ladders
   ii Around any steps into the pool
   iii Underneath the skimmers
   iv Corners of the pool

4 Checking:

   i Checking the water level and topping it up by filling water when needed
   ii Checking the pH level and adjusting the water pH level to be maintained
   iii Checking pool chemicals and ordering any necessary replacements
   iv Checking safety valves of heaters on a regular basis

5 Servicing equipment and appliances
6 Running the pump, a maximum of 8 hours a day, checking for any leakage.

References

[1] Pool Troopers (2019) "Swimming pool: draining and refilling" Retrieved from https://pooltroopers.com/blog/swimming-pool-draining-and-refilling/#:~:text=Most%20pools%20need%20to%20be,to%20create%20a%20healthy%20environment on 03 August 2024.
[2] Designing Buildings Ltd. (2025) "Swimming pool construction" Retrieved from https://www.designingbuildings.co.uk/wiki/Swimming_pool_construction on 16 February 2025.
[3] UP Code (2021) "Skimmer sizing", General Compliance, Chapter 3, Section 315.3.
[4] www.yourpoolplans.com "Public pool recirculation systems standards" Retrieved from https://yourpoolplans.com/public-pool-recirculation-systems-standards/ on 15 August 2024.
[5] www.buildingcalculators.com "Building calculators" Retrieved from https://www.buildingcalculators.com/swimming-pool-calculator#google_vignette on 15 August 2024.
[6] The Pool Butler (2023) "Single speed, two speed, and variable speed pool pumps: which is best?" Retrieved from https://www.thepoolbutler.net/resource/single-speed-two-speed-and-variable-speed-pool-pumps-which-is-best/ on 29 July 2025.
[7] Government of Nova Scotia (UD) "Recirculation and filtration" Section 7, Retrieved from https://novascotia.ca/nse/environmental-health/docs/Aquatic-Guidelines-Section-7.pdf on 23 March 2025.
[8] S.A. Fluidra (2021) "Water turnover rate: a critical factor in pool maintenance" Retrieved from https://www.fluidra.com/projects/turnover-rate/ on 13 March 2025.
[9] In the Swim (2025) "Size and selection of swimming pool pumps" Retrieved from https://intheswim.com/blog/whats-the-best-pool-pump.html on 23 March 2025.
[10] In The Swim (2025) "Swimming pool feet of head calculations" Retrieved from https://intheswim.com/blog/swimming-pool-feet-of-head-calculations.html on 23 March 2025.

[11] S. A. Fluidra (2021) "Water turnover rate: a critical factor in pool maintenance" Retrieved from https://www.fluidra.com/projects/turnover-rate/ on 13 March 2025.
[12] www.water.mecc.edu (UD) "Filtration". Lesson 7, Retrieved from https://water.mecc.edu/courses/ENV115/Lesson7_print.htm on 16 February 2025.
[13] Marcio Mesquitaa, Fábio Ponciano de Deusb; Roberto Testezlafc, Adriano Valentim Diottob (2019) "Removal efficiency of pressurized sand filters during the filtration process" Desalination and Water Treatment, www.deswater.com doi:10.5004/dwt.2019.24285, Retrieved from https://www.deswater.com/DWT_articles/vol_161_papers/161_2019_132.pdf on 16 February 2025.
[14] Swimming Pool Water Treatment "Pressure sand filter" Retrieved from https://www.swimmingpoolwatertreatment.com/pressure_sand_filter.htm#:~:text=Pressure%20sand%20filters%20use%20multi,is%2030%2035%20m%2Fhr.]uju on 15 August 2024.
[15] www.envites.cz "Sand filtration" Retrieved from https://envites.cz/en/sand-filtration on 15 August 2024.
[16] AquaCal (2021) "Top 6 cost-efficient ways to heat your pool" Retrieved from https://www.aquacal.com/top-6-cost-efficient-ways-heat-pool/ on 9 March 2025.
[17] River Pools (2025) "Swimming pool heaters: types, costs, pros & cons" by Jason Hughes Retrieved from https://www.riverpoolsandspas.com/blog/swimming-pool-heater-cost-types on 9 March 2025.
[18] Patrick Paroline (2015) "How much does my pool heater cost to run? A complete guide" Retrieved from https://diy.inyopools.com/article/how-much-does-it-cost-to-run-a-pool-heater/ on 30 July 2025.
[19] B. V. Lenntech "Conditions for water disinfection" Retrieved from https://www.lenntech.com/processes/disinfection/swimming-pool-/swimming-pool-disinfection.htm on 12 March 2025.
[20] UP Code (2021) "General compliance" Chapter 3 Section 313, Subsection 313.4, Retrieved from https://up.codes/s/return-inlets on 30 July 2025.
[21] The Pool Butler "When to drain a swimming pool" Retrieved from https://www.thepoolbutler.net/resource/when-to-drain-a-swimming-pool/ on 13 March 2025.
[22] Hawthorn Woods (UD) "Building regulations" Chapter 3 Swimming Pools, Code of Ordinances. Retrieved from https://codelibrary.amlegal.com/codes/hawthornwoodsil/latest/hawthornwoods_il/0-0-0-5273 on 13 March 2025.

Appendices

Appendix A1

Table A1.1 Technical specifications of domestic type air-cooled water chiller

Nominal cooling capacity		*ton/h*	*1.5*	*2*	*3*	*4*	*5*
Power supply		V/Hz/Ph	220–240/50/1			380–415/50/3	
Cooling (1):	Cooling capacity	W/h	6100	8350	12200	16000	19600
A35/24°C	Power consumption	W	1540	2110	3110	4160	5100
W45/25°C	EER	–	3.96	3.96	3.92	3.85	3.84
Cooling (2):	Cooling capacity	W/h	5200	7200	10600	14000	17500
A46/24°C	Power consumption	W	1770	2440	3670	4950	6040
W45/25°C	EER	–	2.94	2.95	2.89	2.83	2.90
Ambient temp. range		°C			20–53		
Outlet water temp. range		°C			15–35		
Compressor type			Rotary			Scroll	
Water connection (inlet/outlet)		mm	25	25	25	25	25
Rated water flow rate		Cum/h	0.9	1.2	1.8	2.4	3
Noise level		dB(A)	56	57	63	63	63
Net dimension	W	mm	660	660	790	790	790
	D		620	620	750	750	750
	H		655	655	650	850	850

Source: https://www.blueway-heatpump.com/air-cooled-water-chiller.html

Notes:

1. Conditions of "Cooling (1)": Ambient air temperature DB/WB: 35/24°C, Inlet/Outlet water temperature: W45/25°C;
2. Conditions of "Cooling (2)": Ambient air temperature DB/WB: 46/24°C, Inlet/Outlet water temperature: W45/25°C.

Appendix A2

Table A2.1 Pipe or tube materials and dimensions

Materials and dimensions	*Standard*
1. **Ferrous Piping (Welded and Seamless)**	
Standard Specification for Black and Hot-Dipped Zinc-Coated (Galvanized) Welded and Seamless Steel Pipe for Fire Protection Use	ASTM A 795/A 795M
Standard Specification for Pipe, Steel Black and Hot-Dipped, Zinc-Coated	ASTM A53/A53M
2. Materials and dimensions	
Welded and Seamless Wrought Steel Pipe	ASM EB36.10M
Standard Specification for Electric-Resistant-Welded Steel Pipe	ASTM A135/A 135M
3. **Copper Tube (Drawn, Seamless)**	
Standard Specification for Seamless Copper Tube	ASTM B75/B75.M
Standard Specification for Seamless Copper Water Tube	ASTM B88
Standard Specification for general requirement for Wrought Seamless Copper and Copper-alloy Tube	ASTM B251
Standard Specification for Liquid and Paste Fluxes for Soldering of Copper and Copper-alloy Tube	ASTM B813
Specification for Filler Metals for Brazing and Braze Welding	AWS A5.8M/A5.8
Standard Specification for Solder Metal, Section 1: Solder Alloys Containing Less Than 0.2% Lead and Solidus Temperature less than 400°F	ASTM B32
Alloy Material	ASTM B446
4. **CPVC Pipe**	
Standard Specification for Chlorinated Poly Vinyl Chloride (CPVC) Plastic Pipe (SDR-PR)	ASTM F442/F442M

Source: NFPA 13. Available at https://blog.qrfs.com/223-fire-sprinkler-system-pipe-material-nfpa-requirements-and-the-pros-and-cons-of-steel/

Appendix A3

Table A3.1 Friction loss in meter-head, per 30-meter fire hose

Sl. no	*Flow (lpm)*	*Hose size (mm)*				
		38	*45*	*63*	*75*	*100*
1	454.60	16.88	10.90	1.41		
2	568.26	26.37	17.02	2.18	0.84	
3	909.22		43.60	5.63	2.25	
4	1136.52		68.07	8.79	3.52	
5	1363.83			12.66	5.06	1.27
6	1591.13			17.23	6.89	1.69
7	1818.44			22.50	9.00	2.25
8	2045.74				11.39	2.88
9	2273.04				14.06	3.52
10	2727.65				20.25	5.06
11	3182.26				27.56	6.89
12	3636.87				36.01	9.00
13	4091.48				45.57	11.39
14	4546.09					14.06
15	5000.70					17.02
16	5455.31					20.25
17	5909.92					23.77
18	6364.53					27.57
19	6819.13					31.65

1 psi = 0.703 meter head.

Table A3.2 Solid stream flow through nozzles

Sl. no	*Tip size (mm)*	*Flow (lpm) at pressure*	
		345 kPa	*552 kPa*
1	13	197	
2	16	310	
3	19	447	
4	22	610	
5	24	697	
6	25	795	
7	29	1007	
8	32	1242	1571
9	35		1900.
10	39		2260
11	41		2654
12	44		3081
13	48		3532
14	50		4024

Source: https://www.uvu.edu/ufra/docs/certification/standards/ado_standard_1010_nov2025.pdf

Appendix A4

Table A4.1 Water supply requirements for sprinkler systems

Sl. no	*Occupancy classification*	*Sprinkler system*		*Hose stream allowance (lpm)*	*Duration of supply (minutes)*
		Design density (lpm/sqm)	*Design area (sqm)*		
1	Light hazard	4.1	280	950	60
2	Ordinary hazard group 1	6.1	280	1900	60
3	Ordinary hazard group 2	8.2	280	1900	90
4	Extra hazard group 1	12.2	280	2840	120
5	Extra hazard group 2	16.3	280	2840	120

Source: NFPA, 1999. Available at: https://files.thunderheadeng.com/femtc/2014_d2-14-franco-paper.pdf

Table A4.2 Pressure losses in water meters (bar)

Meter size (mm)	*Flow (lpm)*					
	68 or less	*87*	*98*	*117*	*148*	*197*
15 mm	0.67 bar	0.97 bar	1.2 bar	1.8 bar	2.6 bar	*
20 mm	0.48 bar	0. 76 bar	1.5 bar	1.5 bar	2.4 bar	*
25 mm	0.14 bar	0.21 bar	0.21 bar	0.28 bar	0.41 bar	0.69 bar
40 mm	0.07 bar	0.07 bar	0.14 bar	0.14 bar	0.28 bar	0.48 bar
50 mm	0.07 bar	0.07 bar	0.07 bar	0.07 bar	0.14 bar	0.21 bar

* Above maximum rated flow of commonly available meters.

Source: NFPA. Available at: https://www.awwa.org/Portals/0/AWWA/ETS/Resources/ResidentialFireSprinklerSystems.pdf

Appendix A5

Table A5.1 Fuel and gas properties

Property	*Units*	*Natural Gas*	*Propane*	*Butane*
Chemical formula		CH_4	C_3H_8	C_4H_{10}
Molecular weight	kg/k/mole	16	44.10	58.12
Boiling point of liquid at atmospheric pressure	°C	−161.6	−42.1	−0.5
Specific gravity of vapor (Air=1)		0.6	1.53	2.09
Specific gravity of liquid (Water = 1)		0.6	0.512	0.58
Density of vapor			1.5	1.95
Density of liquid at 20°C			0.505<	0.575
Volume occupied/mass of liquid	liter/ton		1957	1743
Calorific value at 20°C	MJ/M^3	38.7		
	MJ/liter			
	MJ/kg			
Latent heat of vaporization	kJ/kg		358	372
Wobbe index		50.13	81.12	
Specific heat–liquid			0.60	0.58
Liquid weight	kg/l			
Vapor volume from 1 liter of liquid at 20°C	M^3			
Vapor volume from 1 kg of liquid at 20°C	M^3			
Vapor pressure at 20°C	kPa g			
Vapor pressure at 38°C	kPa g			
Combustible limits	% gas in air	5–15	2.2–9.6	1.9–8.6
Amount of air to burn 1 M^3 of gas	M^3	9.78		
Ignition temperature in air	°C	612–622		
Maximum flame temperature in air	°C	1964		
Auto ignition temperature	°C		470	410
Critical temperature	°C		96.7	152.1
Critical pressure	kPa g		42.5	38.0
Octane number		100	Over 100	92

Source: https://www.airgas.com.au/fuel-gas-properties/

Appendix A6

Table A6.1 Display height, flow rate and pressure for geyser nozzles of various diameter

Display height	*Nozzle diameter*														
	55 mm			*80 mm*			*105 mm*			*125 mm*			*150 mm*		
Meters	*TDH*	*lpm*	*Bar*	*TDH*	*lpm*	*Bar*	*TDH*	*lpm*	*Bar*	*TDH*	*lpm*	*Bar*	*TDH*	*lpm*	*Bar*
0.50	6.7	30.4	0.20	9.0	82.7	0.27	8.0	128.4	0.24						
0.75	8.3	33.9	0.25	10.3	89.1	0.31	9.7	140.3	0.29						
1.00	10.0	37.2	0.30	11.7	95.0	0.35	11.3	151.3	0.34	12.3	242.0	0.37			
1.50	13.0	42.9	0.39	14.7	105.9	0.44	14.3	171.2	0.43	15.7	273.3	0.47			
2.00	16.3	47.9	0.49	17.3	115.8	0.52	17.7	189.0	0.53	19.0	301.3	0.57	18.7	459.4	0.56
2.50	19.7	52.5	0.59	20.3	124.9	0.61	20.7	205.3	0.62	22.3	327.0	0.67	22.0	497.0	0.66
3.00	23.0	56.7	0.69	23.0	133.4	0.69	24.0	220.4	0.72	25.7	350.8	0.77	25.0	532.0	0.75
4.00	29.7	64.3	0.89	28.7	149.0	0.86	30.3	247.9	0.91	32.3	394.3	0.97	31.3	596.0	0.94
5.00	36.4	71.1	1.09	34.4	163.1	1.03	36.7	272.8	1.10	39.4	433.5	1.18	37.7	653.9	1.13
6.00	42.7	77.3	1.28	40.0	176.2	1.20	43.0	295.6	1.29	46.0	469.6	1.38	44.4	707.3	1.33
7.00				46.0	188.4	1.38	52.4	316.9	1.57	53.0	503.2	1.59	50.7	757.0	1.52
8.00				51.7	199.8	1.55	55.7	336.8	1.67	59.7	534.8	1.79	57.0	803.8	1.71
10.00							68.7	373.7	2.06	73.7	593.1	2.21	70.0	890.5	2.10
12.00							81.4	407.5	2.44	87.4	646.5	2.62	83.0	969.8	2.49
14.00										101.4	696.1	3.04	96.4	1043.6	2.89
16.00										115.4	742.7	3.46	109.4	1112.8	3.28
18.00										129.4	786.7	3.88	122.7	1178.3	3.68
20.00										143.4	828.7	4.30	136.1	1240.8	4.08
25.00													169.8	1386.0	5.09
30.00													204.1	1519.1	6.12

Source: https://cdn.shopify.com/s/files/1/2394/4493/files/Oase_Geyser_Nozzles_Specifications_Sheet_PDF.pdf?v=1643092608

Appendix A7

Table A7.1 Properties of saturated steam (part-1)

Absolute pressure (kPa, kN/sqm)	*Evaporation temperature (°C)*	*Specific volume (cum/kg)*	*Density - ρ - (kg/cum)*	*Specific enthalpy of*			*Specific entropy of steam - s - (kJ/kgK)*
				Liquid - hl - *(kJ/kg)*	*Evaporation* - he - *(kJ/kg)*	*Steam* - hs - *(kJ/kg)*	
0.8	3.8	160	0.00626	15.8	2493	2509	9.058
2.0	17.5	67.0	0.0149	73.5	2460	2534	8.725
5.0	32.9	28.2	0.0354	137.8	2424	2562	8.396
10.0	45.8	14.7	0.0682	191.8	2393	2585	8.151
20.0	60.1	7.65	0.131	251.5	2358	2610	7.909
28	67.5	5.58	0.179	282.7	2340	2623	7.793
35	72.7	4.53	0.221	304.3	2327	2632	7.717
45	78.7	3.58	0.279	329.6	2312	2642	7.631
55	83.7	2.96	0.338	350.6	2299	2650	7.562
65	88.0	2.53	0.395	368.6	2288	2657	7.506
75	91.8	2.22	0.450	384.5	2279	2663	7.457
85	95.2	1.97	0.507	398.6	2270	2668	7.415
95	98.2	1.78	0.563	411.5	2262	2673	7.377
100	99.6	1.69	0.590	417.5	2258	2675	7.360
101.331	100	1.67	0.598	419.1	2257	2676	7.355
110	102.3	1.55	0.646	428.8	2251	2680	7.328
130	107.1	1.33	0.755	449.2	2238	2687	7.271
150	111.4	1.16	0.863	467.1	2226	2698	7.223
170	115.2	1.03	0.970	483.2	2216	2699	7.181
190	118.6	0.929	1.08	497.8	2206	2704	7.144
220	123.3	0.810	1.23	517.6	2193	2711	7.095
260	128.7	0.693	1.44	540.9	2177	2718	7.039
280	131.2	0.646	1.55	551.4	2170	2722	7.014
320	135.8	0.570	1.75	570.9	2157	2728	6.969
360	139.9	0.510	1.96	588.5	2144	2733	6.930
400	143.1	0.462	2.16	604.7	2133	2738	6.894
440	147.1	0.423	2.36	619.6	2122	2742	6.862
480	150.3	0.389	2.57	633.5	2112	2746	6.833
500	151.8	0.375	2.67	640.1	2107	2748	6.819
550	155.5	0.342	2.92	655.8	2096	2752	6.787
600	158.8	0.315	3.175	670.4	2085	2756	6.758
650	162.0	0.292	3.425	684.1	2075	2759	6.730
700	165.0	0.273	3.66	697.1	2065	2762	6.705
750	167.8	0.255	3.915	709.3	2056	2765	6.682
800	170.4	0.240	4.16	720.9	2047	2768	6.660
850	172.9	0.229	4.41	732.0	2038	2770	6.639
900	175.4	0.215	4.65	742.6	2030	2772	6.619
950	177.7	0.204	4.90	752.8	2021	2774	6.601
1000	179.9	0.194	5.15	762.6	2014	2776	6.583

Table A7.2 Properties of saturated steam (part-2)

Absolute pressure (kPa, kN/sqm)	*Evaporation temperature (°C)*	*Specific volume (cum/kg)*	*Density - ρ - (kg/cum)*	*Specific enthalpy of*			*Specific entropy of steam - s - (kJ/kgK)*
				Liquid - hl - (kJ/kg)	*Evaporation - he - (kJ/kg)*	*Steam - hs - (kJ/kg)*	
1050	182.0	0.186	5.39	772	2006	2778	6.566
1150	186.0	0.170	5.89	790	1991	2781	6.534
1250	189.8	0.157	6.38	807	1977	2784	6.505
1300	191.6	0.151	6.62	815	1971	2785	6.491
1500	198.3	0.132	7.59	845	1945	2790	6.441
1600	201.4	0.124	8.03	859	1933	2792	6.418
1800	207.1	0.110	9.07	885	1910	2795	6.375
2000	212.4	0.0995	10.01	909	1889	2797	6.337
2100	214.9	0.0945	10.54	920	1878	2798	6.319
2300	219.6	0.0868	11.52	942	1858	2800	6.285
2400	221.8	0.0832	12.02	952	1849	2800	6.269
2600	226.0	0.0769	13.01	972	1830	2801	6.239
2700	228.1	0.0740	13.52	981	1821	2802	6.224
2900	232.0	0.0689	14.52	1000	1803	2802	6.197
3000	233.8	0.0666	15.00	1008	1794	2802	6.184
3200	237.4	0.0624	16.02	1025	1779	2802	6.158
3400	240.9	0.0587	17.04	1042	1760	2802	6.134
3600	244.2	0.0554	18.06	1058	1744	2802	6.112
3800	247.3	0.0524	19.08	1073	1728	2801	6.090
4000	250.3	0.0497	20.09	1087	1713	2800	6.069

Source: https://www.engineeringtoolbox.com/saturated-steam-properties-d_101.html

Appendix A8

Table A8.1 Optimum pump type for specific flow rate and head applications

Flow rate (lpm)	*Pump head (TDH) (m)*	*Approximate HP: Watt range requirement*	*"Best" pump type*
≤189.27	≤30	≤ 1864.25	ESC
≤189.27	30–100	1864.25–5592.75	ESC, SP
≤189.27	≥100	≥5592.75+	SP
189.27–378.54	≤30	≤1864.25–3355.65	ESC
189.27–378.54	30–100	1864.25–9321.28	ESC, SP, VTP
189.27–378.54	≥100	5592.75–9321.28+	SP, VTP
378.54–946.35	≤30	≤3355.65–7457.00	ESC, VTP
378.54–946.35	30–100	7457.00–20133.90	ESC, SP, VTP
378.54–946.35	≥100	8948.40–22371.00+	SP, VTP
946.35–1892.70	≤30	≤7457.00–11931.20	ESC, VTP
946.35–1892.70	30–100	6338.45–37285.00	ESC, SP, VTP
946.35–1892.70	≥100	18642.50–37285.00+	SP, VTP
1892.70–3785.41	≤30	≤11185.50–23862.40	ESC, VTP, MF, AF
1892.70–3785.41	30–100	11185.50–74570.00	ESC,SP, VTP
1892.70–3785.41	≥100	35793.60–74570.00+	SP, VTP
3785.41–9463.53	≤30	≤22371.00–59656.00	ESC, VTP, MF, AF
3785.41–9463.53	30–100	22371.00–171511.00	ESC, SP, VTP, MF
3785.41–9463.53	≥100	70841.50–171511.00	VSP, VTP, MF

(*Continued*)

Table A8.1 (Continued)

Flow rate (lpm)	*Pump head (TDH) (m)*	*Approximate HP: Watt range requirement*	*"Best" pump type*
>9463.53	≤30	≥37285.00	MF, AF, ESC
>9463.53	30–100	≥37285.00–171511.00	MF, ESC, VTP, SP
>9463.53	>100	≥171511.00+	MF, VTP. SP

Source: https://waterwelljournal.com/irrigation-fundamentals-12/

Note: AF, Axial-flow pump (source); ESC, End-suction centrifugal pump (single stage) (source or booster); MF, Mixed-flow pump (source); SP, Staged (Multistage) pump (source or booster); VTP, Vertical turbine pump (source or booster).

Appendix A9

Table A9.1 Medical fixtures water demand and corresponding fixture unit

Fixtures	*Flow rate (lph)*	*Water/hot water SFU*
Sink, clinic	15	3
Sink, plaster	38	3
Sink, service (general)	76	3
Lavatory. public (general)	23	1½
Lavatory, patient (private)	8	1
Lavatory, clinic	30	3
Basin, mop service	76	3
Shower, staff or patient group	284	3
Shower, patient (private)	114	2
Shower, hydrotherapeutic	1520	11
Bathtub	76	3
Bath, whirlpool, arm	130	3
Bath, whirl pool, arm/foot/knee	57	3
Bath, whirlpool, leg	380	7
Bath, body hydrotherapy (hubbard)	2270	17
Bath, Sitz 114 (30) 3		
Processor, X-ray	57	3

Source: https://www.cedengineering.com/userfiles/An%20Intro%20to%20Med%20Plumb%20&%20Gas.pdf

Appendix A10

Table A10.1 Chlorine dose in fluid ounce, to reach desired chlorine level in water

Sl.no	*Chlorine level (ppm)*	*Water volume (cum)*				
		3.79	*18.93*	*37.85*	*75.71*	*189.27*
1	1	1	5	10	20	50
2	3	3	15	30	50	150
3	5	5	25	50	100	250
4	10	10	50	100	200	500

Source: https://www.wildwestpoolsupplies.com/blog/pool-chemical-water-chemistry-guide/

Table A10.2 Super chlorination ounce (Amount needed to introduce 10 ppm)

Sl.no	*Percentage chlorine (%)*	*Water volume (cum)*				
		3.79	*18.93*	*37.85*	*75.71*	*189.27*
1	Granular 65%	1.6	3.0	6.0	9.0	15.0
2	Di Chloro 56%	0.70	3.3	6.6	10.0	16.5
3	Sodium Hypo 12.5%	3.21 fl.oz	1.0 pt	2.0 pt	3.0 pt	5.1 pt

1 fluid ounce = 2 table spoon
1 cup = 8 fluid ounce (fl.oz)
1 quart = 2 pints (pt)
1 gallon = 8 pints

1 fluid ounce = 29.57 ml
1 gallon = 3.9 liter
= 128 fluid ounces

Source: https://www.wildwestpoolsupplies.com/blog/pool-chemical-water-chemistry-guide/

Appendix A11

Table A11.1 Conversion factors, mixed units

Length (L)

mile	*yard*	*ft.*	*in.*	*m.*	*cm.*
1	1760	5280	6.336×10^{4}	1.609×10^{3}	1.609×10^{5}
5.68×10^{-4}	1	3	36	0.9144	91.44
1.894×10^{-4}	0.333	1	12	0.3048	30.48
1.578×10^{-5}	0.028	0.083	1	0.0254	2.54
6.214×10^{-4}	1.094	3.281	39.37	1	100

Area (A)

mile²	acre	yard²	ft²	in²	m²
1	640	3.098×10^{6}	2.788×10^{7}	4.014×10^{9}	2.59×10^{6}
1.563×10^{-3}	1	4840	43,560	6.27×10^{6}	4047
3.228×10^{-7}	2.066×10^{-4}	1	9	1296	0.836
3.587×10^{-8}	2.3×10^{-5}	0.111	1	144	0.093
2.491×10^{-10}	1.59×10^{-7}	7.716×10^{-4}	6.944×10^{-3}	1	6.452×10^{-4}
3.861×10^{-7}	2.5×10^{-4}	1.196	10.764	1550	1

Volume (V)

acre-ft	U.S. Gal	ft³	in.³	L	m³	cm³
1	325,851	43,560	75.3×10^{-6}	1.23×10^{6}	1230	1.23×10^{9}
3.07×10^{-6}	1	0.134	231.6	3.875	3.875×10^{-3}	3875
2.3×10^{-5}	7.481	1	1728	28.317	0.028	28,317
1.33×10^{-8}	4.329×10^{-3}	5.787×10^{-4}	1	0.016	1.639×10^{-5}	16.39
8.1×10^{-7}	0.264	0.035	61.02	1	1×10^{-3}	1000
8.13×10^{-4}	264.2	35.31	6.10×10^{4}	1000	1	10^{6}

(*Continued*)

Table A11.1 (Continued)

Time (T)					
year	month	day	hour	minute	second
1	12	365	8760	525,600	3.1536×10^7

Velocity (L/T)				
ft/s	ft/min	m/s	m/min	cm/s
1	60	0.3048	18.29	30.48
0.017	1	5.08×10^{-3}	0.3048	0.5080
3.281	196.8	1	60	100
0.055	3.28	0.017	1	1.70
0.032	1.969	0.01	0.588	1

Discharge (L^3/T)					
mgd	gpm	ft³/s	ft³/min	L/s	m³/d
1	694.4	1.547	92.82	43.75	3.78×10^3
1.44×10^{-3}	1	2.228×10^{-3}	0.134	0.063	5.45
0.646	448.9	1	60	28.32	2447
0.011	7.481	0.017	1	0.472	40.78
0.023	15.85	0.035	2.119	1	86.41
2.64×10^{-4}	0.183	4.09×10^{-4}	0.025	0.012	1

Mass (M)					
ton	lb_m	grain	ounce (oz)	kg	g
1	2000	1.4×10^7	32,000	907.2	907,185
0.0005	1	7000	16	0.454	454
7.14×10^{-8}	1.429×10^{-4}	1	2.29×10^{-3}	6.48×10^{-5}	0.065
3.125×10^{-5}	0.0625	437.6	1	0.028	28.35
1.10×10^{-3}	2.205	1.54×10^4	35.27	1	1000
1.10×10^{-6}	2.20×10^{-3}	15.43	0.035	10^{-3}	1

Temperature (T)			
°F	°C	°K	°R
°F	5/9(°F−32)	5/9°F + 255.38	°F+459.69
9/5°C + 32	°C	°C + 273.16	9/5°C + 491.69
9/5°K−459.69	°K−273.16	°K	9/5°K
°R−459.69	5/9°R−273.16	5/9°R	°R

(*Continued*)

Table A11.1 (Continued)

Density (M/L^3)

lb/ft^3	lb/gal(U.S)	kg/m^3	kg/L	g/cm^3
1	0.1337	16.019	0.01602	0.01602
7.48	1	119.8	0.1198	0.1198
0.0624	8.345×10^{-3}	1	0.001	0.001
62.43	8.345	1000	1	1

Pressure (F/L^2)

lb/in^2	ft water	in Hg	atm	mm Hg	kg/cm^2	N/m^2
1	2.307	2.036	0.068	51.71	0.0703	6895
0.4335	1	0.8825	0.0295	22.41	0.0305	2989
0.4912	1.133	1	0.033	25.40	0.035	3386
14.70	33.93	29.92	1	760	1.033	1.013×10^5
0.019	0.045	0.039	1.30×10^{-3}	1	1.36×10^{-3}	133.3
14.23	32.78	28.96	0.968	744.7	1	98,070
1.45×10^{-4}	3.35×10^{-4}	2.96×10^{-4}	9.87×10^{-6}	7.50×10^{-3}	1.02×10^{-5}	1

Viscosity

Dynamic absolute viscosity (μ)

cp	Ibf.ft2	Ib_m/ft.s	g/cm.s	$N.s/m^2$	kg/m.s	dp
1	2.09×10^{-5}	6.72×10^{-4}	0.01	1×10^{-3}	1×10^{-3}	1×10^{-3}
4.78×10^4	1	32.15	478.5	47.85	47.85	47.85
1488	0.031	1	14.88	1.488	1.488	1.488
100	2.09×10^{-3}	0.672	1	0.10	0.10	0.10
1000	0.021	0.672	10	1	1	1

Kinemate viscosity (V)

centistoke	ft^2/s	cm^2/s	m^2/s	myriastoke
1	1.076×10^{-5}	0.01	1.0×10^{-6}	1.0×10^{-6}
9.29×10^4	1	929.4	0.093	0.093
100	1.076×10^{-3}	1	1.0×10^{-4}	1.0×10^{-4}
10^6	10.76	10^4	1	1

Force (F)

Ib_f	N	dyne
1	4.448	4.448×10^5
0.225	1	10^5
2.25×10^{-6}	10^{-5}	1

(*Continued*)

Table A11.1 (Continued)

Energy (E)

kW.h	hp.h	Btu	J	kj	calories
1	1.341	3412	3.6×10^6	3600	8.6×10^5
0.7457	1	2545	2.684×10^6	2685	6.4×10^5
2.930×10^{-6}	3.929×10^{-4}	1	1055	1.055	252
2.778×10^{-7}	3.72×10^{-7}	9.48×10^{-4}	1	0.001	0.239
2.778×10^{-4}	3.72×10^{-4}	0.948	1000	1	239
1.16×10^{-6}	1.56×10^{-6}	3.97×10^{-3}	4.186	4.18×10^{-3}	1

Power (P)

kW	Btu/min	hp	ft.b/s	kg.m/s	cal/min
1	56.89	1.341	737.6	102	14,330
0.018	1	0.024	12.97	1.793	252
0.746	42.44	1	550	76.09	10,690
1.35×10^{-3}	0.077	1.82×10^{-3}	1	0.138	19.43
9.76×10^{-3}	0.558	0.013	7.233	1	137.6
6.98×10^{-5}	3.97×10^{-3}	9.355×10^{-5}	0.0514	7.12×10^{-3}	1

Source: Quasim (2002).

Index

Pages in *italics* refer to figures and pages in **bold** refer to tables.

accelerated sheet waterfall 140, 142
accessibility 55, 85, 320
acidic-waste neutralization **317**
acid-waste drainage 315
acid-waste metering detail 316–318
aerated waterfalls 139, 142–143
air break 319, *320*
air-conditioning 7–8, 10, *11*, 19–21, 28, 210, 228–229
air-cooled chiller 8, 10, *11*, 13–14, 21, *33*
air-cooled condenser 15
air gap 48, 319, *320*, 345
air pre-heater 192, 197–198
air valve 80–81, 212
air vents 32, *33*, 185, 212–214
alarms 88, 96–97, 100–101, 196, 207, 269, 271, 278, 283
alarm system 5, 264, 271, 283, 316
allowable friction loss 276–278
allowable pressure drop 38, 224, 252–253, 256, 279
ambient wet bulb 23–24
amusement xxxiii, 1, 3–4, 124
anesthesia workroom 304
anti-vortex plate 62
appurtenances 4, 6, 12, 24–29, 56, 69, 80–82, 97, 153, 165–167, 232, 268, 314
argon 261, 264
atmospheric vacuum breaker 166, 298–299
automatic dry standpipe systems 57–58
automatic expansion valve 20
automatic or constant pressure nozzle 76
automatic wet standpipe systems 57
autopsy room 305, 314

back flow 46–48, 81–82
backflow hazard ratings 47
back-flow preventer 44, 48, 65, 165–166, 291, 305, 314
back-siphonage 46–47, 49, 299, 305
backwashing 332, 336–337, 339–340, 344, 349
balancing tank 323, 327–328, 331, 347, *348*
balancing valve 24–27, 34, 39
ball-lock coupling 79
barbecue 228, **254**
basin 3, 24, 43, 121–122, 134, 137–138, 141, 143–147, 149–150, 161, 287, 290, 301, 316, 323–326, 344–346
bathing enclosure or stall 215
bayonet coupling 80
bedpan washer 287–289, 300, 302–304
bell fountain 128
bell or mushroom jet fountain 130, 134
below-edge pool 323
bent tube boiler 188
bib-nose type landing valve 72
birthing room 304
blow off 192
blow-off valve 193–194
boiler burner 189, 201
boiler combustion chamber 189, 197
boiler exhaust stack 189–190
boiler heat exchanger 185, 189, 197
boiler shell 186, 194, 202
boiler working principle 185
boil-over waterfall 142–143
borosilicate glass pipe 315
branch and drop pipe size 278
break tank 34
bubbler 35, 159, 161–162
bubble tester 236–237, *238*
buffer tank sizing 36–37
building gas service pipe 231
building gas supply 230–231, 241
built-in oven **254**
butterfly valve 24

cage frame 103
cam-lock coupling 80
capillary tube 20–21

carbon dioxide 190, 227, 229, 260–261, 263, 268, 272, **278**, 292, 341
cascade waterfall 143
cast iron radiator 212
central gas supply system 235–236
centralized piped medical gas supply and vacuum system 263–264
central vacuum plant 266
centrifugal compressor 17–18
centrifugal pump 3, 28, 59, 62, 64, 91, 123, 154–157, 174, 206–207, 329
ceramic radiator 212
change of gas pressure corresponding to height 257–258
changeover valve or regulator 235, 237
check valve 4, 48–49, 62, 65, 81–82, 97, 138, 153, 163, 165–166, 267, 345–346
chilled water xxix, 1–2, 7–10, 12–13, 16, 27–36, 38–39, 293, 306, 309
chilled water buffer tank 25, 36
chilled water circulation pump 28–29
chilled water distribution pipe sizing 36, 38
chilled water distribution piping 29–32
chilled water flow rate 29, 36–37
chilled water supply piping material 35
chilled-water supply system xxxi, 2, 7–39, 309
chilled water system for drinking 32–35
chiller xxxi, 2, 7–9, 11–22, 25–27, *28*, 29–37, 293, *294*
class-I standpipe 69–70, 114
class-II standpipe 69–70
class-III standpipe 69–70, 114
clear-sheet waterfall 139–140, *141*
closed chilled water supply system 9–11, 27
closed circuit cooling tower 22–23
closed circuit tower 293
closed fire sprinkler head 104
clothes dryer 228, **254**
coal fired boiler 185
color coding usage of medical gases 276
column jet fountain 128–129
combined firefighting system 113–114
combined standpipe and sprinkler system 55
compact boiler 186
compressor 2, 12, 16–19, 21, 27, 261–262, 264–265, 267–268, 270, 283, 296–297, 351
concealed pendent head 105–107
concrete 24, 43, 108, 129, 155, 178, 216–217, 252, 322, 325, 335, 347
condensate pump 206–207
condensate recovery system 185
condenser 8–10, 12–17, 20–23, 29, 208, 295, 310, 319
condenser water circulation pump 29
constant or fixed orifice nozzle 75
constant-rate chilled water distribution 29
constant rate chilled water distribution system 29–30
cooking 2, 4, 228
cooking appliances 228, **254**
cooling load 20, 36–37
cooling tower 2, 8–9, 12–15, 22–24, 29, 39, 43, 295
cooling tower sizing 23–24, 36, 39
coriolis flow meter 249–250
coupler 153, 178
cross-connection 45–46, 48–51, 263
crown jet fountain 130
crystalline fountain jet 128, 130–131
curtain. fountain 128
cyclone-fired boiler 188
cyclone steam separator 197
cylinder bundle 235–237
cylinder pressure regulator 273
cystoscopic room 302
compressed air 261–262, 264, 267–268, 270, 285, 296–297, 305

dancing fountain 124–125
dead weight safety valve 199
deflector 102–106, 297
deluge and 101
deluge sprinkler valve *100*
demineralization 309–312
dental lavatory 296–297
dialysis machine 290–292, 310
diaphragm meter 244–245
diatom filter 336–337
diesel engine 63–64, 67, 154, 156–157
differential pressure 26, 32, 34, 48–49
discharge to sewer 316
disinfecting pool water 343
disinfection 4–5, 147, 313, 335, 343
disinfector 5, 288–289, 324, 331
distillation 293, *294*, 309–310
diversity factor 37, 253, 257, 276–277, 279
divider plate steam separator 197
double check valve assembly 48–49
double-containment waste piping 316
down-feed or gravitational distribution system 35
drain cock 62, 81, 88, 167, 182, 192, 216
drain suction inlet 324, 346–347
drain tempering valve 295, 298–300
drift eliminator 23
drinking fountain 2, 7, 28, 32, 34–35, 306
drop sheet waterfall 140
dry fountain 126–127
dry pipe 57–58, 91, 96–98
dry standpipe system 56–58
dry steam 180–181, 187, 197–198
dual media filter and 336–337
dual-speed pump 330

dual water supply system xxix, xxxi–xxxii, 2, 41–51
dual water supply system in a building 44, *45*
dust filter 262, 267–268
drain pipe 145–146, 149, 216, 285, 287, 290, 295, 297–298, 316, 319

early suppression fast response sprinklers 91, 96, 103
economizer 192, 197
electrically fired boiler 185
electric motor 18–19, 28, 61, 63, 154, 157, 329
electron microscope 293, 309
electronic pressure switch 202
emergency shower 286–287
emergency (triage) room 301
epoxy-based sealants 326
equivalent pipe length 176, 256, 276–280
evaporator 8, 12, 16–17, 20–21, 30, 293
examination and treatment room 301–302
expansion tank 198, 202, *203*, 268
expansion valve 12, 15, 17, 20–21
extra hazard 91–92, 108, **110**, 115–116, 118, **353**
eye and face-wash fountain 286–287

fan fountain 28
fan jet fountain *132*
fast response **103**
feed check valve 192, 194
feed pump 191–192, 194, 206–207, 221
feed pump take-off 191
feed water pump 206
feed-water tank 190–191, 221, 223
fiberglass 215, 300–301, 316, 322, 325, 335
fiber reinforced plastics 338
film fountain *135*
film spray fountain 134
filter 122, 138, 147, 185, 189, 195–196, 252, 264, 267, 291–292, 310, 322–323, 326, 329, 331–332, 335–340, 343, 347, 349
filtering pool water 335–336
filtration 3–5, 25, 39, 122, 138–139, 147, 152, 166, 169, 309–310, 312, 322, 328, 332, 335–339, 344
finger jet fountain *133*
fire xxxii, 3–4, 47, 53–54, 57–58, 63, 67–68, 70, 75, 77–78, 82, 85, 88, 96, 99–100, 102–103, 106–107, 113, 115, 118, 198, 202, 229
fire brigade connections 56, 81
fire hose nozzle 68, 75
fire hose reel 68–69, *71*, 73, 83
fire hoses 73–74, 78
fire hydrant xxix, 2, 53–56, 59–60, 62, 64–68, 70, 72, 78–80, **81**, 82, 84–85, 87–88
fire pump controller 66–67
fire pump driver 63–64
fire pump performance curve 94
fire pumps 55, 59–63, 65–66, 84–85, 91, 95, 117, 119
fire pump testing 95
fire sprinkler system xxxii, 55, 87–119
fire standpipe system xxxii, 53–85, 87
fire station hose cabinet 69
fire suppressing 54–55
fire suppression 1–3, 53, 70, 78, 87, 96, 118
fire-tube boiler 185–187, **189**
fire water reservoir 81
firewater reservoir *90*
fixed irrigation sprinkler heads *160*
fixed sprinkler irrigation system 151
flexible bellow 25
float and thermostatic 196
floating fountain 126–127, 134
float type low-water cutoff switch 200–201
float valve 345
floor drain 35, 119, 216, 286, 289–290, 298, 302, 305–306
flow control valve 99, 268
flow through a nozzle 3, 77
flow velocity *225*, 245, 247–249, 258, 332
fluid cooling tower 22–23
foam column fountain 134
foam fountain 136
fog fountain 136
forced circulation boiler 186, 188
fountain basin 138, 145
fountain nozzle 124, 131, 149
fountain planning 137–139
fountain pool 121–150
fountain pump *139*, 150
fracture room 304
fragrance pump 215
friction loss of fire hose 74
fuel gas xxxii, 1–2, 4, 227, 237, 244, 257, **258**, 284, 306, 341
funnel floor drain 290, 295
fusible plug 192, 198, 202

gage glass 192
gas bank manifold 268–271
gas cylinder 233–235, 237, 264, 268, 270, 273
gas cylinder head assembly *234*
gas fired boiler 185, 341
gas fireplace **254**
gas flow 240, 244–250, 252, 254, 258, 268, 271, 276, 279
gaslight **254**
gas log **254**
gas meter 231–232, 236, 241, 244–252
gas pipe installation guidelines 238–239
gas pool heater 340–341
gas service line sizing 252
gas supply pipe 235, 237, 252, 273, 278
gas supply pipe sizing 252–258

gas supply pressure loss 256
gate valve 24, 153, 165, 169, 178, 193, 240, 346
geyser jet fountain 131, 134, *135*
glass or glass-lined pipe 313
globe valve 25, 70, 165–166, 241
gravity displacement type autoclave *219*
gravity filters 335–336
gravity sand filters 336
grid piping layout 111–112
guidelines for a dual water supply system 45–46
gunite 325

hand-washing clinical sink 285
hazard classification 88–89, 107, 109, 114, *116*
head of pump 332–334
health care 1, 5, 228, 260, 283–284, 306, **307**, 314, 320
healthcare plumbing xxxiii, 5, 284–321
heater 248, 269, 285, 292, 309, 324, 331, 341–343
heating 4, 10, 47, 180–181, 185, 188, 202–204, 208–211, *213–214*, 216, 218, 221–222, 224, 228, 235, 252, **254**, 330–331, 335, 340–342, 344
heating pool water 340–343
heat transfer efficiency 16
helium 229, 260–261, **276**
high-pressure boiler 186, 204
high-silicon cast iron 315
horizontal boiler 186
horizontal flash tank 203
horizontal return tubular boiler 186
horizontal straight tube boiler 188
horizontal tubular 186
horizontal type landing valve 72
hot water boiler 185
hot water supply system 308–309
hydronic boiler 254

independent remote 192
indirect sensing method 192
individual cylinder gas supply system 235
individual cylinder system 263
individual gas source 230
indoor fountains 123, 125–126
indoor humidification 218
infinity edge pool 323
inlet and outlet pipe connections 268
in-line chlorination 343–344
in-rack sprinklers 96
inspection opening 192
installation of gas meter 251–252
installing a pressure-reducing valve 80, 184
intensive-care room 301
inverted bucket 196
irrigation xxix, xxxii, 1–3, 43, 49, 51, 151–155, 158–159, 161, 163, 165–169, 171–172, 174–177
irrigation duration 171
irrigation sprinkler head 159–163

jockey pump 64–66, 84, 95, 117
jumping jet fountain 124

kitchen 4, 43, 210, 235, 300, 305–306, 308

laboratory glass washer 285, *286*
lance jet fountain *129*, 130–131
landing valves 56, 69, 72, 85
laundry 4, 228, 306, 308
lay flat hose 73
light hazard 91, 108, **110**, **113**, **115**, **118**, **353**
lighting a fireplace 229
line pressure regulator 270, 273
liquefied petroleum gas 227–228
liquid fuel fired boiler 185
loop piping layout and 111–112
low-pressure boiler 186, 204
low-water cutoff switch 198–200

maintainability 320
manifold room 269–270
manual dry standpipe systems 57–58
manual wet standpipe systems 57
maximum gas demand 253
mechanical pressure switch 202
medical compressed air 261–262, 264
medical compressed air plant 264, 267
medical compressed air supply pipe sizing 280
medical gas 2, 4–5, 260, 268, 272–273, 275, 278, **280**, 284, 286
medical gas demand 262–263
medical gas pipeline 264–273
medical gas pipe sizing 276–277, **281–282**
medical gas supply piping 271–272
medical gas supply pressure regulator 273
medical gas supply system testing 283
medical nitrous oxide plant 264
medical oxygen plant 264–265
medical vacuum 261–262, 266–267, 274, **278**
medium-pressure boiler 186, 204
metal pipe 35, 112, 147, 190, 218, 313
micro irrigation sprinkler head 159, *160–162*, *168*
mild or carbon steel 338
mist fountain 137
misting irrigation sprinkler head 159–160
mobile sprinkler irrigation 151–152, 159
monitor 6, 25, 66, 90, 146, 166–167, 268, 271, 273

monitoring and alarm system 264
mop-service basin 287
multi-directional drives 122
multi-grade media filters 336, 338
multi jet fountain 130
multipurpose nozzle 75, 77
multi-tube boiler 186
mushroom fountain 134, *134*
musical fountain 124
multiport valves 345

natural circulation boiler 186, 188
natural fuel gas 227–228, 258
natural gas supply pressure 232–233
nitrogen 101, 190, 202, 227, 229, 260–262, 264–265, 268, 272, 274, **278**, **280**
nitrous oxide 260–261, 265, 268, 272, 274, **276**, **278**, 279, **280**, 296
nitrous oxide supply pipe sizing 279
Nomograph for Hazen William's formula *333*
non-potable water xxxii, 1–2, 41–46, 50–51, 306, 308, 314
non-return valve 25, 51, 100, 276, 328
normal waterfall xxxi, 1–2, 35, 43, 45, 194, 285
nourishment station 302–303
nuclear fired. boiler 185
nurse station 301–302

oblique type landing valve 72
occupancy hazard 91, *92*, 108
one-pipe steam distribution system 212
open chilled water supply system 9–10
open circuit cooling tower 22
open-circuit tower 22–23
open fire sprinkler head 104, *105*
operating-room 5, 277, 292–293, 303, 308
ordinary hazard 91, 107–108, 113, **115**, **118**, **353**
orifice meter 245–246
outdoor fountain 123, 126–131, 149–150
overflow armature 122, *146*
oxygen 11, 53, 127, 190, 260–266, 268, *269*, 270, 272, 274, **276**, **278**, **280**, 292
oxygen supply pipe sizing 279

piping of swimming pool 348
pool capacity and time rise 342–343
palm tree jet fountain *132*
panels 56, 124, 148, 278, 283, 341
pathology lab 304–305
pathology laboratory 304–305
patient room 300–302
pendent sprinkler head 105
permanent cross-connections 50–51
pharmacy and drug room 305
physical-therapy treatment room 302
pilot-operated safety valve 199
pin-lock coupling (claw coupling) 80
plastic pipe 35, 112, 147, 173–174, *177*, 237, 252, **253**, 299, 313, 315, 348, **352**
plate economizer 198
pool pump 323, 329–330, 332, 334, 339–340, 346–347
pool water flow 330–331
pop-up irrigation sprinkler head *162*
potable water xxxi, 2, 41–46, 48, 50–51, 166, 286, 306–308, 314, 344–345
power of pump 334
pre-action 57–58, 91, 96, 101–102
pressure and temperature gauge 25
pressure control panel 236
pressure filter 147, 312, 335–337, 339
pressure gauge 26, 153, 165–166, 192, 196, 218, 233–234, 273
pressure-regulating type landing valve 72
pressure-regulating valve 80, 153, 166, 236, 273
pressure regulator 5, 26, 233–234, 241–242, 244, 264, 268, 270–271, 273, 309
pressure release valve 268
pressure switches 66, 198, 202, 283
pressurized tank 27, 89–90
primary and secondary chilled water system 30
primary-secondary chilled water distribution 30, *31*
probe low-water cutoff switch 201
propeller pump 154, 156–157
psychrometer 24
pump 3, 9, 11, 28–29, 32, 34, 47, 58, 60–62, 64–65, 82, 89–91, 93–94, 109, 111, 117, 119, 121, 124, 127, 130, 134, 137, 141, 146–147, 149, 152, 154, 157–159, 166, 175–176, 178, 188–189, 206, 210, 220, 245, 266–267, 291–292, 295–297, 328–330, 332, 337, 339–340, 346
pump flow 84, 109, 119, 148, 221, 332
pure-water supply system 309

quality parameters of non-potable water 42–43
quality parameters of potable water 42–43
quartzoid bulb 103–104
quick coupler 165, 167
quick coupling 79–80, 178

Ram strike 181
range xxx, 7, 9, 18, 21–24, 60, 72, 75–76, 83, 93, 104, 108, 115, 123, 154, 158, **159**, 161, 185–186, 191, 201, 204, 216, 221, 224, 227, 232, 244–245, 257–258, 261, 266–267, 274, 285, 287, 309, 312, 315, 339, **351**, **357**–**358**
reciprocating compressor 17–18
recovery room 304

reduced pressure principle assembly 48–49
reduced pressure zone device 51
redundancy 5, 320
refrigerant gas 8, 15–19, 22
refrigerator **254**
regenerative economizer 198
reinforced hose 73
reverse osmosis 291, 296, 305–306, 309–310
right-angle type landing valve 72, 74
ring-lock coupling 80
roller-lock coupling 79
rotary irrigation sprinkler heads 160–161
rotary meter 244–245
rotary screw compressor 19
rough wall waterfall 142–143

safety valve 192, 198–199, 202, 216, 218, 220, 258, 268, 349
saturated steam 180–181, *184*, 205, 218, 225, **356–357**
scroll compressor 18–19
selectable or adjustable nozzle 75
semiautomatic dry standpipe systems 57–58
semi-mobile sprinkler irrigation system 151–152
sequencing fountain 123–124
serviceability 320–321
short firebox boiler 186
shutoff valve 48, 240, 264, 307
sidewall sprinkler head 105–106
single jet fountain 128, 131
single-screw compressor 19
single-speed pump 330
single-stage pressure regulator 241, 273
single-tube boiler 186
six-way valve 338–339
sizing a chiller 36
sizing of gas meter 250
skimmer 323–324, 326–327, 331–332, **334**, 335, 345–347, *348*, 349
skimmer number 327
smooth-bore nozzle 75–77
space heating 4, 180–181, 204, 210–214, 218, 228, **254**
space heating by steam 209–211
specialized plumbing system xxxi, 1–2, 6
special occupancy hazard 91, 108
specialties of specialized plumbing 6
specific heat of substance 219
splash-free floor drain 290
spout fountain 128–129, *130*
spring-loaded safety valve 48, 199, *200*
sprinkler heads 3, 47, 87–88, 90, 92–93, 95–97, 99–107, 109–115, 118, 154, 158–163, 168–170, 172–173, 178
sprinkler irrigation xxix, xxxii, 3, 151–179
sprinkler piping 102, 110–113, 116, 118
square sprinkler arrangement pattern 165
stable fountain 123
stainless steel 24, 27, 35, 62, 69, 72, 129, 136, 142, 195, 206, 212, 216, 218, 271, 291, 301–305, 313, 316, 329, 338
stair-step waterfall 142–143
standard response 103, 107
stand pipes and branch pipes 56, 80
steam 2, 4, 77, 180–225, 265, 269, 289, 293, 295, 306, 310, 313, 319, **356–357**
steam bath generator 215–217
steam boiler 47, 184–186, 189–192, 197, 199, 201–202, 204–205, 208–212, 219–220
steam boiler efficiency 220
steam circulating pump 206–208
steam distribution 182–185, 212–214, 225
steam distribution piping 182–183, *184*, 215
steam dryness 181
steam filter 195–196
steam for sterilization 218
steam injection 191
steam pressure management system 184
steam radiator 210–212, 218, 221–222
steam separator 192, 197
steam shower head 215, 217
steam stop valve 192–193
steam strainer 195
steam trap 185, 192, 195–197, 205
steel radiator 212
sterilizer 218, 289, 295–296, 303, 310, 319
still 122, 293–295
stop valve 236, 240
storage of non-potable water 46
strainer 25, 39, 138, 147, 150, **177**, 185, 195–196, 252, 329, 335
submersible pump 138, 145, 147, 154, 157–158
super-heater 192, 198, *199*
supply of non-potable water in a building 46–49
surface area and ambient air temperature 341–342
swimming xxxiii, 1, 5, 144, 322, 324, 334
swimming pool xxxiii, 5, 43, 51, 322–331, **332**, 334–336, 338–340, 342–349
swimming pool drainage 346–347
swimming pool maintenance 349
swimming pool plumbing xxix, xxxiii, 5, 322–349
swimming pool plumbing principles 347–348
switchboard with controller 267
system xxxi–xxxii, 1–7, 26–27, 32–35, 37, 39, 43, 46–47, 50, 53, 60, 64–69, 80–82, 85, 93–94, 97, 99–100, 113, 115, 117–119, 121–122, 125, 145, 147, 150–151, 163, 165–169, 178, 180–225, 227–258, 260, 262, 265, 270, 276, 278, 283–284, 292, 295, 297, 305, 308, 310, 313–323, 328, 330, 341, 344–347
service pipes installation 297–298

tapping points at the various points of use 236
temperature gauge 191
temperature rating 101, 103–104
temperature sensor and control 215
temporary cross-connections 50–51
terminal unit 264, 274–275
terminal units 30, 32, 268, 273–275, 277
test connections 88
thermal mass flow meter 245, 248
thermal sensitivity and 103
thermodynamic 12, 196
thermostatic 26, 196, 291, 299, 301–302, 304
thermostatic expansion valve 20–21
three-way valve 24–25, 345–346
total flow rate 38, 149, 276–277
trap primer 286, 290–291, 298
tree piping layout 111
triangular sprinkler arrangement pattern *164*, 165
tube economizer 198
tubing hose 73
tulip fountain 128
turbine meter 244–246, 250
turnover 147, 331–332
twin-screw compressor 19
two-pipe steam distribution system 214
two-stage pressure regulator 241–242, *243*, 267
two-way valve 25, 345–346

ultrasonic flow meter 245–248
under-water lighting 122
unplasticized polysulfide sealants 326
up-feed or pressurized distribution system 35
upright sprinkler head 105–106
urban dual-water supply system 43–44

vacuum inlet unit 276
vacuum pipe sizing 279
vacuum plant 264, 266–267
vacuum steam 181
vacuum system 5, 262–263, 267, 296–297
valve assembly 85, 97, 103, 163, 233
variable primary flow chilled water distribution 29, 31–32
variable primary flow chilled water system 29, 31–32
variable speed compressor 18
variable-speed pump 31–32, 330
various qualities of waters supplied 306
velocity of gas flow 258
velocity of steam flow 205
venting chilled water supply piping 32
vertical boiler 186
vertical flash receiver 203
vertical tubular 186
vertical turbine or diffuser pump 154
vinyl 325
vortex flow meter 245, 248–249
vortex type steam separator 197

water service connection 306–307
water source 3, 43–44, 60, 121, 127, 155, 291, 307
water storage 61, 82–83, 89, 107, 113–114, 118, 208, 307
ward room 301
warm air furnace 254
water-cooled chiller 9–10, 13–14, 24
water-cooled condenser 15, 293
water-fall xxxii, 3, 122, 139–145, 147
water fountain xxxii, 121–124, 128, 130–131, 137, 144–145
water inlets 324, 345, 347
water jet fountain 128–131
water level indicator 192, *193*
water meter 45, 153, **164**, 165, 167
water recirculation 345
water separator 267
water supply xxix–xxxii, 1–3, 5–11, 14, 24–30, 32–35, 38–39, 41, 43–48, 50–51, 53–54, 57, 66–67, 87, 89, 93–94, 96, 99, 110, 115, 117, 121, 127, 165–166, 284–286, 291, 296, 298, 301–304, 306–310, 312–314, 330, 344–346, 348, **353**
water-tube boiler 185, 187–189
wet bulb temperature 23–24
wet cooling tower 22–23
wet pipe 57, 96–97
wet standpipe system 56–57, 67
wet steam 181, 184, 205
wind effect 171

zero-edge pool 323

For Product Safety Concerns and Information please contact our
EU representative GPSR@taylorandfrancis.com Taylor & Francis
Verlag GmbH, Kaufingerstraße 24, 80331 München, Germany